Malgorzata Lekka, Daniel Navajas, Manfred Radmacher and Alessandro Podestà (Eds.)
Mechanics of Cells and Tissues in Diseases

Also of Interest

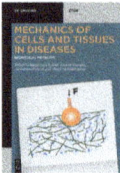

Mechanics of Cells and Tissues in Diseases.
Biomedical Methods
Edited by Malgorzata Lekka, Daniel Navajas, Manfred Radmacher,
Alessandro Podestà, 2022
ISBN 978-3-11-064059-5, e-ISBN (PDF) 978-3-11-064063-2

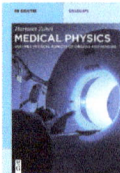

Medical Physics.
Volume 1: Physical Aspects of Organs and Imaging
Hartmut Zabel, 2017
ISBN 978-3-11-037281-6, e-ISBN (PDF) 978-3-11-037283-0

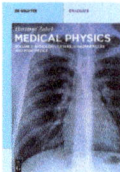

Medical Physics.
Volume 2: Radiology, Lasers, Nanoparticles and Prosthetics
Hartmut Zabel, 2017
ISBN 978-3-11-055310-9, e-ISBN (PDF) 978-3-11-055311-6

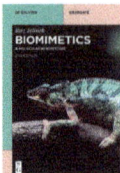

Biomimetics.
A Molecular Perspective
Raz Jelinek, 2021
ISBN 978-3-11-070944-5, e-ISBN (PDF) 978-3-11-070949-0

Quantum Electrodynamics of Photosynthesis.
Mathematical Description of Light, Life and Matter
Artur Braun, 2020
ISBN 978-3-11-062692-6, e-ISBN (PDF) 978-3-11-062994-1

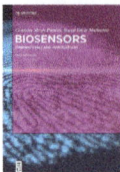

Biosensors.
Fundamentals and Applications
Chandra Mouli Pandey und Bansi Dhar Malhotra, 2019
ISBN 978-3-11-063780-9, e-ISBN (PDF) 978-3-11-064108-0

Mechanics of Cells and Tissues in Diseases

Biomedical Applications

Edited by
Malgorzata Lekka, Daniel Navajas, Manfred Radmacher
and Alessandro Podestà

Volume 2

DE GRUYTER

Editors

Prof. Dr. Malgorzata Lekka
Department of Biophysical Microstructures
Institute of Nuclear Physics
Polish Academy of Sciences
ul. Radzikowskiego 152
31-342 Kraków
Poland
malgorzata.lekka@ifj.edu.pl

Prof. Dr. Daniel Navajas
Unitat de Biofísica i Bioenginyeria
Facultat de Medicina i Ciències de la Salut
Universitat de Barcelona
Institute for Bioengineering of Catalonia
Barcelona. Spain
Spain
dnavajas@ub.edu

Prof. Dr. Manfred Radmacher
Institute of Biophysics
University of Bremen
Otto-Hahn-Allee 1
28359 Bremen
radmacher@uni-bremen.de

Prof. Dr. Alessandro Podestà
Department of Physics "Aldo Pontremoli" and
CIMaINa
Università degli Studi di Milano
via Celoria 16
20133 Milano
Italy
alessandro.podesta@mi.infn.it

ISBN 978-3-11-099972-3
e-ISBN (PDF) 978-3-11-098938-0
e-ISBN (EPUB) 978-3-11-098943-4

Library of Congress Control Number: 2022941163

Bibliographic information published by the Deutsche Nationalbibliothek
The Deutsche Nationalbibliothek lists this publication in the Deutsche Nationalbibliografie;
detailed bibliographic data are available on the Internet at http://dnb.dnb.de.

© 2023 Walter de Gruyter GmbH, Berlin/Boston
Cover image: background: Olga Kurbatova/iStock/Getty Images Plus, drawing: Malgorzata Lekka
Typesetting: Integra Software Services Pvt. Ltd.
Printing and binding: CPI books GmbH, Leck

www.degruyter.com

Preface

The mechanical properties of cells can be used to distinguish pathological from normal cells and tissues in many diseases, not only those where the relation between mechanics and physiology of the disease is obvious, like infarcted heart tissue, but also those where this relation is less obvious or still unknown, like cancer. This book outlines the physics behind cell and tissue mechanics, describes the methods, which can be used to determine the mechanical properties of single cells and tissues, and presents various diseases, in which a mechanical fingerprint could be established. Cell mechanics has the potential to serve as an assay, which could be widely used in the future. This book aims to introduce this topic to researchers from backgrounds as varied as biophysics, biomedical engineering, biotechnology, as well as graduate students from biology to medicine to introduce this novel and exciting concept to the community. In this book, we introduce to several aspects of cell biology, emphasizing the importance of the cytoskeleton, the cell membrane and glycocalyx, and the extracellular matrix. One chapter introduces the physics of continuum mechanics and its application to cells, including viscoelastic measurements. Then, various methods for measuring the mechanical properties of cells and tissues are discussed. Finally, evidence on the mechanical fingerprint of diseases is presented, discussing the properties of pathological cells from cancer, muscular dystrophy to diabetes, to name just a few here.

The first volume presents a comprehensive description of the basic concepts of soft matter mechanics and of the nano- and microscale *biomedical methods* that characterize the mechanical properties of cells and tissues.

The second volume is dedicated to discussing several *biomedical applications* of the mechanical phenotyping of cells and tissues to specific disease models. The topical chapters on mechanics in disease are preceded by chapters describing cell and tissue structure and their relationship with the biomechanical properties, as well as by describing dedicated sample preparation methods for the nano- and microscale mechanical measurements.

This book has been written for the primary benefit of young researchers but also of senior scientists, involved in interdisciplinary studies at the boundary of Physics, Biology and Medicine, and committed to transforming academic scientific and technological knowledge into useful diagnostic tools in the clinic.

We like to thank all authors of the various chapters for their valuable contributions. We appreciate very much your efforts and your continuing support over the time needed to create this work.

https://doi.org/10.1515/9783110989380-202

Acknowledgment

We acknowledge the support of the European Union's Horizon 2020 research and innovation program under the Marie Skłodowska-Curie grant agreement no. 812772, project Phys2Biomed.

https://doi.org/10.1515/9783110989380-203

Table of Contents of Volume 2 – Biomedical Applications

Table of Contents Volume 1 – Biomedical Methods

Soft Matter Mechanics

Instruments and Methods

Contributing Authors

Yara Abidine
University Grenoble Alpes, CNRS, LIPhy
Grenoble, France
Current address: Department of Clinical
Microbiology, Faculty of Medicine
Wallenberg Centre for Molecular Medicine
Umeå University, Sweden
yara.abidine@umu.se

Charles T. Anderson
Department of Biology and Center for
Lignocellulose Structure and Formation
The Pennsylvania State University
University Park, PA, USA
cta3@psu.edu

Massimo Alfano
Division of Experimental Oncology/Unit of
Urology, URI, IRCCS Ospedale San Raffaele
Milan, Italy
alfano.massimo@hsr.it

Nelda Antonovaite
Optics 11 Life,
Amsterdam, Netherlands
nelda373@gmail.com

Manuela Brás
i3S Instituto de Investigação e
Inovação em Saúde, Universidade do
Porto, Portugal
INEB – Instituto de Engenharia Biomédica
Porto, Portugal
FEUP – Faculdade de Engenharia da
Universidade do Porto, Portugal
mbras@i3s.up.pt

Kristian Brat
Department of Respiratory Diseases
University Hospital Brno, Brno
Czech Republic
Brat.Kristian@fnbrno.cz

Massimiliano Berardi
Optics 11 Life,
Amsterdam, Netherlands
massimiliano.berardi@optics11.com

Kevin Bielawski
Optics 11 Life,
Amsterdam, Netherlands
kevin.bielawski@optics11.com

Ignacio Casuso
Aix-Marseille Univ, CNRS, INSERM, LAI, Turing
centre for biological systems, Marseille
France

Shu-wen W. Chen
Institut de Biologie Structurale, Univ.
Grenoble Alpes, CEA, CNRS, F-38000
Grenoble, France

Matteo Chighizola
CIMaINa and Dipartimento di Fisica "Aldo
Pontremoli", Università degli Studi di Milano
Milan, Italy
matteo.chighizola@unimi.it

Thomas Decaens
Institute for Advanced Biosciences, Grenoble-
Alpes University, Inserm U1209 – CNRS UMR
5309, and Hepatology Department, University
Hospital of Grenoble Alpes, Grenoble, France
tdecaens@chu-grenoble.fr

Thierry Desnos
Aix Marseille Université, CNRS, CEA, Institut
de Biosciences et Biotechnologies Aix-
Marseille, Equipe Bioénergies et
Microalgues, CEA Cadarache, Saint-Paul-lez-
Durance, France
thierry.desnos@cea.fr

https://doi.org/10.1515/9783110989380-205

Simone Dinarelli
Institute for the Structure of Matter, CNR
Rome, Italy
simone.dinarelli@ism.cnr.it

Peter Dvorak
Department of Biology, Faculty of Medicine
Masaryk University, Brno, Czech Republic
ICRC, St. Anne's University Hospital, Brno
Czech Republic
dvorak.josefov@icloud.com

Vincent Dupres
Cellular Microbiology and Physics of Infection
Group, Univ. Lille, CNRS, Inserm, CHU Lille
Institut Pasteur Lille, Center for Infection and
Immunity of Lille, Lille, France
vincent.dupres@ibl.cnrs.fr

Allen Ehrlicher
Department of Bioengineering, McGill
University, Montreal, Canada
aje.mcgill@gmail.com

Ramon Farré
Unitat de Biofísica i Bioenginyeria
Facultat de Medicina i Ciències de la Salut
Universitat de Barcelona, Barcelona
Spain CIBER de Enfermedades Respiratorias
Madrid, Spain
Institut d'Investigacions Biomediques August
Pi Sunyer, Barcelona, Spain
rfarre@ub.edu

Conor Fields
School of Chemistry and Conway Institute for
Biomolecular and Biomedical Science
University College Dublin, Dublin
Republic of Ireland

Dorota Gil
Chair of Medical Biochemistry, Jagiellonian
University Medical College, Kraków, Poland
dorotabeata.gil@uj.edu.pl

Marco Girasole
Institute for the Structure of Matter, CNR,
Rome, Italy
marco.girasole@ism.cnr.it

Christian Godon
Aix Marseille Université, CNRS, CEA, Institut
de Biosciences et Biotechnologies Aix-
Marseille, Laboratoire de Signalisation pour
l'adaptation des végétaux à leur
environnement, CEA Cadarache, Saint-Paul-
lez-Durance, France
godon@cea.fr

Wolfgang Goldmann
Department of Physics, Biophysics Group
Friedrich-Alexander-University Erlangen-
Nuremberg, Erlangen, Germany
wolfgang.goldmann@fau.de

Hatice Holuigue
CIMaINa and Dipartimento di Fisica "Aldo
Pontremoli", Università degli Studi di Milano
Milan, Italy
hatice.holuigue@unimi.it

Sébastien Janel
Cellular Microbiology and Physics of Infection
Group, Univ. Lille, CNRS, Inserm, CHU Lille,
Institut Pasteur Lille, Center for Infection and
Immunity of Lille, Lille, France
sebastien.janel@cnrs.fr

Tae-Hyung Kim
University of California, Los Angeles, CA, USA
Current address: Department of Pathology at
the University of New Mexico, Albuquerque
NM, USA
takim@salud.unm.edu

Harinderbir Kaur
Univ. Grenoble Alpes, CEA, CNRS, Institut de
Biologie Structurale, Grenoble, France
harinderbir.kaur@ibs.fr

Prem Kumar Viji Babu
Institute of Biophysics, University of Bremen
Bremen, Germany
Current address: NanoLSI, Kanazawa
University, Kanazawa, Japan
pk@biophysik.uni-bremen.de

Leda Lacaria
Aix-Marseille Univ, INSERM, CNRS, LAI, Turing
Centre for Living Systems, Marseille, France
leda.lacaria@inserm.fr

Frank Lafont
Cellular Microbiology and Physics of Infection
Group, Univ. Lille, CNRS, Inserm, CHU Lille
Institut Pasteur Lille, Center for Infection and
Immunity of Lille, Lille, France
frank.lafont@pasteur-lille.fr

Piotr Laidler
Chair of Medical Biochemistry, Jagiellonian
University Medical College, Kraków, Poland
piotr.laidler@uj.edu.pl

Gil Lee
School of Chemistry and Conway Institute for
Biomolecular and Biomedical Science
University College Dublin, Dublin
Republic of Ireland
gil.lee@ucd.ie

Peng Li
School of Chemistry and Conway Institute for
Biomolecular and Biomedical Science,
University College Dublin, Dublin, Republic of
Ireland

Malgorzata Lekka
Department of Biophysical Microstructures
Institute of Nuclear Physics, Polish Academy
of Sciences, Kraków, Poland
malgorzata.lekka@ifj.edu.pl

Ewelina Lorenc
CIMaINa and Dipartimento di Fisica "Aldo
Pontremoli", Università degli Studi di Milano
Milan, Italy
ewelina.lorenc@unimi.it

Chau Ly
University of California, Los Angeles, CA, USA

Arnaud Millet
Institute for Advanced Biosciences, Grenoble-
Alpes University, Inserm and Research
Department University Hospital of Grenoble
Alpes, Grenoble, France
arnaud.millet@inserm.fr

Daniel Navajas
Unitat de Biofísica i Bioenginyeria
Facultat de Medicina i Ciències de la Salut
Universitat de Barcelona
Institute for Bioengineering of Catalonia
Barcelona. Spain
dnavajas@ub.edu

Hans Oberleithner
Thaur, Austria
oberlei@gmx.at

Jordi Otero
Unitat de Biofísica i Bioenginyeria, Facultat
de Medicina i Ciències de la Salut
Universitat de Barcelona, Barcelona, Spain
CIBER de Enfermedades Respiratorias
Madrid, Spain
jorge.otero@ub.edu

Jean-Luc Pellequer
Univ. Grenoble Alpes, CEA, CNRS, Institut de
Biologie Structurale, Grenoble, France
jean-luc.pellequer@ibs.fr

Martin Pesl
Department of Biology, Faculty of Medicine
Masaryk University, Brno, Czech Republic
ICRC, St. Anne's University Hospital, Brno
Czech Republic
First Department of Internal Medicine, Cardio-
Angiology, Faculty of Medicine, Masaryk
University, Brno, Czech Republic
martin.pesl@fnusa.cz

Alessandro Podestà
Dept. of Physics "Aldo Pontremoli" and
CIMAINA, University of Milano, Milan, Italy
alessandro.podesta@mi.infn.it

Jan Pribyl
CEITEC, Masaryk University, Brno, Czech
Republic

Manfred Radmacher
Institute of Biophysics, University of Bremen
Bremen, Germany
radmacher@uni-bremen.de

Lorena Redondo-Morata
Cellular Microbiology and Physics of Infection
Group, Univ. Lille, CNRS, Inserm, CHU Lille
Institut Pasteur Lille, Center for Infection and
Immunity of Lille, Lille, France
lorena.redondo-morata@inserm.fr

Carmela Rianna
Institute of Biophysics, University of Bremen
Bremen, Germany
Current address: Institute of Applied Physics
University of Tübingen, Tübingen, Germany
carmela.rianna@uni-tuebingen.de

Felix Rico
Aix-Marseille Univ, INSERM, CNRS, LAI, Turing
Centre for Living Systems, Marseille, France
felix.rico@inserm.fr

Jorge Rodriguez-Ramos
Aix-Marseille Univ, CNRS, INSERM, LAI, Turing
Centre for Living Systems, Marseille, France
jorge.rodriguez-ramos@inserm.fr

Vladimir Rotrekl
Department of Biology, Faculty of Medicine
Masaryk University, Brno, Czech Republic
ICRC, St. Anne's University Hospital, Brno
Czech Republic
vrotrekl@med.muni.cz

Niek Rijnveld
Optics 11 Life,
Amsterdam, Netherlands
niek.rijnveld@optics11.com

Amy C. Rowat
University of California, Los Angeles, CA, USA
rowat@ucla.edu

Carsten Schulte
CIMaINa and Dipartimento di Fisica "Aldo
Pontremoli", Università degli Studi di Milano
Milan, Italy
carsten.schulte@unimi.it

Petr Skladal
Department of Biochemistry, Faculty of Science
Masaryk University, Brno, Czech Republic
petr.skladal@ceitec.muni.cz

Zdenek Starek
ICRC, St. Anne's University Hospital, Brno
Czech Republic
44278@mail.muni.cz

Marta Targosz-Korecka
Center for Nanometer-Scale Science and
Advanced Materials, NANOSAM, Faculty of
Physics, Astronomy and Applied Computer
Science, Jagiellonian University, Kraków
Poland
marta.targosz-korecka@uj.edu.pl

Jean-Marie Teulon
Univ. Grenoble Alpes, CEA, CNRS, Institut de
Biologie Structurale, Grenoble, France
jean-marie.teulon@cea.fr

Pouria Tirgar
Department of Bioengineering, McGill
University, Montreal, QC, Canada
pouria.tirgarbahnamiri@mail.mcgill.ca

Anita Wdowicz
School of Chemistry and Conway Institute for
Biomolecular and Biomedical Science
University College Dublin, Dublin
Republic of Ireland

Claude Verdier
University Grenoble Alpes, CNRS, LIPhy
Grenoble, France
claude.verdier@univ-grenoble-alpes.fr

Ellen Zelinsky
Department of Biology and Center for
Lignocellulose Structure and Formation, The
Pennsylvania State University, University
Park, PA, USA

Joanna Zemla
Institute of Nuclear Physics, Polish Academy
of Sciences, Kraków, Poland
Joanna.Zemla@ifj.edu.pl

Marta Zarzycka
Chair of Medical Biochemistry, Jagiellonian
University Medical College, Kraków, Poland
marta.lydka@uj.edu.pl

Cell and Tissue Structure

Małgorzata Lekka

4.1 Cell Structure: An Overview

In living organisms, molecules subject to all the physical laws form spatial complex (bio)chemical structures capable of extracting energy from their environment and using it to build and maintain their internal structure. Each component of a living organism has specific functions at organ and cell levels that maintain cells in a steady state of internal physical and chemical conditions (homeostasis). Diseases occur due to many reasons. Some of them are linked with spontaneous alterations in the ability of a cell to proliferate, while others result from changes generated by external stimuli from the cell microenvironment. Regardless of the cause of diseases, cellular homeostasis undergoes severe alterations to which cells must adapt to survive. Otherwise, they can die. Recent studies on the role of biomechanics in maintaining cells and tissue homeostasis in various pathologies show that it is extremely important to link physical and chemical phenomena with the alterations in the structure of living cells or tissue. Accordingly, in this chapter, basic structural elements are described.

A cell is an individual unit containing various organelles used to maintain all living functions (Lodish et al., 2004). An example of the simplest cell is a bacterium. In bacteria, all cellular processes are carried out within a single cell body. In multicellular organisms, different kinds of cells perform different functions. Cells embedded within their microenvironment (the extracellular matrix, or ECM) assemble in highly specialized tissues (connective, muscle, nervous, and epithelial) as the basis for organ formation. Despite the high level of cellular specialization, most of the animal cells possess similar cellular structures (Figure 4.1.1).

A major component of the cell is the nucleus. The nucleus is a highly specialized organelle that contains genetic information encoded in DNA strands. It is surrounded by a double-layer phospholipidic membrane (called the nuclear envelope) that separates it from other regions present inside the cells. The nuclear membrane contains holes (called nuclear pores) that regulate the passage of molecules to and from the nucleus. A semifluid matrix found inside the nucleus is called nucleoplasm. Within it, most of the nuclear material consists of chromatin, the less condensed form of the cell's DNA that organizes to form chromosomes during mitosis or cell division. The nucleus also contains one or more nucleoli, which are membraneless organelles that manufacture ribosomes – the cell's protein-producing structures.

Close to the cell nucleus, an endoplasmic reticulum with associated ribosomes is located. This organelle is responsible for protein and lipid synthesis. Newly synthesized proteins and lipids are sorted in the Golgi apparatus, from which they are distributed to other cellular compartments or membranes.

Małgorzata Lekka, Institute of Nuclear Physics, Polish Academy of Sciences, Kraków, Poland

https://doi.org/10.1515/9783110989380-001

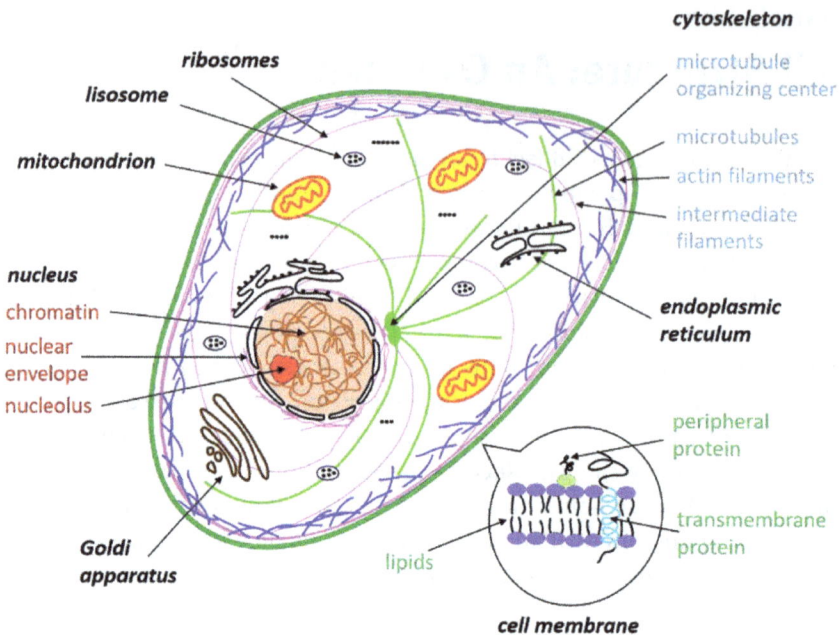

Figure 4.1.1: Schematic structure of an animal cell.

The mitochondria are organelles where energy is stored. They contain two major membranes: the outer and the inner membranes. The inner membrane has restricted permeability, and it is loaded with proteins involved in electron transport and ATP (adenosine triphosphate) synthesis, used for energy production. The outer membrane has many protein-based pores that enable the transport of ions and small molecules.

The lysosomes are specialized organelles that function as the digestive system inside cells and are responsible for the degradation of material taken in from outside the cell and for the digestion of obsolete cellular components. Lysosomes contain arrays of enzymes capable of breaking down any type of biological polymers – proteins, nucleic acids, carbohydrates, and lipids.

Within the cellular space, multiple types of various vesicles (e.g., endosomes) are required for the molecular transport within the cell and between the cell and its environment.

Each cell is surrounded by a cell membrane that separates the cell interior from the surrounding microenvironment. It is not only a structural scaffold within which cells are embedded but also contains various proteins, proteoglycans, and other molecules that participate in distinct cellular functions like adhesion or migration. The cell membrane consists of a double layer of phospholipids in which proteins are embedded. The interaction of the cell with the ECM mainly happens through the action of integral (going across the cell membrane) and peripheral (attached to the outer side of the cell membrane) proteins regulating the transport of substances to and from the cell.

All intracellular organelles are embedded in the cytoplasm filling the cell interior. The cytoplasm contains two elements, that is, the cytosol (a liquid fraction) and the cytoskeleton (a network of protein filaments).

The cytosol is the intracellular fluid comprised of water, dissolved ions, large water-soluble molecules, smaller molecules, and proteins. Within it, multiple levels of organization can be found. These include concentration gradients of small molecules such as calcium, large complexes of enzymes that act together to carry out metabolic pathways, and protein complexes such as proteasomes that enclose and separate parts of the cytosol.

The cytoskeleton is a mesh-like structure composed of various filamentous proteins. Apart from its structural functions related to maintaining cellular shape and providing the tool for organelles' arrangements, the cytoskeleton participates in various processes through interactions with other proteins, such as muscle contraction, cell division, migration, adhesion, and intracellular transport. The cytoskeleton helps establish regularity within the cytoplasm and, together with the plasma membrane, determines the mechanical stability of the cell. The cytoskeleton comprises three main elements – actin, intermediate filaments, and microtubules. A mesh-like structure composed of actin filaments is located beneath the cell membrane. Intermediate filaments form a ring around the cell nucleus and span over the whole cell volume. Microtubules have one end located at the microtubule-organizing center (a centrosome) close to the cell nucleus and the other in the cell membrane.

In the following chapters, detailed descriptions of cell structural components are presented.

Reference

Lodish, H., A. Berk, P. Matsudaira, C. A. Kaiser, M. Krieger, M. P. Scott, L. Zipursky and J. Darnell (2004). "Molecular Cell Biology."

Wolfgang H. Goldmann
4.2 The Cytoskeleton

Adherent cells are anchored via focal adhesions to the extracellular matrix, which is essential for force transduction, cell spreading, and migration. Focal adhesions consist of clusters of transmembrane adhesion proteins of the integrin family and numerous intracellular proteins, including talin and vinculin. They link integrins to actin filaments and are key players of focal adhesions that build up a strong physical connection for transmitting forces between the cytoskeleton and the extracellular matrix. These proteins consist of a globular head and a tail domain that undergo conformational changes from a closed, autoinhibited conformation in the cytoplasm to an open, active conformation in focal adhesions, which is regulated by phosphorylation.

4.2.1 Actin Cytoskeleton

Over the years, much research has provided information on the cellular function of the cytoskeleton, which has helped in understanding the many aspects of cell behavior. Components of the cytoskeletal network are major regulators of processes as diverse as establishing and maintaining gross cell morphology, polarity, transduction of force, motility, and adhesion to matrix components and cells. The cytoskeleton has long been proposed to be involved in the organization/reorganization of reporters in the plasma membrane. It is, therefore, critical to cell recognition mechanisms for many types of associations. These can range from tissue formation to the immune killing of foreign cells. Hence, the association of cytoskeletal elements with membrane components became a paradigm for signal transduction to the cytoplasm from the cell surface and vice versa. Interaction sites for membrane proteins with the interior of the cell are also key integration sites for transmitting signals to several pathways, eliciting pleiotypic responses of cells to signals. Thus, membrane–cytoskeletal complexes are mediators of crosstalk between receptors. Cell surface receptors for diverse ligands, including growth factors and hormones, and cell–matrix and cell–cell adhesion proteins, are transmembrane linked to microfilaments, which in turn interact with both microtubules and intermediate filaments (IFs). These interactive systems of membranes, with all of the cytoskeletal arrays, can elicit the global responses of cells to ligands such as mitogens, which evoke major morphological perturbations (Carraway and Carraway, 2000).

Acknowledgment: The author thanks Ms. Ceila Marshall (MA) for proofreading the manuscript.

Wolfgang H. Goldmann, Department of Physics, Biophysics Group, Friedrich-Alexander-University Erlangen-Nuremberg, Germany

https://doi.org/10.1515/9783110989380-002

The cytoskeleton is a highly dynamic, multifunctional network that connects all compartments of the cell in a three-dimensional space. This intracellular network provides eukaryotic cells with structural support to maintain cell shape and directional locomotion. At the same time, it provides the opportunity for active, directed transport, such as organelles or the separation of chromosomes in mitosis. In addition to actin fibers, the cytoskeleton consists of two other types of protein filaments, microtubules, and intermediate filaments (IFs). All three comprise dynamic protein components that polymerize into spiral-shaped fiber bundles (Figure 4.2.1).

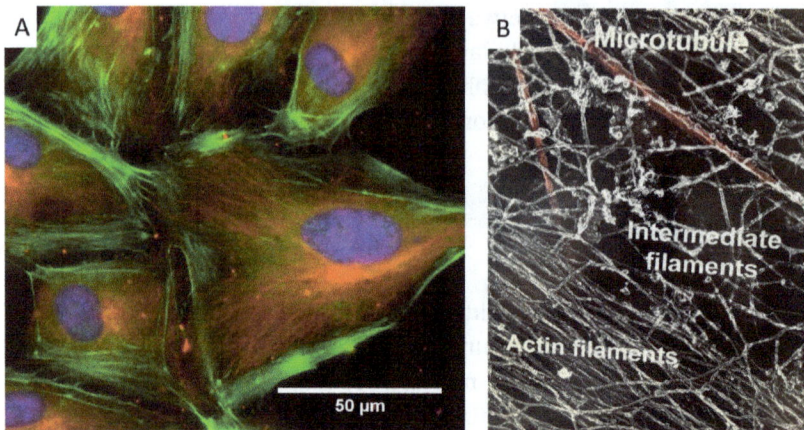

Figure 4.2.1: Filament types of the cell cytoskeleton. (A) HUVEC (human umbilical vein endothelial cells) taken by an 60x oil immersion objective. Green, F-actin (LifeAct-TagGFP2 protein); Red, Tubulin (Monoclonal anti-alpha-Tubulin); Blue, Nuclei (DAPI staining using ibidi mounting medium) adapted and taken from www.ibidi.com with permission. (B) Electron micrograph of the three filament types from a permeabilized cell. After freezing and sublimation of water, the structures were coated with platinum. Microtubules were highlighted in red (adapted and taken from Pollard and Cooper (2009) with permission).

Actin filaments (F-actin), with their flexible, double-helical structure of polymerized globular monomers (G-actin) have a diameter of 7–9 nm. They are found below the plasma membrane as a network (cortical actin) and also in the cytoplasm as discrete fiber bundles (stress fibers) starting from adhesion complexes to the membrane. This type of filament also shows orientation, as polymerization takes place at both ends, but at different rates. The slower growing *pointed end* (minus end) points toward the interior of the cell and the *barbed end* (plus end) polymerizes faster toward the plasma membrane. The actin monomers follow the so-called *treadmill* mechanism (Pollard and Mooseker, 1981), as ATP-bound G-actin attaches preferentially to the plus end through weak non-covalent bonds, while monomers bound to dephosphorylated ADP detach at the minus end of the filament (Carlier and Pantaloni, 1997). At constant G-actin concentrations in the cell, dynamic restructuring of filaments takes place by this mechanism, while the length remains constant. However, some toxins found in sponges and fungi affect the dynamics of actin fibers and are therefore very useful in the study of cellular functions of the actin cytoskeleton. For example, phalloidin, which is commonly used for immunofluorescence, binds and stabilizes F-actin (Cooper, 1987). Substances such as latrunculin A and cytochalasin D, on the other hand, promote depolymerization of the filaments, either by forming a complex with actin monomers or by blocking the *barbed end* of the filament through their attachment. Both the growth and branching of F-actin are precisely regulated by several actin binding proteins (Revenu et al., 2004). Capping proteins bind filament ends and thereby vary the length of the filament (e.g., tropomodulin binds and blocks the minus end) by promoting depolymerization (e.g., cofilin binds G-actin), preventing repolymerization (e.g., gelsolin binds to the plus end), or promoting polymerization (e.g., profilin catalyzes the exchange of actin-bound ADP to ATP) (Paavilainen et al., 2004). Other actin-binding proteins, such as filamin, generate flexible actin gels by linking multiple filaments (van der Flier and Sonnenberg, 2001). An important role in cross-linking F-actin is played by the Arp2/3 complex, which binds laterally to an existing filament and serves as a starting point for the polymerization of another filament at a 70° angle (Krause and Gautreau, 2014). Parallel actin fibers, in comparison, are formed into rigid bundles by binding proteins such as α-actinin or fascin (Sjoblom et al., 2008). Over 50 classes of different actin-binding proteins are now known (Edwards et al., 2014). The dynamic structure of actin filaments is regulated by a large number of factors and can, therefore, be quickly adapted to respective cellular needs (Tseng et al., 2005). As a result, some actin structures are the same in all cell types, while others fulfill a very specific function only in individual cell types. In tissues, for example, actin structures are responsible for the polarity of the cells and the cohesion of the epithelial cells or serve as mechanical support for microvilli on the cell surface. During cell division, actin is used in the form of contractile rings to cut off daughter cells from each other. Apart from the contractile apparatus in muscle cells, the actin cytoskeleton plays a major role in cell movement. The assembly and disassembly of actin regulate filopodia and lamellipodia at the cell front of migrating cells, and forces are generated by ATP hydrolysis of the myosin motor proteins at actin fibers.

As the name suggests, microtubules form a hollow, tubelike structure, with a diameter of approx. 25 nm, consisting of 12–17 laterally attached protofilaments. The protofilaments are composed of dimers, which, in turn, are formed of globular α- and β-tubulins. Microtubules originate from the centrosome and grow by polymerization at the plus end toward the cell periphery. During mitosis, they form the spindle apparatus through which the chromosomes are distributed in the daughter cells. In addition, the transport of organelles or vesicles along the microtubules takes place with the help of motor proteins (kinesin, dynein, etc.). The microtubules are among the most rigid elements in animal cells and contribute to the cell's resistance to shear forces through their structural design (Nogales, 2000).

IFs are flexible, stable, and durable protein fibers with a diameter of 10–12 nm, and, in contrast to the other fiber types, do not exhibit polarity. They additionally connect actin filaments and microtubules with each other, whereby their main purpose is the support function; and through the associated protofilaments, the IFs offer high tensile strength. Therefore, they are mainly found in areas of high mechanical force, such as epithelial cells and long-living structures such as hair. They also line the inner nuclear envelope and stabilize the axons of nerve cells. IFs comprise a heterogeneous group of proteins as the fibers are composed of different proteins, depending on the cell type. A distinction is made between type 1 IF made of acidic and type 2 IF made of basic keratins in epithelial cells, and type 3 IF made of vimentin in mesenchymal cells or desmin in muscle cells. Type 4 IF are the neurofilaments of nerve cells, and type 5 IF are the lamins of the cell nuclear envelope (Herrmann et al., 2007).

4.2.2 Integrins: Adhesion Receptor for the Cytoskeleton

Integrins belong to a family of transmembrane glycoproteins and are each composed of an α-subunit and a β-subunit. In vertebrates, 24 different αβ-heterodimers are found, consisting of one of 18 known α- and one of 8 β-subunits (Hynes, 2002). Figure 4.2.2 gives an overview of the possible combinations of α- and β-subunits and their ligands. Each subunit of the heterodimer has a large extracellular, single transmembrane and small intracellular domain (except for β_4-integrin).

The possible combinations of the two extracellular domains specify ligand binding; these are primarily extracellular matrix proteins. Despite the presence of large ligand proteins such as collagen, laminin, and fibronectin, many integrins recognize only short peptide sequences, such as the three amino acids RGD (Arg–Gly–Asp) found in fibronectin and vitronectin. While some integrins recognize only one specific protein (e.g., $\alpha_5\beta_1$ as fibronectin receptor), others have a variety of different binding partners (e.g., $\alpha_v\beta_3$ with laminin, collagen, fibronectin, von Willebrand factor, and fibrinogen) (Kuhn and Eble, 1994). In addition to the expression of different integrin subunits, the specificity can be further increased by alternative splicing of the cytoplasmic domains; thus, the intracellular function of the integrins can be adapted to the respective tissue (Aplin et al., 1998). Both cytoplasmic domains fulfill important tasks with regard to cytoskeletal attachment and signal transduction. Conserved sequences near the plasma membrane keep the two subunits together and in an inactive state, presumably via salt bridges (Wegener and Campbell, 2008). Ligand binding in the cytoplasm (*inside-out* signaling) or from the extracellular domains (*outside-in* signaling) can cause a conformational change so that the two subunits swing apart, and the heterodimer is activated. In this process, the angled, closed conformation of the extracellular domain changes to an upright, open form (Xiong et al., 2001). Signal

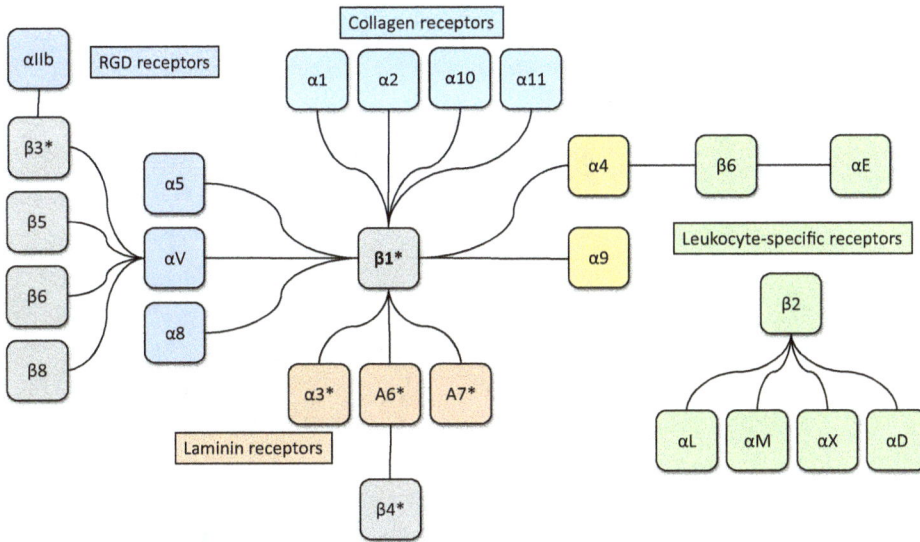

Figure 4.2.2: The integrin family with the various combinations of α- and β-subunits. Large ligand diversity is shown by the frequently occurring integrin heterodimers with β_1- and β_3-subunits. They form receptors for the RGD sequence in fibronectin and vitronectin. β_1Dimers also connect to collagen and laminin. In the basement membrane, $\alpha_6\beta_4$ integrins couple the laminin to intermediate filaments, and heterodimers with β_2- or β_7-subunits are found in cell–cell adhesions of leukocytes. Drawn by Lovis Schween (MSc); Information taken from Hynes (2002).

transduction by integrin molecules can occur in both directions across the plasma membrane. Activation by extracellular ligands often leads to a conformational change that allows cytoplasmic proteins to bind to the intracellular part of the transmembrane proteins, triggering local restructuring of the actin cytoskeleton or activating signaling cascades (Campbell and Humphries, 2011). In contrast, when integrin heterodimers are activated by the interaction of cytoplasmic proteins (e.g., talin), the conformational change of the extracellular domains stimulate binding to matrix proteins and "*clustering*," that is, a local accumulation of integrin molecules in the membrane can occur. This opens up binding sites for extracellular ligands and increases cell adhesion (Liddington and Ginsberg, 2002). Clustering is supported by the lateral homo-oligomerization of the activated α- and β-subunits (Li et al., 2003).

4.2.3 Integrin-Associated Focal Protein Complex

To fulfill the function of chemical and mechanical signal transmission in focal adhesions, the integrins are linked to a multimolecular protein complex on the intracellular side (Calderwood et al., 2003). Over 50 different proteins have already been identified in focal adhesions, which is partly due to cell-specific integrin interactions and partly due to the complexity of the control processes of these numerous proteins (Bershadsky et al., 2003) (Figure 4.2.3).

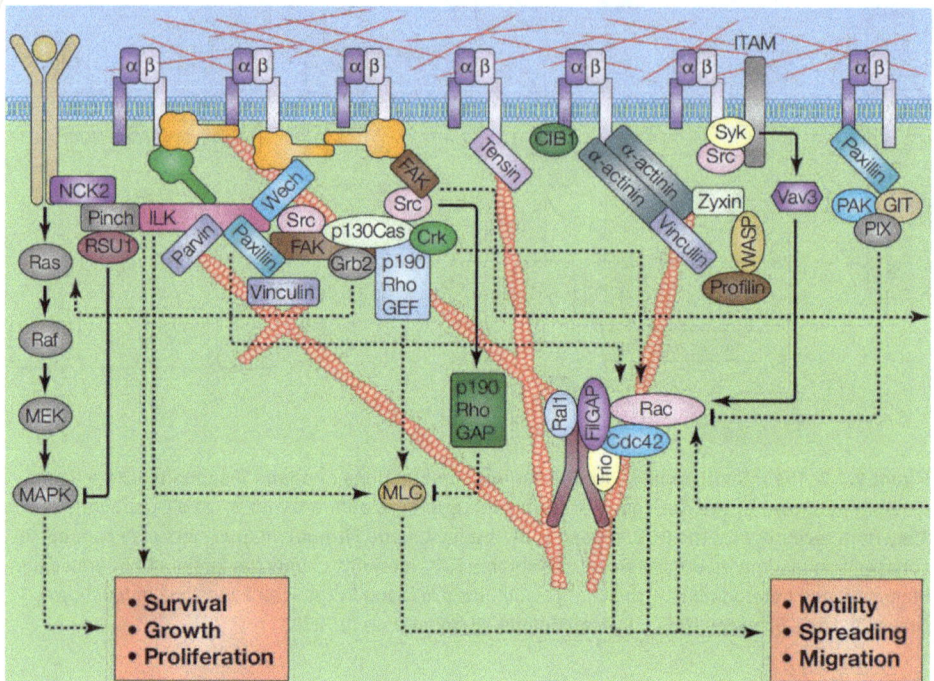

Figure 4.2.3: Proteins involved in the assembly and function of focal adhesions. Through their cytoplasmic domain, integrin heterodimers bind to proteins, such as talin (orange), which, in turn, interact with other focal adhesion proteins (e.g., FAK and vinculin). The entire protein complex then interacts with the actin network. The focal adhesions regulate the actin network via mechanical and biochemical signaling cascades to control the morphodynamics and gene expression of the cell. Taken from Harburger and Calderwood (2009) with permission.

The obligatory cytoplasmic focal contact proteins include talin, paxillin, vinculin, FAK, p130Cas, and α-actinin. In this process, individual proteins such as talin and α-actinin bind directly to integrins and link to other focal proteins such as paxillin and vinculin, ultimately resulting in the recruitment of actin filaments (Brakebusch and Fässler, 2003). The linkage of ECM proteins to the actin cytoskeleton via the integrins and the focal adhesion complex enable bidirectional force transmission (Hynes, 2002). Talin is one of the first proteins involved in the formation of focal contacts and can initiate the activation of integrins. It consists of two polypeptides that form an antiparallel homodimer (Rees et al., 1990). With the N-terminal head domain, talin binds to β_1- or β_3-integrins, as well as to focal adhesion proteins such as FAK or PIP2 (Seelig et al., 2000). PIP2-dependent binding of the FERM domain of talin to an NPXY motif of the β-subunit is a critical step in integrin activation (Nayal et al., 2004). Meanwhile, the larger domain at the C-terminus of talin interacts with F-actin, as well as with other cytoplasmic binding partners, such as vinculin (Bass et al., 1999). For vinculin, there are three known binding sites (VBS) in the talin protein (VBS1: AS 498–636; VBS2: AS 727–965; VBS3: AS 1943–2157), all of which associate with the same region in the vinculin head domain. Vinculin stabilizes the binding of talin to the actin cytoskeleton, providing a direct mechanical coupling of the force-generating apparatus to the integrins (Giannone et al., 2003). Auernheimer et al. (2015) examined the structure and function of vinculin in focal adhesion protein. Calpain-induced proteolysis of talin can restore the connection between integrins and actin fibers and promotes the dissociation of focal adhesions. Like talin, α-actinin also binds to integrins as well as to actin filaments, thus fulfilling a force-transmitting function (Otey and Carpen, 2004). The actin-bundling protein localizes mainly in mature focal adhesions at the attachment site of contractile stress fibers. That integrins not only serve for attachment to the substrate but are significantly involved in signal transduction is shown, for example, by the integrin-specific increase in phosphorylated proteins in cells adhering to the fibronectin-coated surface. In the focal adhesions, in addition to the stabilizing adapter proteins, numerous proteins involved in signaling are found, such as paxillin, FAK, and p130Cas (Schlaepfer and Hunter, 1998). Phosphorylation (MAP kinases, PKC, Src, FAK) and the concomitant recruitment of paxillin to the focal adhesion complex, in turn, activate additional groups of signaling proteins (Brown and Turner, 2004). As a result, Rho-GTPases are mobilized, and the actin cytoskeleton is reorganized. RhoA, in particular, regulates myosin II activity, whereupon, the motor protein, together with actin filaments, can generate intracellular contractile forces in response to mechanical stimuli (Chrzanowska-Wodnicka and Burridge, 1996). Signaling proteins such as the GTPases Rho and Rac also regulate the kinases that control phosphorylation and thus the function of various focal adhesion proteins. When considering a large number of proteins involved and their different tasks, which are as yet poorly understood and may vary from cell type to cell type, it becomes obvious that focal adhesions are dynamic structures with changing size and composition. Due to mechanical coupling and signaling, focal adhesions regulate the structure of the cytoskeleton, mechanotransduction, migration, proliferation, differentiation, and apoptosis of the cell (Goldmann, 2014).

4.2.4 Phosphorylation

Reversible phosphorylation of proteins is one of the most important post-translational modifications and the most common mechanism for regulating protein function and signal transduction. Approximately one-third of the human proteome is phosphorylated at any one time, and it contains an estimated 500 kinases (Manning et al., 2002). Protein kinases are enzymes that catalyze the transfer of the terminal phosphate group from adenosine triphosphate (ATP) to the hydroxyl group of one of the amino acids: serine, threonine, or tyrosine. The opposite reaction, that is, the hydrolysis and release of phosphate, is carried out by protein phosphatases. Since kinases recognize not only the target amino acid of their substrate but also the surrounding consensus sequence, some kinases act very specifically on individual proteins, while others phosphorylate multiple substrates (Pawson and Nash, 2003). The effect of phosphorylation on the respective protein is very diverse; for example, the three-dimensional protein conformation can be changed, an enzyme activity can be regulated, or the interaction of proteins with each other can be enabled. Tyrosine kinases are important components of cell proliferation, differentiation, and migration. Many signal transduction cascades rely on the recruitment of cytoplasmic proteins to the membrane, where they bind to phosphorylated receptors or become phosphorylated, themselves. The class of receptor tyrosine kinases (e.g., EGF or insulin receptors) has a transmembrane domain with an extracellular ligand-binding site (receptor) and the intracellular catalytic center (tyrosine kinase). Receptor kinases are activated by ligand binding; they form dimers and can stabilize their active form by autophosphorylation of cytoplasmic tyrosines as well as providing binding sites for other proteins in the signaling chain. The recruited proteins have conserved binding domains for specific amino acids. For example, the domains, SH2 (Src homology 2) and PTB (phosphotyrosine binding) recognize specific phosphotyrosine motifs (pY). Tyrosine kinases without an extracellular ligand-binding receptor domain include the Src, Abl, and FAK kinase families. These cytoplasmic kinases are activated by hormones, neurotransmitters, cytokines, or growth factors. This activation often begins with the phosphorylation of a tyrosine residue. The c-Src kinase is one of nine members of the Src family, which is found in many different cell types and different cell areas. The protein structure of Src kinases, for example, comprises four domains: a catalytic domain SH1, a SH2, and a SH3 domain, a N-terminal membrane localization sequence with a myristic acid residue, and a subsequent specific region for the respective kinase (Boggon and Eck, 2004). Also important is the short C-terminal tail of Src kinase, with the tyrosine residue it contains at position 527. The kinase can be regulated by multiple extracellular signals, including ECM-integrin contacts and, for example, growth factors such as EGF (Parsons and Parsons, 1997). Transient activation occurs through a conformational change by releasing the intramolecular binding of the SH2 domain to pY527 in the C-terminus and exposing the kinase domain. In addition, autophosphorylation of Y416 in the kinase domain is required to achieve full functionality. During the adhesion of fibroblasts to fibronectin, c-Src is dephosphorylated and localizes in focal adhesions (Kaplan et al., 1994). The binding of the SH2 and SH3 domains to p130CAS might play a role in localization or stabilize the open, active conformation of c-Src. Through the same domains, Src kinase can also bind phosphorylated paxillin or focal adhesion kinase. The tyrosine kinase FAK (focal adhesion kinase) is a 125 kDa protein with a central kinase domain and two proline-rich sequences in the C-terminus. Through the FAT sequence, FAK localizes to focal adhesions (Polte and Hanks, 1995). In adherent cells, integrin signaling and the presence of Src kinase cause an increase in phosphory-lated tyrosines in the FAK protein, resulting in increased activity of the kinase, whereas, in detached cells, the protein is again dephosphorylated (Calalb et al., 1995). Src kinase binds to the autophosphorylated FAK protein and thereby, in turn, pro-motes the association of the adaptor protein p130Cas into the complex, as well as its phosphorylation by FAK (Schlaepfer et al., 1997). The extent of phosphorylation regulates various interactions of FAK, which, in addition to p130Cas, paxillin and talin, binds to a variety of proteins containing an SH2 or SH3 domain (Chen et al., 1995). Thus, FAK also functions as a cross-linking binding partner in the assembly of focal adhesions.

It is clear that the focal adhesion components Src, FAK, p130Cas, and paxillin form a functional unit and are essential for the structure and signaling in adhesions. The activation of tyrosine kinases represents a crucial process of the integrin-mediated signaling cascade, even though it is still unclear how their activation actually occurs. Src, FAK, and other kinases, as well as the antagonistic phosphatases, play an impor-tant role in numerous cellular processes, that is, cell growth, migration, apoptosis, gene transcription, the immune response, or neuronal development (Burke, 1994). Re-versible phosphorylation, thus, transmits and amplifies signals, so that the cell is able to respond quickly to intra- or extracellular stimuli.

4.2.5 Dynamics and Force Generation via Focal Adhesions

Reversible phosphorylation of proteins is one of the most important post-translational modifications, and the most common mechanism for regulating dynamic cell move-ment takes place not only in the course of embryogenesis but also in the adult organ-ism, within the tissues. Migration is particularly evident in wound healing, when fibroblasts migrate in, or in metastasis, when individual cells migrate out of the pri-mary tumor and resettle in another part of the body. Although cells are in vivo sur-rounded by a three-dimensional network of ECM proteins that strongly influence their migration behavior, the basic processes of adhesion and cytoskeleton dynamics can be studied well on two-dimensional substrates. Only the coordinated interplay of force generation and force transmission to the substrate enables the movement and, thus, the response of the cell to external stimuli. For a cell to migrate, it must first adopt a polarized shape, which determines the direction of migration (Figure 4.2.4).

From the actin network at the cell front, broad lamellipodia or single filopodia with long parallel actin fiber bundles are projected toward the membrane by polymeri-zation (Pollard and Borisy, 2003). By protrusion, that is, pushing the membrane for-ward by local actin polymerization, the cell opens up new territory. The assembly and disassembly of filopodia take place within a few minutes. In order to stabilize a formed

1)

2)

3)

4)

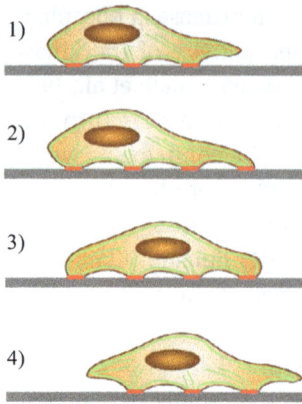

Figure 4.2.4: Four stages of cell migration. (1) Actin polymerization (green) causes the cell to form dynamic protrusions. (2) Certain protrusions are anchored to the substrate by new focal contacts (red). (3) By contraction of the actin cytoskeleton the focal contacts are stabilized. (4) After detachment of existing focal adhesions in the posterior part of the cell, contraction shifts the cell body toward the new adhesions. Taken from commons.wikimedia.org with permission.

filopodium, the actin filaments must be anchored to the extracellular matrix via focal contacts. Tensile forces are established via these new anchors in the cell front by actin-myosin contraction (controlled by Rho kinases) (Beningo et al., 2001). The forces transmitted to the substrate stabilize the new focal adhesions and place the cell under tension (Pasapera et al., 2010). To transform the contraction forces into an efficient forward movement, the adhesions in the posterior part of the cell must detach from the substrate. Once the adhesive structures in the cell rear end have been dissolved (mechanically or biochemically), the cell body moves toward the cell according to the traction forces. Consequently, the spatial distribution of the protrusions and adhesions of different strengths defines the direction of migration of a cell. It is known that cells migrate along gradients of chemical or structure-bound signaling substances (e.g., chemotaxis, haptotaxis). Consequently, signal transduction of the external stimuli and translation into coordinated control of the contractile and adhesive structures must take place. The adhesion process begins with small, punctate, highly dynamic attachments to the substrate, in which, initially, talin establishes a connection between the integrins and the actin filaments (Möhl et al., 2009). This early stage in the cell periphery is also referred to as nascent focal contact. Focal contacts are thought to play a role in the mechanical sensing of the cell as it senses stiffness, geometry, and its environment (Discher et al., 2005, Vogel and Sheetz, 2006, Geiger et al., 2009); their number and size are highly dependent on the properties of the substrate.

Many of these early contacts dissolve within a short time, while others mature into so-called *focal adhesions* through the recruitment of further proteins and the bundling of actin fibers into stress fibers and accompanying force generation (Riveline et al., 2001). The applied forces from the environment, as well as the tensile forces exerted by the actin cytoskeleton from inside the cell, cause a locally enhanced accumulation of integrins in the membrane (clustering) (Choquet et al., 1997) and the accumulation of further proteins, especially vinculin, in the focal complex (Galbraith et al., 2002). In this way, the junction is further stabilized, and there is a growth in the size of the focal adhesions (Golji et al., 2011). In addition to the composition of the protein complex, the degree of phosphorylation of the proteins also changes. This means that kinases are among the first recruited or activated proteins in the complex (Obergfell et al., 2002). Phosphorylation and dephosphorylation can control the dynamics and maturation stage of focal adhesions (Lele et al., 2008). It is conceivable that an increase in dynamics may lead to destabilization and, in combination with the applied actin traction forces, eventually to the dissolution of the focal adhesion (Wolfenson et al., 2011). The dissociation of focal adhesions must be regulated by diverse signaling pathways in addition to the force exerted, which are thought to involve diverse GTPases, FAK, and also Src kinases (Carragher and Frame, 2004). The tensile forces of the cytoskeleton arise from the interaction of myosin motor proteins and actin filaments. Myosin II induces contraction forces through the lateral displacement of actin fibers relative to each other, similar to the sarcomere of muscle cells.

In general, mechanical signaling pathways rely on a signal being transmitted to biomolecules in the form of mechanical forces, such as tensile forces or shear stress. Often, the applied force induces a conformational change in the protein, exposing functional domains (Del Rio et al., 2009). This can be the trigger for cytoskeletal remodeling, cell shape, or modified gene expression (Chiquet et al., 2009). Mechanical stimuli are often transmitted more rapidly than is the case with the perception of chemical signals (Na et al., 2008). As another example, mechanical traction forces acting externally on the cell have been observed to cause calcium influx across the membrane, which, in turn, causes intracellular force generation and protein recruitment to focal adhesions. Stabilization of focal adhesions and force transmission to the substrate are significantly regulated by proteins such as vinculin (Gallant et al., 2005). Consequently, it is essential to decipher the regulatory mechanisms of vinculin recruitment to understand the signaling pathways and control of focal adhesion formation and dynamics (Goldmann et al., 2013).

4.2.6 Conclusions

Adherent cells are in contact with the extracellular matrix via focal adhesions, a connection that is crucial for many cellular processes. To understand how cells perceive their environment and respond to different stimuli, it is essential to learn more about the regulation and functioning of focal adhesions and the proteins involved. Proteins such as vinculin and talin play a central role in the assembly and disassembly of focal complexes; they stabilize the binding of transmembrane integrins to the actin cytoskeleton of the cell and are, thus, crucial in cellular force transmission. Although intensive research has been conducted for years on focal adhesion proteins and many details about the protein structure and interaction partners are now known, it is still unclear exactly how the activation of the molecules and, thus, their exact function are regulated.

References

Aplin, A. E., A. Howe, S. K. Alahari and R. L. Juliano (1998). "Signal transduction and signal modulation by cell adhesion receptors: The role of integrins, cadherins, immunoglobulin- cell adhesion molecules, and selectins." Pharmacological Reviews **50**: 197–263.

Auernheimer, V., L. A. Lautscham, M. Leidenberger, O. Friedrich, B. Kappes, B. Fabry and W. H. Goldmann (2015). "Vinculin phosphorylation at residues Y100 and Y1065 is required for cellular force transmission." Journal of Cell Science **128**: 3435–3443.

Bass, M. D., B. J. Smith, S. Prigent and D. R. Critchley (1999). "Talin contains three similar vinculin binding-sites predicted to form an amphipathic helix." The Biochemical Journal **341**: 257–263.

Beningo, K. A., M. Dembo, I. Kaverina, J. V. Small and Y. L. Wang (2001). "Nascent focal adhesions are responsible for the generation of strong propulsive forces in migrating fibroblasts." The Journal of Cell Biology **153**: 881–888.

Bershadsky, A. D., N. Q. Balaban and B. Geiger (2003). "Adhesion-dependent cell mechanosensitivity." Annual Review of Cell and Developmental Biology **19**: 677–695.

Boggon, T. J. and M. J. Eck (2004). "Structure and regulation of Src family kinases." Oncogene **23**: 7918–7927.

Brakebusch, C. and R. Fässler (2003). "The integrin-actin connection, an eternal love affair." The EMBO Journal **22**: 2324–2333.

Brown, M. C. and C. E. Turner (2004). "Paxillin: Adapting to change." Physiological Reviews **84**: 1315–1339.

Burke, T. R. Jr. (1994). "Protein-tyrosine kinases: Potential targets for anticancer drug development." Stem Cells **12**: 1–6.

Calalb, M. B., T. R. Polte and S. K. Hanks (1995). "Tyrosine phosphorylation of focal adhesion kinase at sites in the catalytic domain regulates kinase activity: A role for Src family kinases." Molecular and Cellular Biology **15**: 954–963.

Calderwood, D. A., Y. Fujioka, J. M. de Pereda, B. Garcia-Alvarez, T. Nakamoto, B. Margolis, C. J. McGlade, R. C. Liddington and M. H. Ginsberg (2003). "Integrin beta cytoplasmic domain interactions with phosphotyrosine-binding domains: A structural prototype for diversity in integrin signaling." Proceedings of the National Academy of Sciences **100**: 2272–2277.

Carlier, M. F. and D. Pantaloni (1997). "Control of actin dynamics in cell motility." Journal of Molecular Biology **269**: 459–467.

Carragher, N. O. and M. C. Frame (2004). "Focal adhesion and actin dynamics: A place where kinases and proteases meet to promote invasion." Trends in Cell Biology **14**: 241–249.

Carraway, K. L. and C. A. C. Carraway (2000). Cytoskeleton: Signaling and cell regulation. A Practical approach. Hames, B. D., ed. 1st edition, Oxford. UK, Oxford University Press, 1–287.

Campbell, I. D. and M. J. Humphries (2011). "Integrin structure, activation, and interactions." Cold Spring Harbor Perspective Biology **3**: a004994.

Chen, H. C., P. A. Appeddu, J. T. Parsons, J. D. Hildebrand, M. D. Schaller and J. L. Guan (1995). "Interaction of focal adhesion kinase with cytoskeletal protein talin." The Journal of Biological Chemistry **270**: 16995–16999.

Chiquet, M., L. Gelman, R. Lutz and S. Maier (2009). "From mechanotransduction to extracellular matrix gene expression in fibroblasts." Biochimica Et Biophysica Acta **1793**: 911–920.

Choquet, D., D. P. Felsenfeld and M. P. Sheetz (1997). "Extracellular matrix rigidity causes strengthening of integrin-cytoskeleton linkages." Cell **88**: 39–48.

Chrzanowska-Wodnicka, M. and K. Burridge (1996). "Rho-stimulated contractility drives the formation of stress fibers and focal adhesions." The Journal of Cell Biology **133**: 1403–1415.

Cooper, J. A. (1987). "Effects of cytochalasin and phalloidin on actin." Journal of Cell Biology **105**: 1473–1478.

Del Rio, A., R. Perez-Jimenez, R. Liu, P. Roca-Cusachs, J. M. Fernandez and M. P. Sheetz (2009). "Stretching single talin rod molecules activates vinculin binding." Science **323**: 638–641.

Discher, D. E., P. Janmey and Y. L. Wang (2005). "Tissue cells feel and respond to the stiffness of their substrate." Science **310**: 1139–1143.

Edwards, M., A. Zwolak, D. A. Schafer, D. Sept, R. Dominguez and J. A. Cooper (2014). "Capping protein regulators fine-tune actin assembly dynamics." Nature Reviews. Molecular Cell Biology **15**: 677–689.

Galbraith, C. G., K. M. Yamada and M. P. Sheetz (2002). "The relationship between force and focal complex development." The Journal of Cell Biology **159**: 695–705.

Gallant, N. D., K. E. Michael and A. J. Garcia (2005). "Cell adhesion strengthening: Contributions of adhesive area, integrin binding, and focal adhesion assembly." Molecular and Cellular Biology **16**: 4329–4340.

Geiger, B., J. P. Spatz and A. D. Bershadsky (2009). "Environmental sensing through focal adhesions." Nature Reviews. Molecular Cell Biology **10**: 21–33.

Giannone, G., G. Jiang, D. H. Sutton, D. R. Critchley and M. P. Sheetz (2003). "Talin-1 is critical for force dependent reinforcement of initial integrin-cytoskeleton bonds but not tyrosine kinase activation." The Journal of Cell Biology **163**: 409–441.

Goldmann, W. H. (2014). "Mechanosensation: A basic cellular process." Progress in Molecular Biology and Translational Science **126**: 75–102.

Goldmann, W. H., V. Auernheimer, I. Thievessen and B. Fabry (2013). "Vinculin, cell mechanics and tumor cell invasion." Cell Biology International **37**: 397–405.

Golji, J., J. Lam and M. R. Mofrad (2011). "Vinculin activation is necessary for complete talin binding." Biophysical Journal **100**: 332–340.

Harburger, S. D. and D. A. Calderwood (2009). "Integrin signaling at a glance." The Journal of Cell Biology **122**: 159–163.

Herrmann, H., H. Bär, L. Kreplak, S. V. Strelkow and U. Aebi (2007). "Intermediate filaments: From cell architecture to nanomechanics." Nature Reviews. Molecular Cell Biology **8**: 562–573.

Hynes, R. O. (2002). "Integrins: Bidirectional, allosteric signaling machines." Cell **110**: 673–687.

Kaplan, K. B., K. B. Bibbins, J. R. Swedlow, M. Arnaud, D. O. Morgan and H. E. Varmus (1994). "Association of the amino-terminal half of c-Src with focal adhesions alters their properties and is regulated by phosphorylation of tyrosine 527." The EMBO Journal **13**: 4745–4756.

Krause, M. and A. Gautreau (2014). "Steering cell migration: Lamellipodium dynamics and the regulation of directional persistence." Nature Reviews. Molecular Cell Biology **15**: 577–590.

Kuhn, K. and J. Eble (1994). "The structural bases of integrin-ligand interactions." Trends in Cell Biology **4**: 256–261.

Lele, T. P., C. K. Thodeti, J. Pendse and D. E. Ingber (2008). "Investigating complexity of protein-protein interactions in focal adhesions." Biochemical and Biophysical Research Communications **369**: 929–934.

Li, R., N. Mitra, H. Gratkowski, G. Vilaire, R. Litvinov, C. Nagasami, J. W. Weisel, J. D. Lear, W. F. DeGrado and J. S. Bennett (2003). "Activation of integrin alphaII beta3 by modulation of transmembrane helix associations." Science **300**: 795–798.

Liddington, R. C. and M. H. Ginsberg (2002). "Integrin activation takes shape." The Journal of Cell Biology **158**: 833–839.

Manning, G., D. B. Whyte, R. Martinez, T. Hunter and S. Sudarsanam (2002). "The protein kinase complement of the human genome." Science **298**: 1912–1934.

Möhl, C., N. Kirchgessner, C. Schäfer, K. Küpper, S. Born, G. Diez, W. H. Goldmann, R. Merkel and B. Hoffmann (2009). "Becoming stable and strong: The interplay between vinculin exchange dynamics and adhesion strength during adhesion site maturation." Cell Mot Cytoskeleton **66**: 350–364.

Na, S., O. Collin, F. Chowdhury, B. Tay, M. Ouyang, Y. Wang and N. Wang (2008). "Rapid signal transduction in living cells is a unique feature of mechanotransduction." Proceedings of the National Academy of Sciences 105: 6626–6631.

Nayal, A., D. J. Webb and A. F. Horwitz (2004). "Talin: An emerging focal point of adhesion dynamics." Current Opinion in Cell Biology 16: 94–98.

Nogales, E. (2000). "Microtubule function." Annual Review of Biochemistry 69: 277–302.

Obergfell, A., K. Eto, A. Mocsai, C. Buensuceso, S. L. Moores, J. S. Brugge, C. A. Lowell and S. J. Shattil (2002). "Coordinate interactions of Csk, Src, and Syk kinases with alphaII beta3 initiate integrin signaling to the cytoskeleton." The Journal of Cell Biology 157: 265–275.

Otey, C. A. and O. Carpen (2004). "Alpha-actinin revisited: A fresh look at an old player." Cell Mot Cytoskeleton 58: 104–111.

Paavilainen, V. O., E. Bertling, S. Falck and P. Lappalainen (2004). "Regulation of cytoskeletal dynamics by actin-monomer-binding proteins." Trends in Cell Biology 14: 386–394.

Parsons, J. T. and S. J. Parsons (1997). "Src family protein tyrosine kinases: Cooperating with growth factor and adhesion signaling pathways." Current Opinion in Cell Biology 9: 187–192.

Pasapera, A. M., I. C. Schneider, E. Rericha, D. D. Schlaepfer and C. M. Waterman (2010). "Myosin II activity regulates vinculin recruitment to focal adhesions through FAK-mediated paxillin phosphorylation." The Journal of Cell Biology 188: 877–890.

Pawson, T. and P. Nash (2003). "Assembly of cell regulatory systems through protein interaction domains." Science 300: 445–452.

Pollard, T. D. and M. S. Mooseker (1981). "Direct measurement of actin polymerization rate constants by electron microscopy of actin filaments nucleated by isolated microvillus cores." The Journal of Cell Biology 88: 654–659.

Pollard, T. D. and G. G. Borisy (2003). "Cellular motility driven by assembly and disassembly of actin filaments." Cell 112: 453–465.

Pollard, T. D. and J. A. Cooper (2009). "Actin, a central player in cell shape and movement." Science 326: 1208–1212.

Polte, T. R. and S. K. Hanks (1995). "Interaction between focal adhesion kinase and Crk associated tyrosine kinase substrate p130Cas." Proceedings of the National Academy of Sciences 92: 10678–10682.

Rees, D. J., S. E. Ades, S. J. Singer and R. O. Hynes (1990). "Sequence and domain structure of talin." Nature 347: 685–689.

Revenu, C., R. Athman, S. Robine and D. Louvard (2004). "The co-workers of actin filaments: From cell structures to signals." Nature Reviews. Molecular Cell Biology 5: 635–646.

Riveline, D., E. Zamir, N. Q. Balaban, U. S. Schwarz, T. Ishizaki, S. Narumiya, Z. Kam, B. Geiger and A. D. Bershadsky (2001). "Focal contacts as mechanosensors: Externally applied local mechanical force induces growth of focal contacts by an mDia1-dependent and ROCK-independent mechanism." The Journal of Cell Biology 153: 1175–1186.

Schlaepfer, D. D., M. A. Broome and T. Hunter (1997). "Fibronectin-stimulated signaling from a focal adhesion kinase-c-Src complex: Involvement of the Grb2, p130cas, and Nck adaptor proteins." Molecular and Cellular Biology 17: 1702–1713.

Schlaepfer, D. D. and T. Hunter (1998). "Integrin signaling and tyrosine phosphorylation: Just the FAKs?." Trends in Cell Biology 8: 151–157.

Seelig, A., X. L. Blatter, A. Frentzel and G. Isenberg (2000). "Phospholipid binding of synthetic talin peptides provides evidence for an intrinsic membrane anchor of talin." The Journal of Biological Chemistry 275: 17954–17961.

Sjoblom, B., A. Salmazo and K. Djinovic-Carugo (2008). "Alpha-actinin structure and regulation." Cellular and Molecular Life Sciences: CMLS 65: 2688–2701.

Tseng, Y., T. P. Kole, J. S. Lee, E. Fedorov, S. C. Almo, B. W. Schafer and D. Wirtz (2005). "How actin crosslinking and bundling proteins cooperate to generate an enhanced cell mechanical response." Biochemical and Biophysical Research Communication **334**: 183–192.

van der Flier, A. and A. Sonnenberg (2001). "Structural and functional aspects of filamins." Biochimica et Biophysica Acta **1538**: 99–117.

Vogel, V. and M. P. Sheetz (2006). "Local force and geometry sensing regulate cell functions." Nature Reviews. Molecular Cell Biology **7**: 265–275.

Wegener, K. L. and I. D. Campbell (2008). "Transmembrane and cytoplasmic domains in integrin activation and protein-protein interactions." Molecular Membrane Biology **25**: 376–387.

Wolfenson, H., A. Bershadsky, Y. I. Henis and B. Geiger (2011). "Actomyosin-generated tension controls the molecular kinetics of focal adhesions." Journal of Cell Science **124**: 1425–1432.

Xiong, J. P., T. Stehle, B. Diefenbach, R. Zhang, R. Dunker, D. L. Scott, A. Joachimiak, S. L. Goodman and M. A. Arnaout (2001). "Crystal structure of the extracellular segment of integrin alphaV beta3." Science **294**: 339–345.

Dorota Gil, Marta Zarzycka, Piotr Laidler

4.3 Cell Membrane and Glycocalyx

The cell membrane is a thin self-organized structure surrounding every living cell. In *eukaryotic* and *prokaryotic* cells that are built out of the same kind of chemical components, the physicochemical interactions at the membrane are also similar. The differences are due to specific lipid, protein, and carbohydrate components. Together with the cytoskeleton, the membrane gives the cell its structure and protects the integrity of its interior by selective transport systems, allowing the movement of specific molecules from one side to the other. Cell membrane serves as a base of attachment for the cytoskeleton in some organisms and the cell wall in others.

Knowledge of cell membrane structure has evolved based on evidence from physicochemical, biochemical, and electron microscopic investigations. Lipids and proteins are two major components of all biological membranes. The fundamental structure of the membrane is the phospholipid bilayer, which is responsible for its basic function as a barrier between two aqueous compartments, so the membrane is impermeable to water-soluble molecules, including ions and most of the polar/hydrophilic solutes. Proteins embedded within the phospholipid bilayer are responsible for the dynamic function of membranes, including selective transport of molecules and cell–cell recognition; they serve as receptors for various signaling molecules (Singer and Nicolson, 1972, Alberts et al., 2002).

4.3.1 Lipids as Components of Cell Membrane

Phospholipids are the most abundant lipids of the cell membrane. There are basically five major phospholipids, four of which are based on glycerol (phosphatidylcholine, phosphatidylethanolamine, phosphatidylserine, and phosphatidylinositol) and one based on sphingosine (sphingomyelin).

Glycerophospholipids are complex lipid molecules made up of glycerol esterified with two fatty acids, one saturated and one usually polyunsaturated, and phosphate that is also esterified with aminoalcohol or alcohol (choline, serine, ethanolamine, and inositol). The latter molecule composes the so-called head groups, which are highly hydrophilic, while fatty acid chains, called tails, are hydrophobic.

Sphingomyelins are built of sphingosine with fatty acid bound by amide bond and phosphocholine. Phospholipid molecules are amphipathic structures, with clearly separated hydrophobic and hydrophilic groups, and sometimes even polar groups;

Dorota Gil, Marta Zarzycka, Piotr Laidler, Chair of Medical Biochemistry Jagiellonian University Medical College, Kraków, Poland

https://doi.org/10.1515/9783110989380-003

for these reasons, they spontaneously form in water a double-layered membrane. When in water or an aqueous solution, the hydrophilic heads of phospholipids orient themselves to be on the outside, facing the ligands, while the hydrophobic tails are on the inside of membrane (Figure 4.3.1).

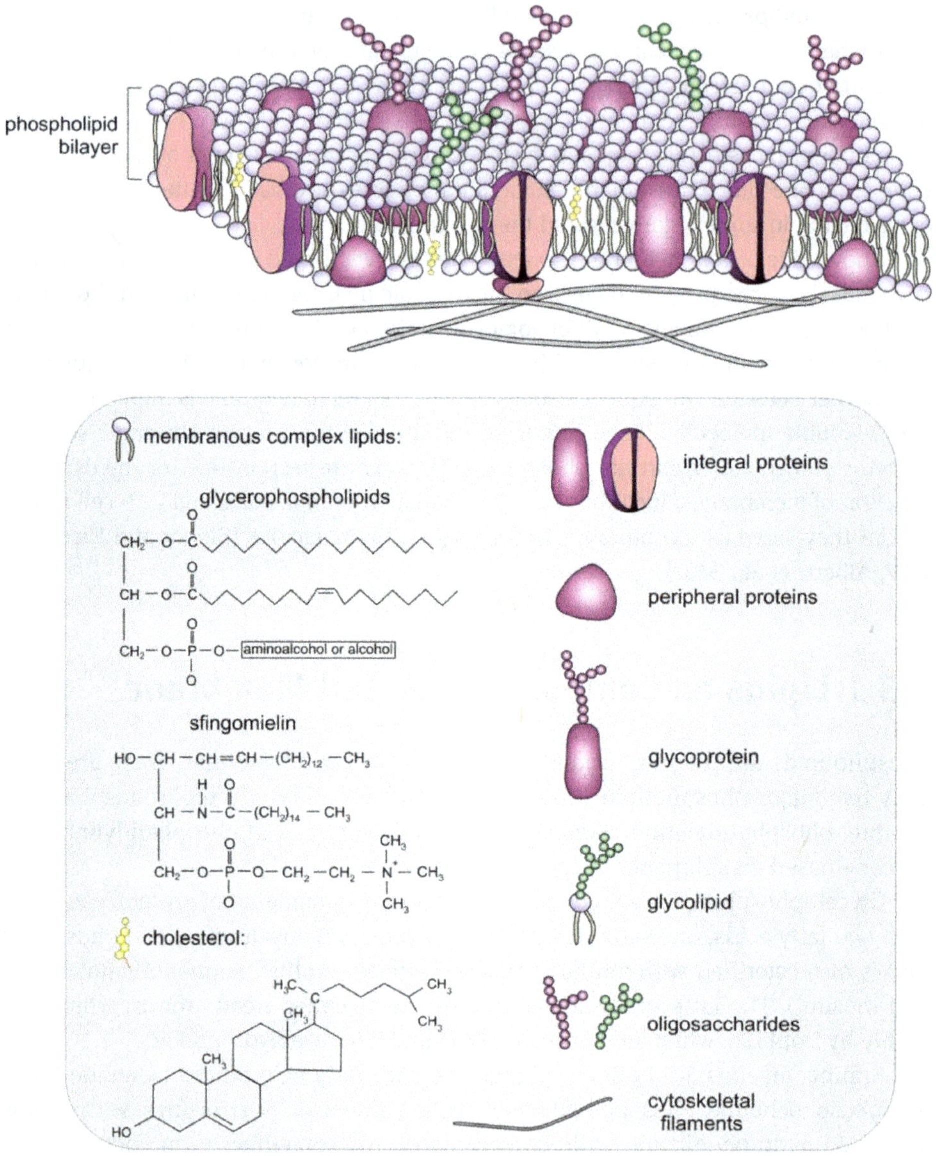

Figure 4.3.1: The schematic view of a biological membrane and chemical structures of selected, most common components.

The phospholipids are asymmetrically distributed between the two halves of the membrane bilayer. The outer leaflet of the plasma membrane consists mainly of phosphatidylcholine and sphingomyelin, whereas phosphatidylethanolamine and phosphatidylserine are the predominant phospholipids of the inner leaflet. A fifth phospholipid, phosphatidylinositol, is also localized on the intracellular side of the plasma membrane. Although phosphatidylinositol is a quantitatively minor membrane component, it plays an important role in cell signaling. The head groups of both phosphatidylserine and phosphatidylinositol are negatively charged, so their predominance in the inner leaflet results in a net negative charge on the cytosolic side of the plasma membrane.

In addition to the phospholipids, the plasma membranes of animal cells contain glycolipids and cholesterol. Glycolipids are located on the outer leaflet of the cell membrane, with their carbohydrate portions exposed at the cell surface. They are relatively minor membrane components, constituting only about 2% of the lipids of most plasma membranes, but often determine the antigenic properties of a cell. On the other hand, cholesterol is an appreciable membrane constituent of animal cells, being present in about the same molar amounts as the phospholipids.

With four fused rings and a branched hydrocarbon chain, cholesterol is a compact, rigid, hydrophobic molecule which plays a distinct role in membrane structure. Cholesterol molecules are dispersed between membrane phospholipids. Cholesterol contains one hydrophilic hydroxyl group facing the extracellular surrounding, while the rest of its rigid, planar hydrophobic structure is embedded in the sea of phospholipid fatty acids. Due to its rigidity and lack of structural elasticity, cholesterol alters the fluidity of the membrane and participates in controlling its microstructure. Depending on the temperature, cholesterol has distinct effects on membrane fluidity. At high temperatures, cholesterol interferes with the movement of the phospholipid fatty acid chains, making the outer part of the membrane-less fluid and reducing its permeability to small molecules. However, at low temperatures, cholesterol has the opposite effect by interfering with interactions between fatty acid chains and preventing membranes from freezing and maintaining membrane fluidity. Membrane fluidity controls membrane-bound enzyme activity and functions such as phagocytosis or cell signaling. Cholesterol is not found in the membranes of plant cells.

4.3.2 Cell Membrane Proteins

Singer and Nicolson (1972) identified two classes of membrane-associated proteins called peripheral and integral membrane proteins, respectively. This classification is based on the ease of removal of a specific protein from an isolated membrane.

Peripheral proteins dissociate from the membrane following treatments with polar reagents, such as solutions of extreme pH or high salt concentration, which do not disrupt the phospholipid bilayer. The names imply a physical localization on the membrane. These proteins are not inserted into the hydrophobic interior of the lipid bilayer, they are rather loosely attached to membranes, some bound to integral proteins. Negatively charged phospholipids interact with the positively charged region of proteins by electrostatic bonds. Some peripheral proteins have hydrophobic sequences at one end of polypeptide chain that serve as an anchor in the membrane lipids. Other kinds of peripheral membrane proteins are anchored only in the inner leaflet of the plasma membrane by covalently attached lipids and play important roles as mediators in transmitting signals from cell surface receptors to intracellular targets. Isolated peripheral proteins are water soluble, and many of them are enzymes.

Isolation of integral protein requires rather drastic treatments, which disrupt the phospholipid bilayer, such as use of detergents or organic solvents. Many integral proteins are transmembrane proteins, which contain sequences of hydrophobic aminoacids in the lipid bilayer and hydrophilic parts exposed on both sides of the membrane. Integral proteins usually contain tightly bound lipids to their hydrophobic domains. The disruption of protein–lipid interaction leads to denaturation of protein and loss of its biological function. Most enzymes that are integral membrane proteins require a membrane lipid for activity. Integral proteins contain different specific domains, for ligand binding, for catalytic activity, and for attachment of lipids or carbohydrates. Most transmembrane proteins of the plasma membrane are glycoproteins, with their oligosaccharides (glycans) exposed at the surface of the cell. These proteins in the cell membrane have many different functions. Some are enzymes that catalyze biochemical reactions, and some are involved in the transport of substances across the membrane. Integral proteins, which contain specific domains, one for external ligand binding and another for catalytic activity, can function as receptors. They can bind corresponding ligands to initiate cellular signaling pathways.

4.3.3 Proteins and Lipids Diffuse in the Membrane

Interaction among lipids and between lipids and proteins is very complex and dynamic. There is a degree of fluidity in the lipid portion of membranes, so both proteins and lipids are able to diffuse laterally through the membrane. In some cases, the mobility of membrane proteins is restricted by their association with the cytoskeleton, but in other cases, the mobility of membrane proteins may be restricted by their associations with other membrane proteins, including proteins on the surface of adjacent cells, or with the extracellular matrix.

Proteins do not move across the membrane. Lipid displacement from one side of the membrane to the other (flip-flop) is a very slow process. The plasma membrane is a mechanosensing structure that transmits environmental stimuli and triggers intracellular mechanisms, therefore mediating in the cellular response to mechanical stimuli and the subsequent biochemical responses. Upon applying external stress, cells have multiple ways of supplying lipids to the plasma membrane, buffering the increase of tension, forming curved structures, or supporting lipid reorganization.

4.3.4 Glycocalyx

The extracellular portions of the plasma membrane proteins are generally glycosylated. Likewise, the carbohydrate portions of glycolipids are exposed on the outer surface of the plasma membrane. Due to their presence, the cell interacts with the environment through numerous glycan–protein structures, characterized by the various proportion of glycan and protein components called proteoglycans and basically acidic polysaccharide glycosaminoglycans (GAG). Therefore, the surface of the cell is covered by a carbohydrate multifunctional coat, known as the glycocalyx, formed by the oligosaccharides of glycolipids, transmembrane glycoproteins, and various associated with them, highly hydrophilic protein–carbohydrate complexes. Proteoglycans and glycoproteins are generic structural components of a glycocalyx, but the precise biochemical composition and structure is determined by the specific cell type and the prevailing mechanical and physicochemical conditions (Lahir, 2016).

The glycocalyx is a highly hydrated fibrous meshwork of carbohydrates, which projects out and covers the outside of many eukaryotic and prokaryotic cells, particularly bacteria.

Plant and bacterial cell possess a stiff layer, and this structural property has protective, functional, and supportive nature. For example, the bacterial glycocalyx mediates cell attachment, retains humidity during exposure to dry environments, protects against molecular and cellular antibacterial agents (antibiotics, surfactants, bacteriophages, phagocytes), and is associated with the ability of the bacteria to initiate an infection.

In animal cells, the glycocalyx is a unique carbohydrate-rich boundary, and this layer functions as a barrier between a cell and its surrounding (Tarbell and Cancel, 2016). The glycocalyx also serves as a mediator for cell–cell interactions, regulates the cell's permeability, and protects a cell membrane from the direct action of physical forces and stresses allowing the membrane to maintain its integrity as well as transmits physical forces to the cytoskeleton. Weinbaum et al. (2003) proposed a model in which core proteins in the glycocalyx serve as lever arms that strengthen and transform physical forces and stresses in the glycocalyx to transfer them to the cytoskeleton (Figure 4.3.2).

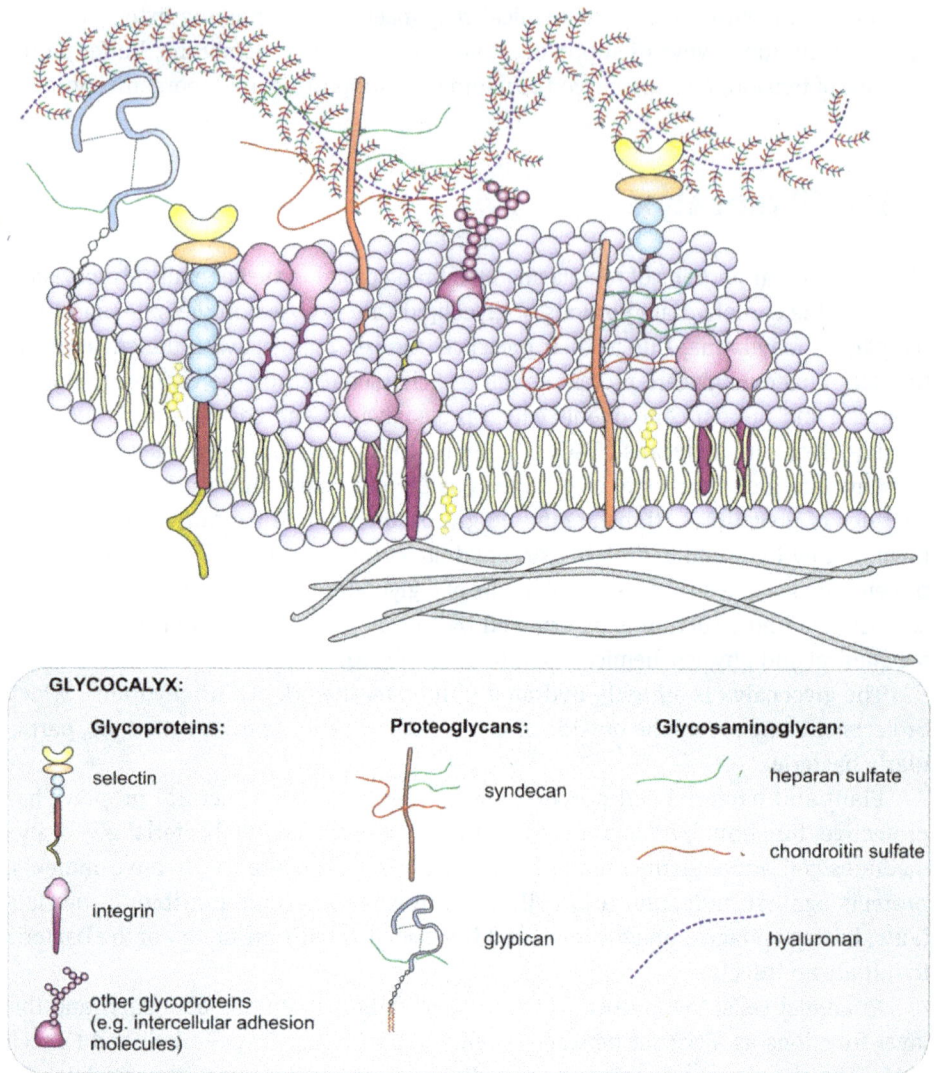

Figure 4.3.2: The localization and interactions of glycocalyx components with membrane counterparts.

4.3.4.1 Structure of Glycocalyx

4.3.4.1.1 Glycoproteins

Backbone molecules connect glycocalyx to the cell membrane. They are glycoconjugates with relatively small (5–12 sugar residues) and branched carbohydrate side chains. The most abundant, functionally important glycoproteins are selectins, integrins, and other adhesion molecules with immunoglobulin domains.

Selectins are transmembrane proteins that recognize specific carbohydrates on the cell surface. There are three subsets of selectins: E – in endothelial cells, P – in platelets and endothelial cells, and L – in granulocytes and monocytes and on most lymphocytes. All three known members of the selectin family (L-, E-, and P-selectin) contain a small cytoplasmic tail, a transmembrane domain, several consensus repeats (2, 6, and 9 for L-, E-, and P-selectin, respectively), epidermal growth factor-like domain, and lectin-like calcium-dependent domain at the N-terminus. Each selectin has a carbohydrate recognition domain that mediates binding to specific glycans on other cells. In a multicellular organism, some oligosaccharides act as antigens. This feature is very common during adhesive interaction important in inflammation and progression of cancer.

Integrins are transmembrane receptors that link the extracellular matrix to the cell. Integrins are heterodimers composed of two noncovalently associated transmembrane glycoprotein subunits called α and β. Both of them have large extracellular domains, single-spanning transmembrane domains, and a short cytoplasmic tail. Integrins function as transmembrane linkers between the extracellular matrix and the actin cytoskeleton and also function as signal transducers, activating various intracellular signaling pathways when activated by matrix components binding (Alberts et al., 2002).

Other glycoproteins forming the glycocalyx are adhesion molecules like intercellular adhesion molecules (endothelial, plate, vascular), glycoproteins acting in coagulation, fibrinolysis, and homeostasis.

4.3.4.1.2 Proteoglycans

The proteoglycan core proteins may be incorporated into the cell membrane by a glycosylphosphatidylinositol anchor (glypicans) or a transmembrane domain that links to the cytoskeleton (syndecans). Proteoglycans are formed by the covalent attachment of a core protein with one or more GAG chains through serine residues (Couchman and Pataki, 2012).

Syndecans are single transmembrane domain proteins capable of carrying three to five heparan sulfate (HS) and chondroitin sulfate (CS) chains. They interact with a large variety of ligands, growth factors including fibroblast growth factors, vascular endothelial growth factor, transforming growth factor-beta, integrins, extracellular proteins such as fibronectin and collagens I, III, V, or thrombospondin, and tenascin. There are four types of syndecans in human cells, namely syndecan-1 to syndecan-4. They are composed of three distinct domains, an N-terminal variable extracellular domain with GAG attachment sites, a single-conserved transmembrane domain, and a short C-terminal cytoplasmic domain with two conserved regions flanking a variable region unique for each syndecans (Couchman and Pataki, 2012). Syndecan ectodomains can be shed from cells and compete for cell surface binding (Tarbell and Cancel, 2016).

Glypicans belong to the family of heparan sulfate proteoglycans, which are linked to the cell surface via glycosylphosphatidylinositol, thus, can be released by phospholipase activity. The modular structure of the glypicans has been highly conserved throughout evolution: N-terminal signal sequence, a likely globular domain containing a characteristic pattern of 14 cysteine residues, a domain with the GAG attachment sites and hydrophobic C-terminal sequence involved in the formation of the anchor. Six glypicans have been identified so far in vertebrates. In general, glypicans are expressed predominantly during development. Their expression levels change in a stage- and tissue-specific manner, suggesting that glypicans are involved in the regulation of morphogenesis (De Cat and David, 2001, Filmus, 2001).

4.3.4.1.3 Glycosaminoglycan

GAGs are long linear, acidic heteropolysaccharides with repeating disaccharide units: D-glucuronic acid, L-iduronic acid, D-galactose or D-N-acetyloglucosamine, D-N-acetylogalactosamine. GAGs differ in length and are modified by sulfation and/or (de)acetylation to a variable extent. GAGs can be divided into the following four major categories: HS /heparin (HP), CS/dermatan sulfate (DS), keratan sulfate (KS), and hyaluronic acid or hyaluronan (HA). Glycocalyx GAGs except HA are covalently linked to core proteoglycans. The point of attachment is a serine (Ser) residue to which the GAG is covalently bound by O-glycosidic bond.

HS is the most abundant in the glycocalyx, accounting for 50–90% of the total GAGs. HS is composed from 50 to 150 unbranched negatively charged disaccharide units consisting of (D-glucuronic or L-iduronic acid) and D-glucosamine or N-acetyl-D-glucosamine. Various degrees of sulfation occur (at the oxygen and/or nitrogen containing groups) on each monosaccharide unit ranging from zero to tri-sulfation. Heparan sulfate is less sulfated than HP. HS facilitates several important biological processes in health and disease, including cell adhesion, regulation of cell growth and proliferation, developmental processes, cell surface binding of lipoprotein lipase and other proteins, angiogenesis, viral invasion, and tumor metastasis (Rabenstein, 2002).

HA is unique among the GAGs; it is an unbranched, non-sulfated GAG that consists of repeating disaccharide units of *N*-acetyl glucosamine and D-glucuronic acid. HA polymers are very large (with molecular weights of 100,000–10,000,000 D). HA is highly hygroscopic, and its aqueous state is highly viscous and elastic. Indeed, the hyaluronans are the largest polysaccharides produced by vertebrate cells. It forms non-covalently linked complexes with proteoglycans in the ECM and it can be bound to the cell membrane by, for example, CD44 antigen, cell-surface glycoprotein involved in cell–cell interactions, cell adhesion, and migration (Tarbell and Cancel, 2016).

CS chains are unbranched polysaccharides of variable length containing two alternating monosaccharides: D-glucuronic acid (GlcA) and *N*-acetyl-D-galactosamine (GalNAc). CS chains are linked to hydroxyl groups on serine residues of certain proteins. CS is an important structural component of cartilage and provides much of its resistance to compression (Baeurle et al., 2009).

DS, also known as CS-B, is composed of linear polysaccharides assembled as disaccharide units containing a hexosamine *N*-acetyl galactosamine (GalNAc) or glucuronic acid (GlcA). DS is defined as a CS by the presence of GalNAc, but GlcA residues are epimerized into L-iduronic acid.

KS is a linear polymer of lactosamine, $3Gal\beta1\text{-}4GlcNac\beta1$, sulfated at the C6 of both hexoses. Cell types that secrete KS are neuronal cells, chondrocytes, and keratinocytes. The class designations are based upon these protein linkage differences. KSI is N-linked to specific asparagine residues via *N*-acetylglucosamine, and KSII is O-linked to specific serine or threonine residues via *N*-acetyl galactosamine. The third type of KS (KSIII) has also been isolated from brain tissue that is O-linked to specific serine or threonine residues via mannose (Funderburgh, 2000).

The glycocalyx covers the cellular structure and is closely associated with cell biology and cellular structure. The disorders in the structure or composition of the glycocalyx play a major role in disease mechanisms. Diseases related to vascular system (hypertension, stroke, atherosclerosis, kidney disease, and sepsis) are associated with a degraded glycocalyx (Weinbaum et al., 2003). Disruption of glycocalyx can lead to the loss of permeability barrier and the decrease in the protection of cells against damage by various biological and chemical factors. By contrast, the glycocalyx on cancer cells is found to be more "robust," promoting a clustering of intergins and increased signaling of growth factor. Also, elevated level of HS and HA is associated with tumor growth and metastasis.

The glycocalyx plays an important role in mechanotransduction. Glycocalyx structures transduce biochemical and mechanical forces into signals leading to cellular responses (Lahir, 2016, Tarbell and Cancel, 2016).

References

Alberts, B., A. Johnson, J. Lewis, M. Raff, K. Roberts and P. Walter (2002). Molecular biology of the cell. New York, Garland Science USA.

Baeurle, S., M. Kiselev, E. Makarova and E. Nogovitsin (2009). "Effect of the counterion behavior on the frictional–compressive properties of chondroitin sulfate solutions." Polymer **50**(7): 1805–1813.

Couchman, J. R. and C. A. Pataki (2012). "An introduction to proteoglycans and their localization." Journal of Histochemistry & Cytochemistry **60**(12): 885–897.

De Cat, B. and G. David (2001). Developmental roles of the glypicans. Seminars in cell & developmental biology. Elsevier.

Filmus, J. (2001). "Glypicans in growth control and cancer." Glycobiology **11**(3): 19R–23R.

Funderburgh, J. L. (2000). "Mini review keratan sulfate: Structure, biosynthesis, and function." Glycobiology **10**(10): 951–958.

Lahir, Y. H. (2016). "Understanding the basic role of glycocalyx during cancer." Journal of Radiation and Cancer Research **7**(3): 79.

Rabenstein, D. L. (2002). "Heparin and heparan sulfate: Structure and function." Natural Product Reports **19**(3): 312–331.

Singer, S. J. and G. L. Nicolson (1972). "The fluid mosaic model of the structure of cell membranes." Science **175**(4023): 720–731.

Tarbell, J. and L. Cancel (2016). "The glycocalyx and its significance in human medicine." Journal of Internal Medicine **280**(1): 97–113.

Weinbaum, S., X. Zhang, Y. Han, H. Vink and S. C. Cowin (2003). "Mechanotransduction and flow across the endothelial glycocalyx." Proceedings of the National Academy of Sciences **100**(13): 7988–7995.

Ellen Zelinsky, Charles T. Anderson

4.4 Functions of Plant Cell Walls in Root Growth and Environmental Interactions

4.4.1 Plant Roots

The roots of plants are less visible than their shoots but are nonetheless just as important. They anchor plants to enable upright growth and allow them to collect vital water and nutrients from the soil. Shoots contain photosynthetic tissues, which generate the energy-rich molecules that drive plant growth and development, but photosynthesis would not be possible without the water, magnesium, iron, sulfur, potassium, and other nutrients that are captured by roots. Roots (see Figure 4.4.1) grow at their tips by the continuous production of new cells from an apical meristem, and these cells elongate and then differentiate into distinct tissue layers. Unlike shoots, which are covered in a waxy cuticle that provides protection from pathogens and excess water loss but also prevents the absorption of water and nutrients, the roots of plants typically lack a cuticle. This feature facilitates the absorption of water and nutrients from the soil but leaves roots open to attacks by herbivores and pathogens and subjects them to physical stresses such as drought. The only barriers between root tissues and the soil are the outer walls of root epidermal cells. These cell walls lend strength to penetrate the soil, provide flexibility for rapid growth, and protect against unwanted stresses.

4.4.2 Plant Cell Walls

Plant cell walls are extracellular matrices made of interacting networks of biopolymers and are categorized into primary walls, which are laid down before and during cell expansion, and secondary walls, which are deposited mainly after growth has ceased. The primary cell wall is composed of three major classes of polysaccharides: cellulose, pectins, and hemicelluloses, plus smaller amounts of structural proteins, enzymes, small molecules, and water. Together, these components act centrally in

Acknowledgment: This work was supported by The Center for Lignocellulose Structure and Formation, an Energy Frontier Research Center funded by the U.S. Department of Energy (DOE), Office of Science, Basic Energy Sciences (BES), under Award # DE-SC0001090.

Ellen Zelinsky, Charles T. Anderson, Department of Biology and Center for Lignocellulose Structure and Formation, The Pennsylvania State University, University Park

https://doi.org/10.1515/9783110989380-004

the coordinated growth of plant cells. The main driving force of growth comes from turgor pressure that accumulates within the plant cell. The internal environment of a plant cell often has a larger negative osmotic potential due to high solute concentrations within the cell. This promotes transport of water into the vacuole of the cell, increasing turgor pressure. Without the presence of the cell wall, the cell would expand in response to this increase in pressure until it reached the maximum volume allowed by the plasma membrane and then burst. To prevent this, the cell wall must have high tensile strength to withstand the turgor pressure generated in the cell. However, if the primary cell wall were merely a rigid cage that stopped cell expansion, then the plant would never be able to grow. This is where the cell wall's second major physical attribute comes in to play. A high-enough turgor pressure within the cell can cause cellulose microfibrils to slip past each other, allowing the wall to yield and the cell to expand irreversibly (Cosgrove, 2018). Not all deformations are irreversible in cell walls. Cross-linking of matrix polysaccharides, pectins and hemicelluloses, are vital to the elasticity of the cell wall, allowing it to expand temporarily and then return to its previous state (Abasolo et al., 2009). This gives plants the flexibility to bend without breaking, a vital trait for organisms that cannot move to take shelter from wind and rain, for example. In primary walls, hemicelluloses (primarily xyloglucan in eudicots or xylans in grasses) and pectins (primarily homogalacturonan (HG)) interact closely with cellulose to maintain the integrity of the cell wall. The points at which cell wall components come together are called biomechanical hotspots, from which the majority of the strength of the wall arises (Cosgrove, 2018).

4.4.3 Cellulose

Cellulose in plants takes the form of microfibrils, which are long, thin structures containing 18–24 hydrogen-bonded chains of β-1,4-linked glucose (Fernandes et al., 2011, Thomas et al., 2013). Cellulose is synthesized by a complex of proteins called cellulose synthases (CESAs). For example, there are ten different genes in *Arabidopsis thaliana* that code for CESAs (Carroll and Specht, 2011). Some *CESA* genes are required for the synthesis of primary cell walls, whereas others are specific to the synthesis of secondary cell walls (Taylor et al., 2003, Desprez et al., 2007, Watanabe et al., 2018). Each CESA protein polymerizes a single glucan chain by forming a repeating unit of two glucose rings with a glycosidic linkage in a beta conformation through the first and fourth carbons of adjacent glucoses. This repeating unit is called cellobiose, and long chains of cellobiose make up the glucan chains of cellulose (Brown Jr, Saxena et al., 1996). CESAs work in conjunction with each other, forming complexes (CESA complexes, CSCs) that appear as six-lobed rosettes. The current model of CSC conformation is as a hexamer of trimers, which would generate 18 glucan chains simultaneously (Nixon et al., 2016). These chains can then interact through van der Waals associations and hydrogen bonds, forming the microfibrils that are a major source of rigidity in the cell wall. Cellulose microfibrils can also bundle together to create even larger fibers in the cell wall, presumably providing extra strength. The synthesis of cellulose appears to take place only at the cell surface. CSCs are integralmembrane complexes and are guided along cortical microtubules, and the physical action of generating cellulose pushes them along the plane of the cell face (Diotallevi and Mulder, 2007, McFarlane et al., 2014). Additional deposition of cellulose reinforces the walls in the direction of cellulose orientation. Cellulose microfibrils have very high tensile strength along the axis of the fibril, but bundles of fibrils can be pulled apart when off-axis forces are applied. Anisotropic cell expansion, or expansion in one direction to a higher degree than another, relies on these physical properties of cellulose microfibrils (Slabaugh et al., 2016). As parallel microfibrils are laid down across the cell face, the turgor pressure in the cell causes the cell to expand in the direction perpendicular to the direction of microfibrils. In this way, plant cells can control their direction of expansion and thus their shape. In cells of the root, where the direction of growth is along the root axis and rapid elongation is important, most of the cellulose is deposited in a transverse orientation (Anderson et al., 2010).

4.4.4 Pectins

Pectins can reversibly form gels in the cell wall and are part of the matrix polysaccharides that surround and interconnect cellulose microfibrils. Pectins may be familiar in everyday use, as they are often used to make jellies and jams. Without the addition of pectins, the jelly would not be able to hold together. In plants, pectins have similar functions, in that they help maintain wall integrity and adhere adjacent cells together. Disrupting the synthesis of pectin can cause major defects in cell–cell adhesion (Daher and Braybrook, 2015). This is important since part of how plant cells coordinate their growth is by maintaining tight physical associations between adjacent cells. Pectins are also thought to help determine wall porosity, which is limited to a few nanometers in many cases (Baron-Epel et al., 1988), and wall hydration status.

There are several types of pectins, the most common of which in primary cell walls is homogalacturonan (HG). All pectins are polymers with backbones containing galacturonic acids (Atmodjo et al., 2013). Additional forms of pectins include rhamnogalacturonan I (RG I) and RG II. These rhamnogalacturonans can contain arabinogalactan and other side chains and are more branched than HG, which is a simple chain of α-1,4-linked galacturonic acids. Pectins are synthesized in the Golgi, like most glycans in the secretory pathway, and delivered to the cell wall via post-Golgi vesicle trafficking. When they are first synthesized, the carboxyl groups of most of the galacturonic acid residues in HG are methyl-esterified. This keeps the chains neutral, so they do not repel each other, and also prevents ionic crosslinking between chains. Proteins called pectin methyl esterases (PMEs) have the ability to remove these methyl ester groups. This action reveals a negatively charged carboxyl group on the galacturonic acid. Calcium ions (Ca^{2+}) present in the cell wall can interact with a galacturonic acid residue from each of two chains, binding them together. This cross-linking of pectin chains makes it so that they cannot move apart from each other easily and is thought to cause the cell wall to become stiffer (Peaucelle et al., 2015, Zhang et al., 2019). Some pectin domains can also be covalently linked to one another, forming a large network of linked pectins throughout the cell wall (Anderson, 2019). Another way pectins can be modified in the cell wall is by pectin/pectate lyases and polygalacturonases (Yang et al., 2018). These are enzymes that can cleave demethyl-esterified HG. Pectin-degrading enzymes often act on pectin that has been randomly demethyl-esterified. Large sections of demethyl-esterified pectins are usually cross-linked with Ca^{2+} and are therefore lessaccessible to degrading enzymes.

4.4.5 Hemicelluloses

Hemicelluloses are a highly varying group of polysaccharides. The exact definition of what constitutes a hemicellulose is also somewhat contested since polysaccharides such as arabinogalactan can be considered to be hemicelluloses despite being part of pectin molecules (Scheller and Ulvskov, 2010). More broadly, some people use "hemicellulose" to refer to any cell wall polysaccharide that is not cellulose or pectin. It has been proposed that only polysaccharides with backbones comprising β-1,4-linked glycans in an equatorial configuration can be considered hemicelluloses (Scheller and Ulvskov, 2010). This includes xyloglucans, xylans, mannans, glucomannans, and β-1-3,1-4-linked glucans.

The functions of hemicelluloses in the cell wall are varied and incompletely understood. The most abundant hemicellulose in the primary cell walls of eudicots is xyloglucan, which was once thought to tether cellulose microfibrils together, preventing them from separating and thus constraining wall expansion (Pauly et al., 1999). This was supported by the observation that auxin-induced elongation of pea shoots also results in an increase in xyloglucan metabolism, as well as other data (Labavitch and Ray, 1974, Park and Cosgrove, 2015). Further work, however, has suggested that xyloglucans may not actually tether microfibrils together since digestion of xyloglucan in the cell wall does not increase expansion (Park and Cosgrove, 2012). It is hypothesized that xyloglucan instead interacts with cellulose at specific biomechanical hotspots that are the targets of expansins, proteins that enhance wall expansion (Wang et al., 2013). In addition, the removal of xyloglucans from the wall has been shown to increase the rate of cellulose tagging by nanogold particles, suggesting that xyloglucan may actually coat cellulose microfibrils (Zheng et al., 2018). Xylans are the most abundant hemicellulose in grass cell walls, but are also prevalent in the secondary walls of eudicots (Rennie and Scheller, 2014). Evidence for a function in secondary walls is seen in mutants affecting xylan function that result in collapsed xylem cells (Yuan et al., 2016). Mannans are much less abundant than either xyloglucan or xylan but are found in both the primary and secondary cell walls of grasses and eudicots (Scheller and Ulvskov, 2010). Mannans have a role in seed storage as evidenced by an embryo lethal mutant in a mannan synthase, but otherwise not much is known about their function (Goubet et al., 2003).

Secondary cell walls contain cellulose and hemicelluloses, but very little pectin. In addition, they often contain lignin, a polyphenolic compound that provides hydrophobicity and mechanical strength to the cell wall. Lignin is the last major component of the cell wall to appear during wall development and polymerizes directly in the apoplast via free radical reactions between monolignols, which can be delivered from the lignifying cell or from neighboring cells (Smith et al., 2013).

4.4.6 Root Growth and Anatomy

Roots are necessary for the uptake of water and nutrients in most plants. For this reason, they need to be very plastic to adjust to varying conditions in the soil or to accommodate for growth of the shoot. In eudicots, root structure consists of a single primary root with lateral roots branching off from it as well as root hairs protruding from the epidermis. Root hairs are important for increasing the surface area of the root to facilitate the uptake of nutrients and water as well as for interacting with microbial symbionts. Depending on the plant species and soil environment, the primary root and lateral roots can be organized in different ways. Some nutrients, such as phosphorous, are more abundant in shallower regions of the soil, whereas water and more soluble nutrients tend to be found deeper in soil, especially in water-limited conditions. Plants that can tolerate soils with low phosphorous content often have large networks of lateral roots close to the soil surface to maximize phosphorus uptake. Drought-tolerant plants, on the other hand, often focus their root growth vertically to explore deeper soil layers and potentially find more water. The ability to control the degree and direction of growth in roots is dependent on the cell wall, which determines the rates at which cells can expand in different regions of the root.

At the tip of the root, both primary and lateral, there is a structure called the root cap. This is a collection of cells which are progressively generated and shed from the root tip by controlled wall degradation (Mravec et al., 2017). The root cap secretes a cuticle that protects the root apical meristem as it moves through the soil and aids in lateral root emergence (Berhin et al., 2019).

The cells of the root are organized radially into distinct cell rows, which make up the different tissue layers of the root. Going inward from the surface, these consist of the epidermis, cortex, endodermis, and stele (Figure 4.4.1). The stele is composed of vascular tissues, xylem and phloem, as well as the pericycle, which is responsible for the initiation of lateral roots (Péret et al., 2009). Each of these tissue layers is initially generated by a group of cells called the root apical meristem located just below the root cap at the tip of the root. Within the meristem, a subset of cells form a quiescent center that is maintained in an indeterminate state and divides slowly. The quiescent center and the cells surrounding it make up the first zone of growth in the root called the meristematic zone. This zone is characterized by undifferentiated cells that arise from the quiescent center and divide rapidly, maintaining a very small cell size. As more cells are generated through these cell divisions, the older cells begin to move into the elongation zone. Here, cell division mostly stops, although some cells may still divide in the younger parts of the elongation zone closer to the root tip. At the same time, elongation drastically increases until the cells reach their final size, many times larger than where they started. This leads into the final growth zone, the maturation zone, in which the majority of root cells reside. At this point, division has ceased, and cells finalize their differentiation into specific cell types. The early maturation zone is most often characterized by the initiation of root hairs although some cells will continue to elongate even after forming root hairs, so the actual definition of this boundary is not always clear. At this stage in development, secondary cell walls may be deposited, and a structure called the Casparian strip forms. Secondary walls begin to be laid down in protoxylem and metaxylem cells in the stele and can adopt helical configurations that allow for continued cell elongation even after they begin to be deposited (Schuetz et al., 2014). The Casparian strip is a specialized layer of cell wall that is deposited in the endodermis and contains suberin and lignin. It restricts the passage of water between the cortex and pericycle to a symplastic route, allowing for the regulated entry and exit of water to and from the stele. Peroxidases have been identified that drive the localized deposition of lignin in the Casparian strip (Naseer et al., 2012).

As the primary regulator of anisotropic growth, the cell wall is vital to maintaining the shape of both the cells and the organs as a whole. The cell wall must be able to expand rapidly in the elongation zone and then cease expansion in the maturation zone. This can be accomplished by a change in cell wall patterning and composition. Part of this is accomplished through passive reorientation of cellulose as the cells progress through the elongation zone (Anderson et al., 2010). In the early elongation zone, cellulose is oriented in a transverse direction, perpendicular to the axis of growth. Due to the expansion of the cell, the same transverse microfibril bundles are reoriented so that their angle relative to the growth axis decreases. This change in angle is hypothesized to increase the mechanical tension of the cell wall in the axial direction, slowing elongation of the cells (Anderson et al., 2010). Other changes in cell wall composition could also take place to cause the slowing of growth toward the maturation zone, such as pectin demethyl-esterification and calcium cross-linking (McCartney et al., 2003).

Auxin has been shown to promote growth in shoots and is involved in the initiation of lateral roots and root hairs. Contradictory to this is its clear inhibition, at physiological concentrations, of growth in the primary root. Auxin-mediated shoot elongation is an action requiring the loosening of the cell wall, and it has been hypothesized that auxin might induce tightening in the cell wall of roots to inhibit growth (Liszkay et al., 2004). This appears not to be the case however because the application of auxin to roots causes a reduction in the size of the elongation zone without an effect on the elemental elongation rate of the root cells (Rahman et al., 2007). This modulation of elongation zone size is likely related to auxin's role in determining developmental patterning in root tissues. Auxin is produced in the quiescent center and moves through polar auxin transporters to different tissues in the root. Gradients of auxin concentration in the root tissues help determine the developmental zones in the root (Di Mambro et al., 2017). Changes in auxin concentration not only cause the transition from the elongation zone to the maturation zone but also the transition from the meristematic zone to the elongation zone. This process involves the same kind of auxin-mediated cell-wall loosening as in shoots. Promotion of wall loosening allows cells to expand, marking the beginning of the elongation zone (Barbez et al., 2017). However, the cell- and tissue-scale effects of auxin and other growth regulatory hormones remain incompletely understood in roots.

4.4.7 Root Cell Walls and Stress

Plants, being mostly sessile organisms, must have ways of coping with changes in their environment, whether good or bad. There are two categories of stresses that plants can experience: abiotic and biotic. Abiotic stress refers to any stress caused by a nonliving part of the environment, including drought, extreme temperature, salinity, and extreme pH. Biotic stress is caused by living organisms and refers mainly to damage by herbivory, pathogens, or parasites.

Among abiotic stresses, drought and soil salinity have a significant impact on the osmotic status of root cells, and since internal turgor pressure is the driving force of cell expansion, stresses that cause a reduction in turgor pressure also damage the plant's ability to grow. In some plant species, drought causes changes in wall composition, potentially modulating the expansibility of the wall and its ability to hold water (Zhu et al., 2007, Leucci et al., 2008). Temperature can be a stress factor at both extreme highs and lows: high temperatures can cause excess water loss via transpiration while the stomata are open, depleting soil water (Wu et al., 2018). The opposite situation is true for low temperatures, where plasma membrane fluidity is altered and reactions slow down. Freezing temperatures are problematic in particular since plant cells are mostly water. If ice crystals form inside or enter cells, their structure can be disrupted, and they will no longer be able to function properly. To avoid damage by low temperatures, cell walls thicken, which slows growth as a result (Takahashi et al., 2019). This is characterized by an increase in pectin content as well as PME activity, both of which lead to a decrease in the cell wall pore size to prevent the spread of ice crystals (Yamada et al., 2002, Sasidharan et al., 2011).

Nutrient toxicity can have subtler effects on root growth and wall composition. While plants require many nutrients to survive, some of these are harmful at high concentrations. Aluminum toxicity is the most common case of this, but iron can also have toxic effects in certain soils. Although iron is essential for processes such as the electron transport chain, too much iron can result in the formation of hydroxyl radicals which are very damaging to the plant. For this reason, plants have mechanisms that limit the uptake of iron and other nutrients (Vigani et al., 2019). Aluminum, on the other hand, has the ability to bind to pectins in the cell wall, replacing Ca^{2+} cross-links and decreasing cell wall extensibility (Schmohl et al., 2000). To alleviate this effect, xyloglucan endotransglucosylase/hydrolases (XTHs) aid in loosening the cell wall to sustain root growth (Osato et al., 2006). Nutrient deficiencies can be another problem for root growth. Boron is important for cell wall structure, as it has been shown to cross-link RG II molecules (O'Neill et al., 2004). Through cross-linking, pectins can form a hydrating gel that helps to protect the root from drought and salinity. This has been shown by studies that found cultivars with increased drought and salinity tolerance also had higher levels of pectins in their walls (Larsen et al., 2011, Tenhaken, 2015). In cases of boron deficiency, ethylene and auxin work in conjunction to inhibit root growth (Růžička et al., 2007, Camacho-Cristóbal et al., 2015).

As mentioned above, plants are anchored in one spot by their roots. They cannot flee from herbivores or pathogens, but this does not mean they are defenseless. Upon encountering a host, pathogens must first contend with the plant cell wall. The same properties that allow the cell wall to withstand internal turgor pressure also act as a barrier to invaders such as nematodes, pathogenic fungi, and parasitic plants. All of these use lytic enzymes to some degree to break down cell wall components. One species of nematode in particular, *Meloidogyne incognita*, has 61 genes encoding cell-wall-degrading enzymes, including cellulases, polygalacturonases, pectate lyases, and arabinanases. In addition, the genome also contains genes encoding expansins, which can be used to loosen interactions between cell wall polymers to allow easier degradation (Abad et al., 2008). Pathogenic fungi also make use of lytic enzymes, although the degree to which they use them depends on the lifestyle of the particular fungus. Necrotrophic fungi survive by destroying the cell wall and taking up nutrients from the remains of the dead plant cells. Unlike most other pathogens, necrotrophic fungi do not need their host to survive and so can use highly damaging methods to break through the cell wall before the plant can activate defense responses (Campion et al., 1997). Biotrophic fungi, on the other hand, infect plants through subtler means and so only do minimal damage to the cell wall so as not to alert the plant to the invasion. Parasitic plants have a similar strategy, and both bio-trophic fungi and parasitic plants often form a specialized structure called a hausto-rium which connects the parasite directly to the vasculature of the host. Formation of

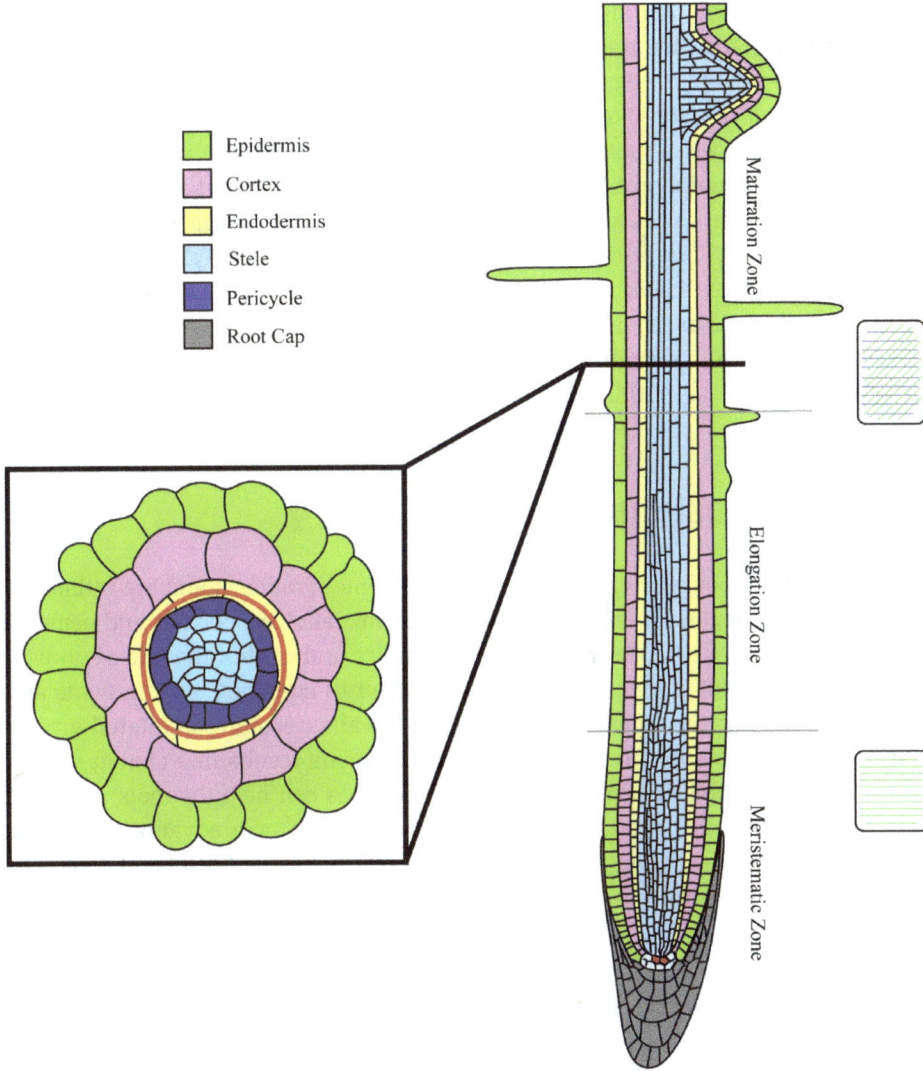

Figure 4.4.1: Anatomy of an *Arabidopsis thaliana* root. Organization of tissue layers and growth zones in a growing root are shown as well as changing cellulose orientations as cells expand (right). Box (left) shows a transverse section from the maturation zone of the root, where secondary cell wall deposition has begun in the vascular cells of the stele and the Casparian strip has formed (red line). Cell cartoons on the right show orientations of cellulose in different zones (adapted from Overvoorde et al., 2010, Somssich et al., 2016).

the haustorium requires the parasite tissue to grow into the host's root, penetrating multiple cell layers until it reaches the vascular tissues (Losner-Goshen et al., 1998, Bleischwitz et al., 2010, Yoshida and Shirasu, 2012).

In addition to being a physical barrier to pathogen invasion, the cell wall has sensing mechanisms that signal to the immune response pathways of the plant. These are triggered when the pathogen's cell-wall-degrading enzymes cause an increase in the amounts of oligosaccharides in the apoplast. For example, it is believed that FERONIA, a receptor-like kinase involved in many cellular processes including mechanical responses in roots (Shih et al., 2014), can bind pectin fragments that are released upon degradation by fungal polygalacturonases (Feng et al., 2018, Voxeur et al., 2019). Pectate lyases and polygalacturonases can only degrade pectins which have been demethyl-esterified and so cell walls with highly methylated pectins will be difficult for a pathogen to degrade. For this reason, many pathogens encode PMEs. Fungal PMEs are more likely to demethyl esterify pectins in a more random manner, conducive to degradation by polygalacturonases and pectate lyases as reviewed in (Lionetti et al., 2012).

4.4.8 Conclusions

The cell walls of root cells change over developmental gradients and are intimately tied to root morphogenesis and tissue differentiation. Cellulose, pectins, and hemicelluloses interact closely to determine wall mechanics and control the rate of growth. The plasticity of root systems allows them to adapt to changing environments and stresses by altering the composition of their cell walls, resulting in altered growth and permeability. Modification of cell wall components results in a change of root growth, either positive or negative. Such modifications can help the roots to defend against environmental stresses, but pathogens can also take advantage of these mechanisms when invading. Responses to such invasions involve a highly complex system of signaling pathways that depend on the integrity of the cell wall.

References

Abad, P., J. Gouzy, J.-M. Aury, P. Castagnone-Sereno, E. G. Danchin, E. Deleury, L. Perfus-Barbeoch, V. Anthouard, F. Artiguenave and V. C. Blok (2008). "Genome sequence of the metazoan plant-parasitic nematode Meloidogyne incognita." Nature Biotechnology **26**(8): 909–915.

Abasolo, W., M. Eder, K. Yamauchi, N. Obel, A. Reinecke, L. Neumetzler, J. W. Dunlop, G. Mouille, M. Pauly and H. Höfte (2009). "Pectin may hinder the unfolding of xyloglucan chains during cell deformation: Implications of the mechanical performance of Arabidopsis hypocotyls with pectin alterations." Molecular Plant **2**(5): 990–999.

Anderson, C. T. (2019). Pectic polysaccharides in plants: Structure, biosynthesis, functions, and applications. Extracellular sugar-based biopolymers matrices. Springer, 487–514.

Anderson, C. T., A. Carroll, L. Akhmetova and C. Somerville (2010). "Real-time imaging of cellulose reorientation during cell wall expansion in Arabidopsis roots." Plant Physiology **152**(2): 787–796.

Atmodjo, M. A., Z. Hao and D. Mohnen (2013). "Evolving views of pectin biosynthesis." Annual Review of Plant Biology **64**.

Barbez, E., K. Dünser, A. Gaidora, T. Lendl and W. Busch (2017). "Auxin steers root cell expansion via apoplastic pH regulation in Arabidopsis thaliana." Proceedings of the National Academy of Sciences **114**(24): E4884–E4893.

Baron-Epel, O., P. K. Gharyal and M. Schindler (1988). "Pectins as mediators of wall porosity in soybean cells." Planta **175**(3): 389–395.

Berhin, A., D. de Bellis, R. B. Franke, R. A. Buono, M. K. Nowack and C. Nawrath (2019). "The root cap cuticle: A cell wall structure for seedling establishment and lateral root formation." Cell **176**(6): 1367–1378.

Bleischwitz, M., M. Albert, H.-L. Fuchsbauer and R. Kaldenhoff (2010). "Significance of cuscutain, a cysteine protease from cuscuta reflexa, in host-parasite interactions." BMC Plant Biology **10**(1): 227.

Brown, R. M. Jr, I. M. Saxena and K. Kudlicka (1996). "Cellulose biosynthesis in higher plants." Trends in Plant Science **1**(5): 149–156.

Camacho-Cristóbal, J. J., E. M. Martín-Rejano, M. B. Herrera-Rodríguez, M. T. Navarro-Gochicoa, J. Rexach and A. González-Fontes (2015). "Boron deficiency inhibits root cell elongation via an ethylene/auxin/ROS-dependent pathway in Arabidopsis seedlings." Journal of Experimental Botany **66**(13): 3831–3840.

Campion, C., P. Massiot and F. Rouxel (1997). "Aggressiveness and production of cell-wall degrading enzymes by Pythium violae, Pythium sulcatum and Pythium ultimum, responsible for cavity spot on carrots." European Journal of Plant Pathology **103**(8): 725–735.

Carroll, A. and C. D. Specht (2011). "Understanding plant cellulose synthases through a comprehensive investigation of the cellulose synthase family sequences." Frontiers in Plant Science **2**: 5.

Cosgrove, D. J. (2018). "Diffuse growth of plant cell walls." Plant Physiology **176**(1): 16–27.

Daher, F. B. and S. A. Braybrook (2015). "How to let go: Pectin and plant cell adhesion." Frontiers in Plant Science **6**: 523.

Desprez, T., M. Juraniec, E. F. Crowell, H. Jouy, Z. Pochylova, F. Parcy, H. Höfte, M. Gonneau and S. Vernhettes (2007). "Organization of cellulose synthase complexes involved in primary cell wall synthesis in Arabidopsis thaliana." Proceedings of the National Academy of Sciences **104**(39): 15572–15577.

Di Mambro, R., M. De Ruvo, E. Pacifici, E. Salvi, R. Sozzani, P. N. Benfey, W. Busch, O. Novak, K. Ljung and L. Di Paola (2017). "Auxin minimum triggers the developmental switch from cell division to cell differentiation in the Arabidopsis root." Proceedings of the National Academy of Sciences **114**(36): E7641–E7649.

Diotallevi, F. and B. Mulder (2007). "The cellulose synthase complex: A polymerization driven supramolecular motor." Biophysical Journal **92**(8): 2666–2673.

Feng, W., D. Kita, A. Peaucelle, H. N. Cartwright, V. Doan, Q. Duan, M.-C. Liu, J. Maman, L. Steinhorst and I. Schmitz-Thom (2018). "The FERONIA receptor kinase maintains cell-wall integrity during salt stress through Ca2+ signaling." Current Biology **28**(5): 666–675.

Fernandes, A. N., L. H. Thomas, C. M. Altaner, P. Callow, V. T. Forsyth, D. C. Apperley, C. J. Kennedy and M. C. Jarvis (2011). "Nanostructure of cellulose microfibrils in spruce wood." Proceedings of the National Academy of Sciences **108**(47): E1195–E1203.

Goubet, F., A. Misrahi, S. K. Park, Z. Zhang, D. Twell and P. Dupree (2003). "AtCSLA7, a cellulose synthase-like putative glycosyltransferase, is important for pollen tube growth and embryogenesis in Arabidopsis." Plant Physiology **131**(2): 547–557.

Labavitch, J. M. and P. M. Ray (1974). "Relationship between promotion of xyloglucan metabolism and induction of elongation by indoleacetic acid." Plant Physiology **54**(4): 499–502.

Larsen, F. H., I. Byg, I. Damager, J. Diaz, S. B. Engelsen and P. Ulvskov (2011). "Residue specific hydration of primary cell wall potato pectin identified by solid-state 13C single-pulse MAS and CP/MAS NMR spectroscopy." Biomacromolecules **12**(5): 1844–1850.

Leucci, M. R., M. S. Lenucci, G. Piro and G. Dalessandro (2008). "Water stress and cell wall polysaccharides in the apical root zone of wheat cultivars varying in drought tolerance." Journal of Plant Physiology **165**(11): 1168–1180.

Lionetti, V., F. Cervone and D. Bellincampi (2012). "Methyl esterification of pectin plays a role during plant–pathogen interactions and affects plant resistance to diseases." Journal of Plant Physiology **169**(16): 1623–1630.

Liszkay, A., E. van der Zalm and P. Schopfer (2004). "Production of reactive oxygen intermediates (O2–, H2O2, and OH) by maize roots and their role in wall loosening and elongation growth." Plant Physiology **136**(2): 3114–3123.

Losner-Goshen, D., V. H. Portnoy, A. M. Mayer and D. M. Joel (1998). "Pectolytic activity by the haustorium of the parasitic plant Orobanche L. (Orobanchaceae) in host roots." Annals of Botany **81**(2): 319–326.

McCartney, L., C. G. Steele-King, E. Jordan and J. P. Knox (2003). "Cell wall pectic (1→ 4)-β-d-galactan marks the acceleration of cell elongation in the Arabidopsis seedling root meristem." The Plant Journal **33**(3): 447–454.

McFarlane, H. E., A. Döring and S. Persson (2014). "The cell biology of cellulose synthesis." Annual Review of Plant Biology **65**: 69–94.

Mravec, J., X. Guo, A. R. Hansen, J. Schückel, S. K. Kračun, M. D. Mikkelsen, M. Mouille, I. E. Johansen, P. Ulvskov and D. S. Domozych (2017). "Pea border cell maturation and release involve complex cell wall structural dynamics." Plant Physiology **174**(2): 1051–1066.

Naseer, S., Y. Lee, C. Lapierre, R. Franke, C. Nawrath and N. Geldner (2012). "Casparian strip diffusion barrier in Arabidopsis is made of a lignin polymer without suberin." Proceedings of the National Academy of Sciences **109**(25): 10101–10106.

Nixon, B. T., K. Mansouri, A. Singh, J. Du, J. K. Davis, J.-G. Lee, E. Slabaugh, V. G. Vandavasi, H. O'Neill and E. M. Roberts (2016). "Comparative structural and computational analysis supports eighteen cellulose synthases in the plant cellulose synthesis complex." Scientific Reports **6**(1): 1–14.

O'Neill, M. A., T. Ishii, P. Albersheim and A. G. Darvill (2004). "Rhamnogalacturonan II: Structure and function of a borate cross-linked cell wall pectic polysaccharide." Annual Review of Plant Biology **55**: 109–139.

Osato, Y., R. Yokoyama and K. Nishitani (2006). "A principal role for AtXTH18 in Arabidopsis thaliana root growth: A functional analysis using RNAi plants." Journal of Plant Research **119**(2): 153–162.

Overvoorde, P., H. Fukaki and T. Beeckman (2010). "Auxin control of root development." Cold Spring Harbor Perspectives in Biology **2**(6): a001537.

Park, Y. B. and D. J. Cosgrove (2012). "A revised architecture of primary cell walls based on biomechanical changes induced by substrate-specific endoglucanases." Plant Physiology **158**(4): 1933–1943.

Park, Y. B. and D. J. Cosgrove (2015). "Xyloglucan and its interactions with other components of the growing cell wall." Plant and Cell Physiology **56**(2): 180–194.

Pauly, M., P. Albersheim, A. Darvill and W. S. York (1999). "Molecular domains of the cellulose/ xyloglucan network in the cell walls of higher plants." The Plant Journal **20**(6): 629–639.

Peaucelle, A., R. Wightman and H. Höfte (2015). "The control of growth symmetry breaking in the Arabidopsis hypocotyl." Current Biology **25**(13): 1746–1752.

Péret, B., A. Larrieu and M. J. Bennett (2009). "Lateral root emergence: A difficult birth." Journal of Experimental Botany **60**(13): 3637–3643.

Rahman, A., A. Bannigan, W. Sulaman, P. Pechter, E. B. Blancaflor and T. I. Baskin (2007). "Auxin, actin and growth of the Arabidopsis thaliana primary root." The Plant Journal **50**(3): 514–528.

Rennie, E. A. and H. V. Scheller (2014). "Xylan biosynthesis." Current Opinion in Biotechnology **26**: 100–107.

Růžička, K., K. Ljung, S. Vanneste, R. Podhorská, T. Beeckman, J. Friml and E. Benková (2007). "Ethylene regulates root growth through effects on auxin biosynthesis and transport-dependent auxin distribution." The Plant Cell **19**(7): 2197–2212.

Sasidharan, R., L. A. Voesenek and R. Pierik (2011). "Cell wall modifying proteins mediate plant acclimatization to biotic and abiotic stresses." Critical Reviews in Plant Sciences **30**(6): 548–562.

Scheller, H. and P. Ulvskov (2010). "Hemicelluloses." Annu Rev Plant Biol **61**: 263–289.

Schmohl, N., J. Pilling, J. Fisahn and W. J. Horst (2000). "Pectin methylesterase modulates aluminium sensitivity in Zea mays and Solanum tuberosum." Physiologia Plantarum **109**(4): 419–427.

Schuetz, M., A. Benske, R. A. Smith, Y. Watanabe, Y. Tobimatsu, J. Ralph, T. Demura, B. Ellis and A. L. Samuels (2014). "Laccases direct lignification in the discrete secondary cell wall domains of protoxylem." Plant Physiology **166**(2): 798–807.

Shih, H.-W., N. D. Miller, C. Dai, E. P. Spalding and G. B. Monshausen (2014). "The receptor-like kinase FERONIA is required for mechanical signal transduction in Arabidopsis seedlings." Current Biology **24**(16): 1887–1892.

Slabaugh, E., T. Scavuzzo-Duggan, A. Chaves, L. Wilson, C. Wilson, J. K. Davis, D. J. Cosgrove, C. T. Anderson, A. W. Roberts and C. H. Haigler (2016). "The valine and lysine residues in the conserved FxVTxK motif are important for the function of phylogenetically distant plant cellulose synthases." Glycobiology **26**(5): 509–519.

Smith, R. A., M. Schuetz, M. Roach, S. D. Mansfield, B. Ellis and L. Samuels (2013). "Neighboring parenchyma cells contribute to Arabidopsis xylem lignification, while lignification of interfascicular fibers is cell autonomous." The Plant Cell **25**(10): 3988–3999.

Somssich, M., G. A. Khan and S. Persson (2016). "Cell Wall heterogeneity in root development of Arabidopsis." Frontiers in Plant Science **7**: 1242.

Takahashi, D., M. Gorka, A. Erban, A. Graf, J. Kopka, E. Zuther and D. K. Hincha (2019). "Both cold and sub-zero acclimation induce cell wall modification and changes in the extracellular proteome in Arabidopsis thaliana." Scientific Reports **9**(1): 1–15.

Taylor, N. G., R. M. Howells, A. K. Huttly, K. Vickers and S. R. Turner (2003). "Interactions among three distinct CesA proteins essential for cellulose synthesis." Proceedings of the National Academy of Sciences **100**(3): 1450–1455.

Tenhaken, R. (2015). "Cell wall remodeling under abiotic stress." Frontiers in Plant Science **5**: 771.

Thomas, L. H., V. T. Forsyth, A. Šturcová, C. J. Kennedy, R. P. May, C. M. Altaner, D. C. Apperley, T. J. Wess and M. C. Jarvis (2013). "Structure of cellulose microfibrils in primary cell walls from collenchyma." Plant Physiology **161**(1): 465–476.

Vigani, G., Á. Solti, S. Thomine and K. Philippar (2019). "Essential and detrimental – an update on intracellular iron trafficking and homeostasis." Plant and Cell Physiology **60**(7): 1420–1439.

Voxeur, A., O. Habrylo, S. Guénin, F. Miart, M.-C. Soulié, C. Rihouey, C. Pau-Roblot, J.-M. Domon, L. Gutierrez and J. Pelloux (2019). "Oligogalacturonide production upon Arabidopsis thaliana–Botrytis cinerea interaction." Proceedings of the National Academy of Sciences **116**(39): 19743–19752.

Wang, T., Y. B. Park, M. A. Caporini, M. Rosay, L. Zhong, D. J. Cosgrove and M. Hong (2013). "Sensitivity-enhanced solid-state NMR detection of expansin's target in plant cell walls." Proceedings of the National Academy of Sciences **110**(41): 16444–16449.

Watanabe, Y., R. Schneider, S. Barkwill, E. Gonzales-Vigil, J. L. Hill, A. L. Samuels, S. Persson and S. D. Mansfield (2018). "Cellulose synthase complexes display distinct dynamic behaviors during xylem transdifferentiation." Proceedings of the National Academy of Sciences **115**(27): E6366–E6374.

Wu, H.-C., V. P. Bulgakov and T.-L. Jinn (2018). "Pectin methylesterases: Cell wall remodeling proteins are required for plant response to heat stress." Frontiers in Plant Science **9**: 1612.

Yamada, T., K. Kuroda, Y. Jitsuyama, D. Takezawa, K. Arakawa and S. Fujikawa (2002). "Roles of the plasma membrane and the cell wall in the responses of plant cells to freezing." Planta **215**(5): 770–778.

Yang, Y., Y. Yu, Y. Liang, C. T. Anderson and J. Cao (2018). "A profusion of molecular scissors for pectins: Classification, expression, and functions of plant polygalacturonases." Frontiers in Plant Science **9**: 1208.

Yoshida, S. and K. Shirasu (2012). "Plants that attack plants: Molecular elucidation of plant parasitism." Current Opinion in Plant Biology **15**(6): 708–713.

Yuan, Y., Q. Teng, R. Zhong, M. Haghighat, E. A. Richardson and Z.-H. Ye (2016). "Mutations of Arabidopsis TBL32 and TBL33 affect xylan acetylation and secondary wall deposition." PLoS One **11**(1).

Zhang, T., H. Tang, D. Vavylonis and D. J. Cosgrove (2019). "Disentangling loosening from softening: Insights into primary cell wall structure." The Plant Journal **100**(6): 1101–1117.

Zheng, Y., X. Wang, Y. Chen, E. Wagner and D. J. Cosgrove (2018). "Xyloglucan in the primary cell wall: Assessment by FESEM, selective enzyme digestions and nanogold affinity tags." The Plant Journal **93**(2): 211–226.

Zhu, J., S. Alvarez, E. L. Marsh, M. E. LeNoble, I.-J. Cho, M. Sivaguru, S. Chen, H. T. Nguyen, Y. Wu and D. P. Schachtman (2007). "Cell wall proteome in the maize primary root elongation zone. II. Region-specific changes in water soluble and lightly ionically bound proteins under water deficit." Plant Physiology **145**(4): 1533–1548.

Dorota Gil, Marta Zarzycka, Piotr Laidler

4.5 Extracellular Matrix

The extracellular matrix (ECM) is an essential part of the microenvironment in which cells live and function. ECM is a non-cellular constituent present in all tissues and organs composed of macromolecules that are synthesized and secreted by cells of a tissue and are arranged into an organized network. The animal ECM includes the interstitial matrix and the basal membrane. The interstitial matrix is present in various animal cells (i.e., in the intercellular spaces). A gelatinous mixture of polysaccharides and fibrous proteins fills the interstitial space. The basement membrane, laying at the same side of all epithelial cells in the layer, separates epithelial cells from connective tissue (Theocharis et al., 2016).

ECM is specialized to perform different functions in different tissues; thus, the composition and structure of ECM vary. For example, ECM adds strength to tendons and attachment in the skin or is involved in filtration in the kidney. However, cell adhesion and cell–cell communication are common functions of the ECM. Generally, ECM is composed of water, space-filling molecules – proteoglycans and fibrous proteins including collagen, elastin, and adhesive proteins including fibronectin and laminin, but the physical nature of ECM varies from tissue to tissue (Figure 4.5.1). ECM provides not only essential physical scaffolds but also regulates many cellular processes like growth, migration, differentiation, survival, homeostasis, and morphogenesis by initiating crucial biomechanical and biochemical pathways. Moreover, the ECM is a highly dynamic structure that constantly undergoes either enzymatic or non-enzymatic remodeling (Frantz et al., 2010).

4.5.1 Composition of the ECM

Collagens are the most abundant protein in the human body and also are the most abundant protein in the ECM. Resistant to shearing forces, these fibrous proteins, which play not only a structural function but also regulate cell adhesion, support chemotaxis and migration. Collagen is synthesized and secreted into ECMs mainly by fibroblasts as well as by endothelial and epithelial cells. It is mostly present in tendons, ligaments, bone, and skin. In bone, collagen fibers are oriented at an angle to other collagen fibers to provide resistance to mechanical shear stress applied from any direction. Bundled collagen in tendons gives strength. In the basement membrane, collagen is dispersed as a gel-like substance and provides support and strength. Thirty

Dorota Gil, Marta Zarzycka, Piotr Laidler, Chair of Medical Biochemistry, Jagiellonian University Medical College, Kraków, Poland

https://doi.org/10.1515/9783110989380-005

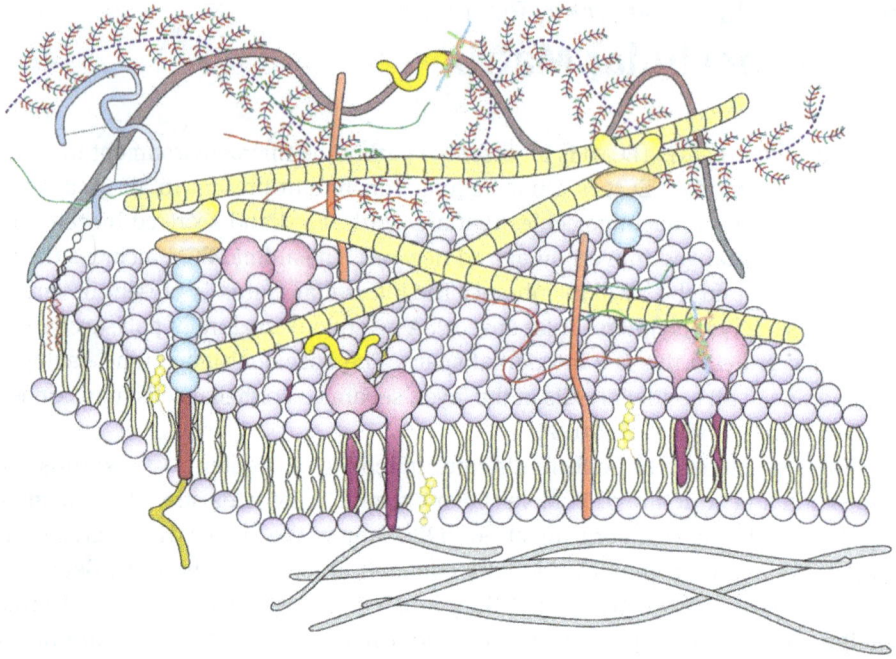

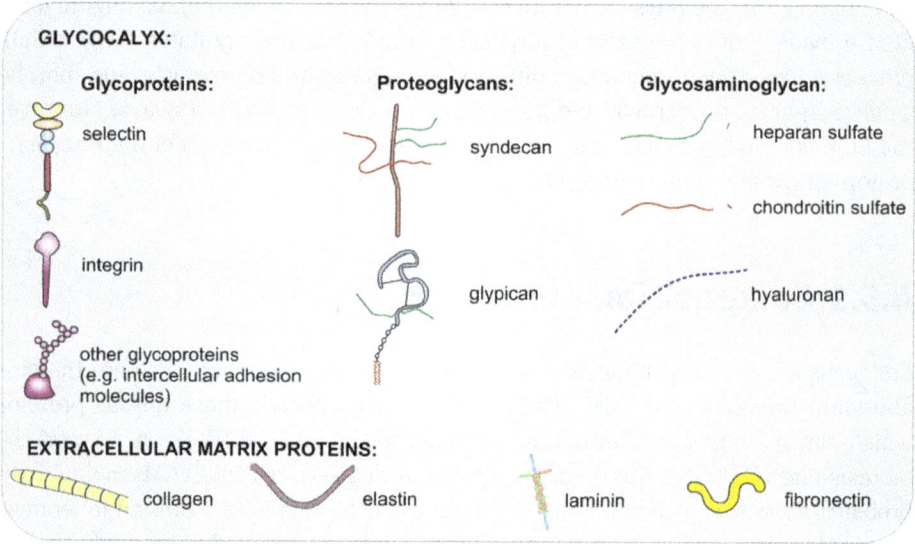

Figure 4.5.1: Extracellular matrix components and their interaction with membrane.

types of collagens have been identified, described, and classified into several groups according to the structure they form. However, over 90% of collagen in the human body is collagen types I, II, III, and IV. Types I, II, and III are fibrillary collagens whose linear polymers form fibrils but type IV is a network-forming collagen that becomes a three-dimensional mesh rather than distinct fibrils. Collagen molecules are composed of three helical α-type chains of amino acids which differ slightly in their chemical composition (α1 or α2) and bind around one another forming a right-handed triple helix called tropocollagen. In each α-chain of tropocollagen, the amino acid sequence is glycine-proline-X and glycine-X-hydroxyproline, where X is any amino acid other than glycine, proline, or hydroxyproline. α-Chains have tendency to form left-handed helices spontaneously due to the presence of high content of proline, 4-hydroxyproline, and glycine and their regular repetitive arrangement. The biosynthesis of collagen is a complex multistep process including a few intermediate forms as procollagen molecules that undergo posttranslational modifications before they reach the final form of tropocollagen. Hydroxylation of proline and lysine is possible in the presence of hydroxylases which require ferrous ions, α-ketoglutarate, O_2, and vitamin C. Hydroxyproline is essential for the formation of intramolecular hydrogen bonds and contributes to the stability of the triple helical conformation. Hydroxylysine residues and lysine are able to form stable intermolecular cross-links between tropocollagen molecules in fibrils after oxidation by copper-dependent lysyl oxidase. These intermolecular cross-links are responsible for the physical and mechanical properties of collagen fibrils. Some of the hydroxylysines are modified by glycosylation. Glucose and galactose residues mediate the interaction with proteoglycans (Marastoni et al., 2008, Frantz et al., 2010, Theocharis et al., 2016). Collagen biosynthesis and structure are markedly modified during aging and are responsible for ECM remodeling in several pathological processes including tumorigenesis. ECM stiffening, caused by increased collagen deposition and cross-linking, disrupts tissue morphogenesis and promotes tumor progression. Paradoxically, in aging tissue, collagen production slows down, but collagen fibers are frequently inappropriately cross-linked through glycation caused by free radicals. Additionally, over time, collagen fibers become rigid. The combination of elevated or inappropriate collagen cross-linking contributes to tissue stiffening; therefore, an aged tissue is mechanically weaker and less elastic but also more rigid than a young one (Kular et al., 2014).

Elastin is another major fibrous protein of ECM and is responsible for the elasticity of tissues. It is found mainly in tissues and organs which undergo repeating stretching forces like skin arteries, lungs, heart, bladder, and elastic cartilage. Elastin gives them the possibility to stretch without unwanted tearing. Elastin is rich in hydrophobic amino acids such as glycine, alanine, and proline and also contains lysine and hydroxyproline only in small amounts, but it is not a glycoprotein. Although elastin has no regular secondary structure, it is classified as a fibrous protein because it contains an unordered coiled structure and is relatively insoluble in water. Elastin is made by linking together several small soluble precursor tropoelastin molecules to make the insoluble polymer. Tropoelastin has a unique composition and tendency to self-associate. The most common interchain cross-link in elastins is the result of the conversion of the amine groups of lysine to reactive aldehydes by lysyl oxidase. This results in the spontaneous formation of desmosine cross-links. Elastin is synthesized and secreted by fibroblasts, smooth muscle cells, chondrocytes, or endothelial cells. Elastin is usually associated with other proteins in elastic fibers. Elastic fibers include scaffold proteins such as elastin or fibrillin which play a structural role and other associated proteins which perform a regulatory role in bridging molecules during the formation of elastic fibers. Fibrillin is a glycoprotein, which contains calcium-binding EGF-like domains, integrin-binding Arg–Gly–Asp (RGD) sequences, and heprin-binding domains. These components suggest that fibrillins are required for binding a specific cell surface receptor and play a crucial role in cell signaling. Other associated proteins are microfibril-associated glycoproteins (MAGPs) which bind both tropoelastin and fibrillins and probably play an important role in linking them during elastic fiber formation. Fibulins, calcium-binding glycoproteins, also participate in the formation of elastic fibers. They interact with every other elastic fiber component (tropoelastin, fibrillin) and with numerous ECM components including fibronectin, proteoglycans, or laminins (Kielty et al., 2002).

Elastin is a very long-living protein, very stable with low or missing turnover. Elastin degradation delivers elastin peptide fragments with significant biological activity which are especially seen in organs with abundance of elastin. For example, elastic peptide fragments, which are chemotactic for monocytes and fibroblasts, contribute to the development and formation of plaque in atherosclerosis. In adults, damaged elastic fibers are often repaired improperly, and in such a case, the integrity of the elastic network is destroyed (Kular et al., 2014).

Fibronectin (FN) is a high-molecular weight glycoprotein of the ECM that contains 4–5% of N-linked or O-linked carbohydrates. FN is produced by a variety of cell types with critical function in development, blood clotting, cell migration/adhesion, and wound healing. FN consists of two nearly identical monomers linked by a pair of disulfide bonds. FN is produced by one gene, but alternative splicing leads to the formation of multiple variants. Based on its solubility, FN is divided into two types: soluble plasma fibronectin, a major protein component of blood plasma secreted by hepatocytes, and insoluble cellular molecules as major component of the ECM. The latter is secreted by various cells, primarily fibroblasts, as a soluble protein dimer and is then assembled into supramolecular fibers. FN contains three different binding domains: collagen-binding domain, cell-binding domain, and proteoglycan-binding domain. The characteristic feature of FN is the peptide sequence RGD present in all binding domains. These sequences allow FN to recognize and bind to cells and other components of ECM. FN interacts mainly with two membrane receptors: integrins or syndecans. Sometimes syndecans are co-receptors with integrins in cell-FN binding and enhance integrin-mediated cell migration and intracellular signaling pathway. FN promotes cell cycle progression and proliferation and is also involved in cell survival and protection from apoptosis of many cell types, including epithelial and endothelial cells and leukocytes (Pankova et al., 2019).

Laminins (LN) are large cross-shaped glycoproteins, with highly conserved structure of the long arm terminated with a large globule involved in cellular interaction, and short arms, each one consisting of rod-like arrays and globular domains. They are made up of three different chains: α, β, and γ. In humans, 11 genes code subunits (5α, 3β, and 3γ), and they are located on chromosomes 1, 3, 6, 7, 9, 18, and 20. Laminin molecules can bind with each other and also interact with another ECM protein by short arms, which allow them to form sheets found in the basement membranes. They can also interact with other components of ECM and cells. Laminins form independent networks and are associated with type IV collagen networks, fibronectin, and they also bind to cell membranes through integrin receptors and other plasma membrane molecules (Aumailley, 2013). A fibrillary laminin matrix is ideally suited to transmit mechanosignals in the form of stretch, and there is evidence that this occurs in lung cells (Jones et al., 2005). The possibility of such interactions and the fact that the distribution of LN isoforms is tissue-specific suggest that laminin is vital for the maintenance and survival of tissues. They contribute to cell attachment and differentiation, cell shape and movement, maintenance of tissue phenotype, and promote tissues survival. LN plays a crucial role in early embryonic development and organogenesis. Similar to fibronectin, LN promotes tumor growth and metastasis (Aumailley, 2013).

ECM is a dynamic, complex structure which undergoes controlled remodeling normally by several matrix proteases. The most important proteases responsible for the catabolism of almost all ECM molecules are metalloproteases. Matrix metalloproteinases (MMPs), are calcium-dependent zinc-containing endopeptidases, which are also known to be involved in the cleavage of cell surface receptors. The MMPs have a common domain structure and are synthesized as inactive zymogens. MMPs are classified into six groups according to substrate specificity, sequence similarity, and domain organization. The breakdown of ECM protein resulting in the formation of a favorable microenvironment for tissue remodeling is associated with various physiological or pathological processes such as morphogenesis, angiogenesis, tissue repair, cirrhosis, arthritis, and metastasis. Tissue homeostasis is maintained by a dynamic dialogue between cells and surrounding ECM. ECM differently modulates cell growth and migration by binding cell receptors and cells in a controlled feedback by the production of ECM proteins and MMPs that regulate ECM structure and composition. Steady-state perturbations occur in pathological conditions, for instance during cancer progression. ECM transmits its signals to the cells through specific receptors. Integrins especially bind to RGD sequences presented in proteins (such as FN, laminin, and vitronectin), and their binding specificity is promoted by residues close to RGD motif. Integrins by linking to the actin cytoskeleton function as mechanotransducers transforming mechanical forces created by ECM proteins such as fibronectin into chemical signals. The syndecan family is another core of proteins, which facilitate interaction of integrins with ECM proteins. In response, intracellular signals are transmitted to outside the cell by influencing integrin affinity for ECM or expression of integrin and synthesis and deposition of ECM proteins. Therefore, integrins have the possibility to regulate many cellular processes such as proliferation, survival, migration, and invasion that can be misregulated in response to changes in the ECM composition. ECM is particularly modified in all pathologies especially in cancer (Marastoni et al., 2008). Growing evidence proves that a healthy microenvironment prevents the cancerous outgrowth of epithelial cells, whereas age-related modification of ECM enables initiation and progression of malignancy. One implication is the increase in stiffness with age. The geometry, rigidity, and other physical properties of the ECM are sensed by the cells and ultimately direct their responses. It is known that matrix stiffness influences stem-cell lineage. Increased ECM stiffness activates integrins and promotes focal adhesion formation and cell motility.

There are several techniques to study ECM architecture in 3D such as atomic force microscopy, confocal microscopy, transmission electron microscopy/scanning electron microscopy, or second-harmonic generation. These techniques could similarly be used to study how the mechanical forces, elasticity of ECM, and its architecture regulate cell behavior (Kular et al., 2014).

References

Aumailley, M. (2013). "The laminin family." Cell Adhesion & Migration 7(1): 48–55.

Frantz, C., K. M. Stewart and V. M. Weaver (2010). "The extracellular matrix at a glance." Journal of Cell Science **123**(24): 4195–4200.

Jones, J. C., K. Lane, S. B. Hopkinson, E. Lecuona, R. C. Geiger, D. A. Dean, E. Correa-Meyer, M. Gonzales, K. Campbell and J. I. Sznajder (2005). "Laminin-6 assembles into multimolecular fibrillar complexes with perlecan and participates in mechanical-signal transduction via a dystroglycan-dependent, integrin-independent mechanism." Journal of Cell Science **118**(12): 2557–2566.

Kielty, C. M., M. J. Sherratt and C. A. Shuttleworth (2002). "Elastic fibres." Journal of Cell Science **115**(14): 2817–2828.

Kular, J. K., S. Basu and R. I. Sharma (2014). "The extracellular matrix: Structure, composition, age-related differences, tools for analysis and applications for tissue engineering." Journal of Tissue Engineering **5**: 2041731414557112.

Marastoni, S., G. Ligresti, E. Lorenzon, A. Colombatti and M. Mongiat (2008). "Extracellular matrix: A matter of life and death." Connective Tissue Research **49**(3–4): 203–206.

Pankova, D., Y. Jiang, M. Chatzifrangkeskou, I. Vendrell, J. Buzzelli, A. Ryan, C. Brown and E. O'Neill (2019). "RASSF1A controls tissue stiffness and cancer stem-like cells in lung adenocarcinoma." The EMBO Journal **38**(13).

Theocharis, A. D., S. S. Skandalis, C. Gialeli and N. K. Karamanos (2016). "Extracellular matrix structure." Advanced Drug Delivery Reviews **97**: 4–27.

Carsten Schulte

4.6 Mechanotransduction

Mechanotransduction refers to the intricate processes and signaling pathways that connect the biophysical cues of the cellular microenvironment with the cell and mediate their reciprocal crosstalk. The (extra)cellular structures described in this chapter, that is, the extracellular matrix (ECM) (see Chapter 4.5), the cell membrane (with the embedded adhesion receptors) and glycocalyx (see Chapter 4.3), and the cytoskeleton (see Chapter 4.2), strongly interact with and influence one another through mechanotransductive actions. The impact of mechanotransduction is not limited to only direct modulations of the cell morphology and mechanics affected by the cytoskeleton but also reaches into the nucleus, where it eventually also regulates cellular decision making, program, and fate. Here, we will give an overview about how the cell senses mechanical and structural cues of its microenvironment and how the events in the cell/microenvironment interface translate the perceived information into appropriate cellular responses by the mechanotransductive machinery and signaling. Also, we will introduce the contribution of aberrations in mechanotransduction and the (extra)cellular structures involved in diseases.

4.6.1 Mechanosensing of Biophysical Cues in the Cell/Microenvironment Interface by Integrin Adhesion Complexes

As outlined in detail in Chapter 4.5, the ECM is built up by self-assembling nanometric building blocks secreted by the cells themselves, often macromolecules consisting of protein and sugar components (called proteoglycans or glycoproteins, depending on the relation between these two components). These building blocks are furthermore highly interlinked with each other, forming a combination of fibrillar (e.g., collagen type I or fibronectin) or meshwork (e.g., collagen type IV, laminins) structures that determine the mechanical and topographical properties of the ECM. The rigidity and structural configuration of ECM of different tissues vary substantially, from the very soft brain matrix with a high content of glycosaminoglycans and water to the very

Acknowledgment: C.S. acknowledges the support of the European Union's Horizon 2020 research and innovation program under the FET Open grant agreement no. 801126, project EDIT.

Carsten Schulte, Interdisciplinary Centre for Nanostructured Materials and Interfaces (C.I.Ma.I.Na.), Università degli Studi di Milano, Milan, Italy

https://doi.org/10.1515/9783110989380-006

rigid, mineralized bone matrix constituted predominantly by collagen type I fibers (Gasiorowski et al., 2013, Young et al., 2016, Leclech et al., 2020, Chighizola et al., 2019).

These mechanical and structural/topographical features are critical biophysical parameters that the cell can perceive and interpret by mechanosensing in the cell/microenvironment interface and subsequent mechanotransductive processes, which will be described in the following. The cellular capacity of microenvironmental mechanosensing is primarily realized by the principal adhesive structures of the cell, the integrin adhesion complexes (IAC) (Gauthier and Roca-Cusachs, 2018, Sun et al., 2019, Kechagia et al., 2019) (Figure 4.6.1). The name for the eponymous integrins was initially chosen because, in 1986, integrins were identified as "integral membrane complex involved in the transmembrane association between the extracellular matrix and the cytoskeleton" (Tamkun et al., 1986). Integrins are heterodimeric receptors that are embedded in the cell membrane, assembling noncovalently in one of 24 possible combinations of always one α- (18 types) and one β-subunit (eight types). Depending on the subunit combination, their extracellular domain can recognize and bind to ligands present in ECM proteins (two prominent examples are the RGD and LDV motifs). The intracellular tails are usually very short (except for the β4-subunit) and can be bound by different IAC proteins under appropriate circumstances, which will be detailed in the following (Gauthier and Roca-Cusachs, 2018, Sun et al., 2019, Kechagia et al., 2019, Chighizola et al., 2019) (Figure 4.6.1A).

How accurate and well-chosen the name "integrin" actually was, became manifest in the decades after its discovery; especially in recent years, it has been disclosed that a major role of IAC is the integration of information coming from microenvironmental biophysical cues into the cell (Gauthier and Roca-Cusachs, 2018, Sun et al., 2019, Kechagia et al., 2019, Chighizola et al., 2019). In 1993, with the help of magnetic tweezer experiments, it was discovered by Wang et al. that integrins are mechanosensitive and capable of transducing mechanical signals to the actin cytoskeleton (Wang et al., 1993). To do so, integrins have to be activated, that is, they change from a bent and closed conformation with low ligand affinity to an extended and open conformation with high affinity for their ligands and separated cytoplasmic tails. This integrin activation can be induced from outside the cell by actual integrin/ligand binding (outside-in signaling), but it can also be triggered intracellularly, for example, by signaling originating from G-protein coupled receptors (inside-out signaling) (Gauthier and Roca-Cusachs, 2018, Sun et al., 2019, Kechagia et al., 2019, Chighizola et al., 2019). Also, other important cell surface receptors can cross-talk with and influence integrin activation and signaling, such as GPI-anchored proteins (Ferraris et al., 2014, Schulte et al., 2016a, Kalappurakkal et al., 2019), receptor tyrosine kinases (Yang et al., 2016), syndecans (Bass et al., 2007, Morgan et al., 2013), and CD44 (Seidlits et al., 2010, Kim and Kumar, 2014). The activated integrin conformation is energetically less favorable than the bent one; however, there are several stabilizing events that can occur to maintain this transition and to increase its lifetime (Sun et al., 2019). Talin (Figure 4.6.1A) and kindlin binding to the integrin tails is a first step and also crucial for integrin activation (Jiang et al., 2003, Theodosiou et al., 2016). The talin rod can furthermore link the integrin to actin filaments (F-actin) and the forces of the retrograde flow generated by actin polymerization and actomyosin contraction (Figure 4.6.1B). This engages the so-called molecular clutch (first hypothesized in 1988 by Mitchison and Kirschner (1988)) in the nascent adhesions enabling force transmission and loading. The extent of force loading in the molecular clutch is determined by critical biophysical parameters of the ECM, in particular, by the rigidity and the spatial organization and dimensionality of the adhesion sites (Oria et al., 2017, Gauthier and Roca-Cusachs, 2018, Sun et al., 2019, Kechagia et al., 2019, Chighizola et al., 2019, Chighizola et al., 2020) (Figure 4.6.1D).

The more rigid the substrate, the stronger is the force transmission and the faster the force loading in the molecular clutch (Elosegui-Artola et al., 2016). Forces in the range of a few pN keep integrins in their open and extended configuration (Strohmeyer et al., 2017, Li and Springer, 2017) and forces in the order of tens of pN can furthermore lead to a catch bond formation between integrin and ligand (demonstrated for α5β1 and αVβ3 integrin) (Kong et al., 2009, Chen et al., 2017). Both events increase the lifetime of the integrin activation and the molecular clutch (Gauthier and Roca-Cusachs, 2018, Sun et al., 2019, Kechagia et al., 2019, Chighizola et al., 2019). In addition, forces from >5 to 25 pN activate talin by stretching the helix bundle domains of its rod domain, which sequentially reveals cryptic binding sites for vinculin. Vinculin, therefore, binds to these disclosed binding sites of the activated talin, first near the integrin and then at higher forces closer to the F-actin (Del Rio et al., 2009, Grashoff et al., 2010, Ciobanasu et al., 2014, Yao et al., 2014, Case et al., 2015a, Elosegui-Artola et al., 2016). During this process, vinculin becomes activated by itself, binds the F-actin, and forms a catch bond with it (Huang et al., 2017), which leads to strong reinforcement of the molecular clutch and the nascent adhesion (Carisey et al., 2013, Case et al., 2015a). Insufficient and very slow force loading causes, instead, a disassembly of the ECM/integrin/talin/F-actin linkage, before the force thresholds that permit reinforcement can be surpassed (Gauthier and Roca-Cusachs, 2018, Sun et al., 2019, Kechagia et al., 2019, Chighizola et al., 2019) (Figure 4.6.1C,D).

The substrate rigidity is not the only factor that regulates force loading and subsequent IAC maturation. The situation is more complex, as there are several spatial and structural determinants influencing the processes in the molecular clutch in a decisive manner, as well (Oria et al., 2017, Kechagia et al., 2019, Chighizola et al., 2019, Chighizola et al., 2020). It is well-established that on rigid substrates too large ligand spacing distances with the threshold being ~ 60–70 nm (Arnold et al., 2004, Liu et al., 2014), impede the formation of mature IAC, which led to the hypothesis that there might be a kind of a "molecular ruler" (potentially talin). Curiously, it has been found recently that on soft substrates increasing the ligand spacing distance instead favors integrin clustering and IAC maturation. This counterintuitive event could again be explained by force loading within the molecular

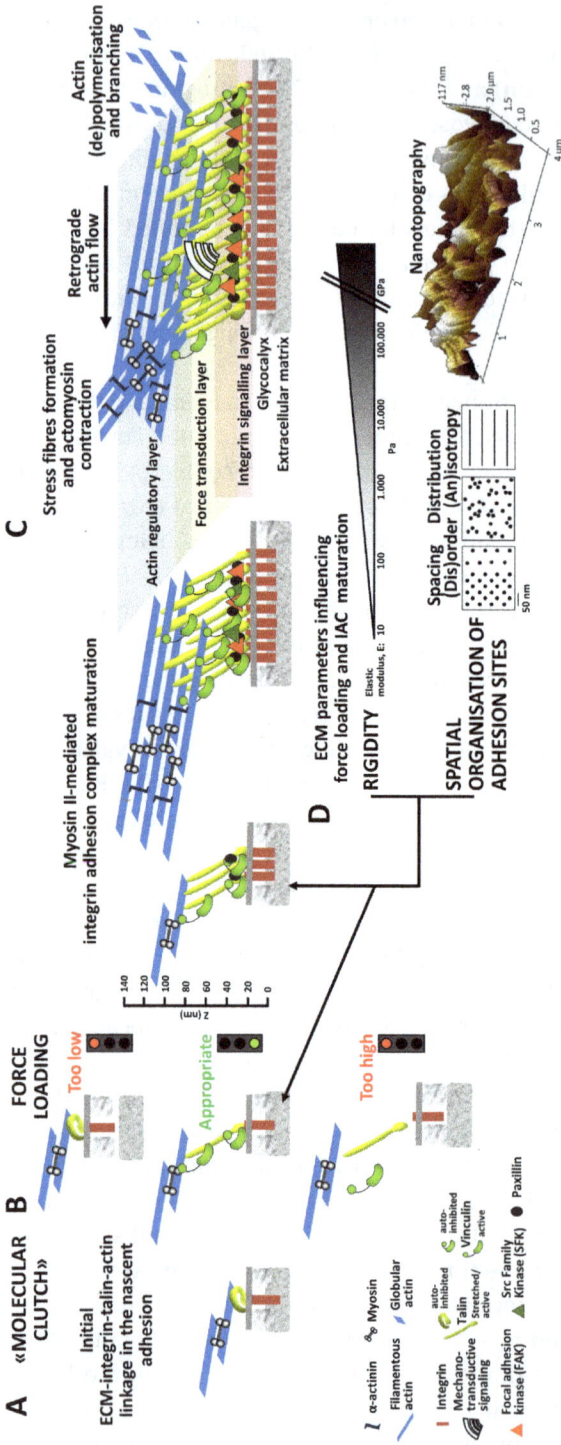

Figure 4.6.1: Integrin-mediated mechanosensing and maturation of integrin adhesion complexes in dependency of biophysical extracellular matrix parameters. (A) The scheme shows the initial ECM–integrin–talin–actin linkage in nascent adhesions, the so-called molecular clutch. (A, B) Whether the nascent adhesions reinforce and mature into (C) integrin adhesion complexes (IAC) with a particular layered nanoarchitecture depends on the force loading within the molecular clutch (D), which is determined by certain indicated biophysical parameters of the ECM. (The nanotopography is represented by an AFM recording of a decelluarized bladder ECM, (courtesy of Prof. Alessandro Podestà, University of Milan). Further details can be found in Section 4.6.1.

clutch. A very strong force loading per integrin as in the case of large ligand spacing on rigid substrates, leads to an adhesion collapse, probably because of the insufficient force distribution due to restrictions in integrin recruitment to the adhesion sites. On the contrary, on soft substrate, the aforementioned deficient force loading per integrin is increased to levels above the necessary thresholds if the ligand spacing is augmented, which then enables the stabilizing events and IAC maturation (Liu et al., 2014, Oria et al., 2017, Kechagia et al., 2019, Chighizola et al., 2019). In addition, IAC maturation seems to be favored if a minimal adhesion unit consisting of a few integrins bound to ligands in adjacency of a few tens of nm can be formed as a core structure that can also attract unligated but activated integrins, forming nanocluster bridges (Changede and Sheetz, 2017, Changede et al., 2019). In line with these observations, it has been demonstrated that the topographical configuration of the cell substrate influences IAC formation and the mechanotransductive sequence (Chighizola et al., 2020, Dalby et al., 2014, Park et al., 2016, Schulte et al., 2016b, Maffioli et al., 2017, Baek et al., 2018, Park et al., 2018, Chighizola et al., 2019). The combination of mechanical properties and local spatial organization of the binding sites encountered by the cell in its microenvironment (also in terms of (dis)order (Huang et al., 2009, Schvartzman et al., 2011) and (an)isotropy (Ferrari et al., 2010, Ray et al., 2017, Baek et al., 2018, Park et al., 2018, Chen et al., 2019, Changede et al., 2019)), thus, determines the mechanotransductive cellular response (Gauthier and Roca-Cusachs, 2018, Sun et al., 2019, Kechagia et al., 2019, Chighizola et al., 2019) (Figure 4.6.1C,D).

As described in Chapter 4.3, a pericellular sugar coat linked to proteoglycans, glycoproteins, and glycolipids forming the glycocalyx is present in the cell/microenvironment interface, and in recent years, its functional involvement in mechanotransduction-related processes has become increasingly evident. The sugar chains of the glycocalyx usually reach out much farther into the extracellular space than the extracellular domains of activated integrins (which have a length of ~20 nm). This structure influences integrin clustering in several ways. Integrins that succeed to bind their ligands cause a crowding/compacting of the adjacent glycocalyx. This creates an upward mechanical loading towards the cell membrane that contributes to keeping the integrins in their extended activated configuration. The compacted glycocalyx leads, furthermore, to the formation of a steric kinetic trap that fosters integrin clustering by restricting the lateral diffusion of integrins (Paszek et al., 2014, Kuo et al., 2018, Gauthier and Roca-Cusachs, 2018, Sun et al., 2019, Kechagia et al., 2019, Chighizola et al., 2019, Chighizola et al., 2021).

If the microenvironmental conditions are suitable, integrins cluster into modules (each module consisting of around 20–50 integrins with a dimension of ~100 nm) (Changede et al., 2015, Changede and Sheetz, 2017). Other IAC proteins are recruited to the reinforced initial integrin/talin/vinculin/F-actin axis, leading also to increasing bundling of actin filaments (up to the formation of stress fibers). This furthers the IAC maturation into multiprotein assemblies of specific nano architecture with three layers: an integrin signaling layer, a force transduction layer, and an actin regulatory layer (Carisey et al., 2013, Case et al., 2015a). The IAC can grow hierarchically into structures of increasing dimensions, first into punctate focal complexes, later into elongated focal adhesions, or, in some cases, even into fibrillary adhesions. There are numerous proteins that have been identified as potential IAC components (>150 proteins, depending on the cell biological context, of which ~60 proteins (many containing LIM domains) were defined as a consensus adhesome). Among them, various adaptor, signaling, and actin cross-linking proteins can be found, such as paxillin, integrin-linked kinase (ILK), LIMS1 (LIM and senescence cell antigen-like-containing domain protein 1)/PINCH, focal adhesion kinase (FAK), src family kinases (SFK), p130Cas, p21 (Rac1)-activated kinase (PAK), zyxin, and α-actinin. These recruitments turn the IAC into sophisticated signaling platforms that can modulate the cellular behavior and state in various ways (Gauthier and Roca-Cusachs, 2018, Sun et al., 2019, Kechagia et al., 2019, Humphries et al., 2019, Green and Brown, 2019, Chighizola et al., 2019) (Figure 4.6.1C).

4.6.2 Mechanotransductive Processes and Signaling in Control of the Cytoskeletal Organization and Cellular Mechanics

Many of the IAC downstream signaling events (Figure 4.6.2) govern the organization of the cytoskeleton, and, as a consequence, the cellular mechanics, which eventually affects mechanosensitive transcription factors and nuclear organization (Figure 4.6.2A, details will be outlined in the following). During IAC lifetime, intricate and intertwined signaling cascades are set in motion that have, for example, a strong impact on Rho guanosine triphosphatases (RhoGTPases) activities (in particular, Rac1, Cdc42, and RhoA, as it was found in 1995 in a seminal work by Nobes and Hall (1995)), which decisively control the spatiotemporal dynamics of the actin cytoskeleton and IAC (dis)assembly (Huveneers and Danen, 2009, Lawson and Ridley, 2018).

RhoGTPases are active when they bind guanosine triphosphate (GTP) and inactive when they bind guanosine diphosphate (GDP). The switch between these two states is controlled in a complex manner by regulatory proteins (Figure 4.6.2B, specific examples below). GTPase-activating proteins (GAPs) stimulate GTP hydrolysis of the RhoGTPases, cycling them to the inactive form. RhoGEFs (guanine nucleotide exchange factor) instead exchange the GDP with GTP, thus activating the RhoGTPases. The inactive form can furthermore be sequestered by Rho guanine nucleotide dissociation inhibitors (RhoGDIs), which inhibit the nucleotide exchange. Active RhoGTPases will trigger effectors, such as Rho-associated kinases (ROCK), Wiskott-Aldrich syndrome protein family verprolin-homologous protein (WAVE) or Wiskott-Aldrich syndrome family (WASp) complexes, thereby regulating versatile cellular responses, in particular, by their impact on the cytoskeletal organization (Hodge et al., 2016).

In the early stages of IAC formation (nascent adhesions to focal complex), FAK recruitment to the IAC (fostered by paxillin (Scheswohl et al., 2008) and talin (Lawson et al., 2012)) triggers Tyr397-autophosphorylation of FAK, which allows SFK to bind the Tyr397-phosphorylated FAK. This, in turn, leads to FAK phosphorylation at other phosphorylation sites by the SFK (Mitra and Schlaepfer, 2006). This activated FAK/SFK complex is the starting point for many mechanotransductive signaling cascades. It recruits and phosphorylates further proteins, for example, p130Cas (Chodniewicz and Klemke, 2004). The SFK-mediated phosphorylation and activation of p130Cas is force-dependent (Sawada et al., 2006) and enables Dock180/ELMO1 binding, which functions as a RhoGEF for Rac1. This SFK/FAK/p130cas/Dock180/ELMO1 signaling sequence thus recruits and activates Rac1. The activated Rac1 then binds the WAVE regulatory complex which, in turn, activates the actin nucleating Arp2/3 complex, leading to actin polymerization and branching in the lamellipodia and membrane protrusion. In addition, FAK/SFK-triggered paxillin phosphorylation enables PAK/GIT-1/β-PIX (the latter is also known as ArhGEF7) recruitment, which activates Cdc42. Activated Cdc42 regulates filopodia formation through the WASp and Arp2/3 complex (Brugnera et al., 2002, Premont et al., 2004, Mitra and Schlaepfer, 2006, Huveneers and Danen, 2009, Vicente-Manzanares and Horwitz, 2011, Lawson and Ridley, 2018) (Figure 4.6.2A).

RhoA activity is instead suppressed in the initial phase of IAC maturation in an SFK/FAK/p120RasGAP/p190RhoGAP-dependent manner (Ren et al., 1999, Arthur et al., 2000, Arthur and Burridge, 2001, Tomar et al., 2009). Depending on the phosphorylation status of the various FAK and paxillin phosphorylation sites (Hamadi et al., 2005, Zaidel-Bar et al., 2007, Miller et al., 2013), at later stages of focal adhesion formation, p115RhoGEF, p190RhoGEF (also known as Rgnef or ArhGEF28), and LARG (leukemia-associated RhoGEF) are recruited, which results in an increasing RhoA activation (Dubash et al., 2007, Lim et al., 2008, Guilluy et al., 2011, Vicente-Manzanares and Horwitz, 2011, Miller et al., 2014, Lawson and Ridley, 2018). RhoA activation leads to higher actomyosin contractility mediated by ROCK-dependent myosin light-chain phosphorylation and activation. PAK- and ROCK/LIMK-dependent cofilin (also called actin depolymerization factor) phosphorylation, and thus inactivation, contributes further to F-actin stabilization and stress fiber formation (Chrzanowska-Wodnicka and Burridge, 1996, Lawson and Ridley, 2018). Moreover, Rho signaling-mediated mechanotransductive processes can lead to a remodeling of the ECM architecture through cytoskeletal forces that are transmitted to the substrate via IAC (Humphrey et al., 2014).

IAC are very dynamic structures, and there is a constant turnover of cellular adhesion sites with the microenvironment, for example, near the leading edge during cell migration or in neuronal growth cones. The IAC life cycle depends on the maturation status they achieve. Nascent adhesions have a short lifetime of less than a minute. If they associate with actin filaments and develop into focal complexes, their lifetime increases to a few minutes. In the case of maturation into focal adhesions, or even fibrillary adhesions, the lifetime of these structures increases up to tens of minutes or hours, respectively (Webb et al., 2004, Choi et al., 2008, Vicente-Manzanares and Horwitz, 2011).

The disassembly mechanism(s) (Figure 4.6.2C) of the later, more mature IAC are not yet completely understood. It involves endo/exocytic processes and probably also microtubule interaction with the focal adhesions (Stehbens and Wittmann, 2012, Noordstra and Akhmanova, 2017, Garcin and Straube, 2019), as well as calpain activity, a proteolytic protein, which has many focal adhesion (such as FAK itself, paxillin, and talin) and cytoskeletal components as substrates (Franco et al., 2004, Franco and Huttenlocher, 2005, Cortesio et al., 2011). Focal adhesions are transiently targeted by microtubule plus-ends in a repeated manner, which induces the demounting of focal adhesions (Kaverina et al., 1999). The microtubule capture is APC (adenomatous polyposis coli)-dependent (Juanes et al., 2019) and realized by microtubule docking to the plasma membrane in the vicinity of focal adhesions via cortical microtubule stabilization complexes that encompass numerous proteins, such as EB (end-binding protein) 1, CLASPs (cytoplasmic linker-associated proteins), KIF21A (kinesin-like protein), MICAL3 (microtubule-associated monoxygenase, calponin, and LIM domain containing), ELKS (protein rich in the amino acids E, L, K, and S), LL5β, and liprins. Liprins link the microtubules to Kank (KN motif and Ankyrin repeat domain-containing protein), which is bound by talin in the focal adhesions. Also ILK can contribute to the microtubule stabilization at focal adhesions via mDia (Diaphanous-related formin)/IQGAP (IQ motif-containing GAPs) (Bouchet et al., 2016, Noordstra and Akhmanova, 2017, LaFlamme et al., 2018, Garcin and Straube, 2019). The microtubule capturing leads to a local sequestering of GEF-H1 (called also ArhGEF2) and, therefore, suppression of Rho/ROCK-mediated myosin II filament assembly (Azoitei et al., 2019, Rafiq et al., 2019). Furthermore, the delivery of factors (e.g., MAP4K4 and matrix metalloproteases) and autophagosomes takes place along microtubules, which promote the disassembly of focal adhesions (Yue et al., 2014, Stehbens et al., 2014, Sharifi et al., 2016, LaFlamme et al., 2018). In addition, regulated by the state of Tyr397 phosphorylation of FAK, integrins are endocytosed at microtubule/focal adhesion interaction sites in a clathrin/dynamin/Rab5/Rab21-dependent (Ezratty et al., 2005, Pellinen et al., 2006, Ezratty et al., 2009, Mendoza et al., 2013) or ILK/caveolin-dependent manner (Shi and Sottile, 2008). FAK is also involved in the recruitment of calpain to the focal adhesions (Carragher et al., 2003). Interestingly, these focal adhesion disassembly elements, that is, calpain- and endocytosis-related proteins, become enriched in focal adhesions upon myosin II activity (Kuo et al., 2011) (Figure 4.6.2C).

The IAC-mediated modulations of the cytoskeletal organization and membrane tension also influence another important group of mechanosensitive elements, that is, mechanically activated channels in the cell membrane, for example, Piezo or

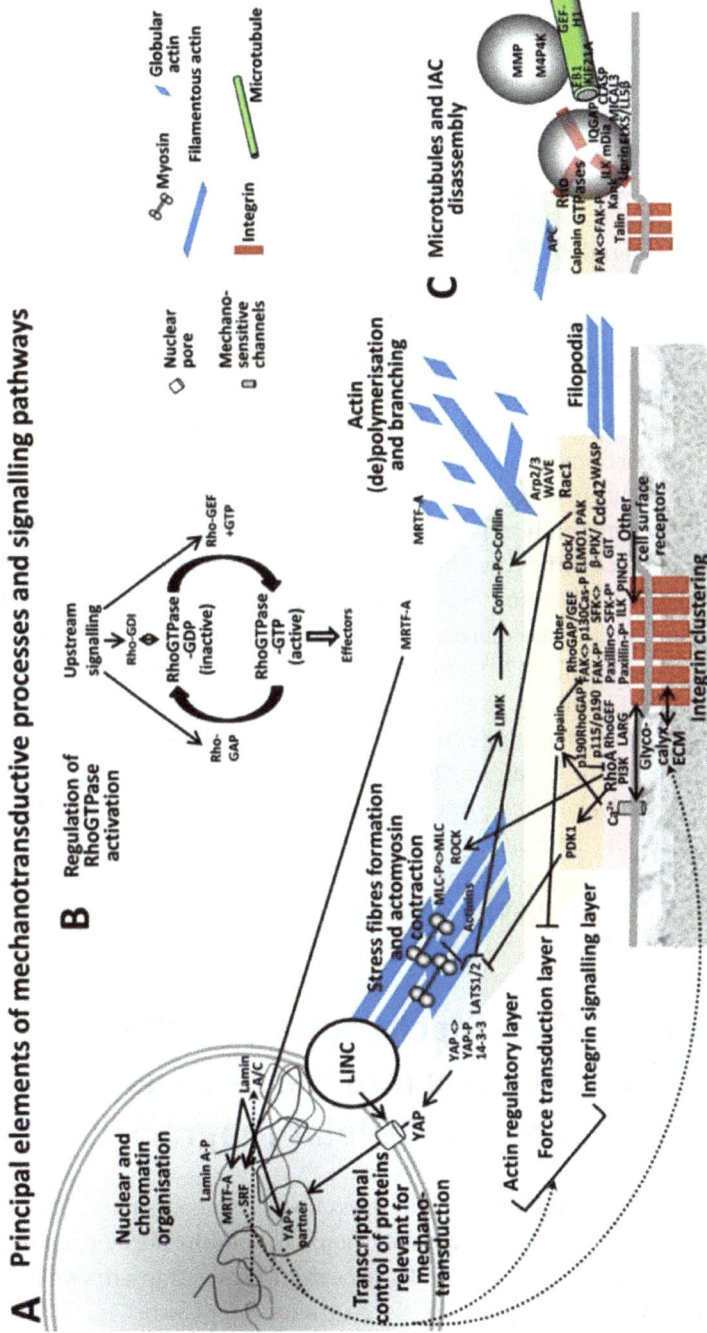

Figure 4.6.2: Principal elements of mechanotransductive processes and signaling pathways. (A) The graphic schematically illustrates different structures and processes along the IAC-mediated mechanotransductive signaling sequence from the cell/microoenvironment to the nucleus, highlighting related prominent proteins and their cross-talk. Further details can be found in the Sections 4.6.2 and 4.6.3. (B) The scheme outlines how RhoGTPase activation is regulated (GDI: GDP dissociation inhibitor; GAP: GTPase-activating protein; GEF: guanine nucleotice exchange factor). (C) The graphic visualizes major processes and proteins involved in integrin adhesion disassembly. Further details can be found in Section 4.6.2.

TRP (transient receptor potential) channels. The Ca^{2+} influx pulses triggered by these mechanosensitive channels can feedback on focal adhesion dynamics, in particular, via the Ca^{2+}-dependent calpain (Pathak et al., 2014, Nourse and Pathak, 2017, Canales et al., 2019, Ridone et al., 2019) (Figure 4.6.2A). Through the actin cytoskeleton, IAC are furthermore connected to cadherin-mediated cell/cell adhesions, which actually share some constituents (e.g., vinculin, ELMO, Dock, Rac1) and organizational features (modular cluster assembly) with focal adhesions. The IAC/cadherin interplay determines the intra- and intercellular forces and tension of cells, that is, their mechanical landscape, which regulates tissue integrity and homeostasis (Mui et al., 2016, Changede and Sheetz, 2017).

In a nutshell, the impact of mechanotransductive processes on the cytoskeleton is realized by complex signaling sequences involving RhoGTPases, -GAPs, -GEFs, and –GDIs, as well as RhoGTPase signaling-related kinases and effectors. Together, their numerous interconnections, which are not always fully understood in-depth and often also depend on the specific cell biological context, precisely regulate and fine-tune the IAC turnover/maturation rates, cytoskeletal organization/mechanics, and, eventually, cell and tissue structure. The correct coordination of these dynamics is essential in many cell biological and developmental processes, for example, cell migration, differentiation, and tissue morphogenesis (Vicente-Manzanares and Horwitz, 2011, Changede and Sheetz, 2017, Lawson and Ridley, 2018, Humphries et al., 2019, Kechagia et al., 2019, Chighizola et al., 2019). Additional details about the cytoskeleton and the concept of cell tensegrity, that is, the idea of considering the cell as a prestressed tensegrity structure (Ingber, 2003b, Ingber, 2003c), can be found in Chapters 2.1 and 4.2, respectively. Changes in the cytoskeletal configuration and tension due to modulations of the mechanotransductive sequence and signaling can furthermore propagate into the nucleus, impacting, eventually, also on the cellular program and fate (Wang et al., 2009, Martino et al., 2018) (Figure 4.6.2A).

4.6.3 Propagation of the Mechanotransductive Processes and Signals into the Nucleus and the Impact on the Cellular Program

The extent of F-actin formation and bundling into stress fibers influences the shuttling of mechanosensitive transcription factors from the cytoplasm into the nucleus, for example, YAP (yes-associated protein) and MRTF-A (myocardin-related transcription factor A). At low F-actin level in the cell, Lats (large tumor suppressor) 1/2 kinase phosphorylates YAP and the phosphorylated YAP interacts with 14-3-3. The YAP/14-3-3 complexes are retained in the cytoplasm and eventually degraded by proteasomes. Upon IAC maturation-induced stress fiber formation, β1-integrin/SFK/FAK/PI3K/PDK1

(Kim and Gumbiner 2015) and β1-integrin/SFK/Rac1/PAK pathways decrease Lats1/2 activity, which thus attenuates YAP phosphorylation. This causes YAP release from 14-3-3 and its translocation into the nucleus, where it exerts a co-transcriptional activity (Panciera et al., 2017, Totaro et al., 2018) (with interacting partners, for example, TEAD1, RUNX, p73, Smad (Kim et al., 2018), potentially also REST (Nardone et al., 2017)). In addition, the force generated by actomyosin contraction of stress fiber contributes to YAP entering the nucleus by acting on the nuclear pores (Elosegui-Artola et al., 2017). MRTF-A is instead affected by the amount of cytoplasmic G-actin to which it binds. If the G-actin pool decreases (as in the case of augmented F-actin formation), free MRTF-A increasingly translocates into the nucleus, where it binds the serum response factor (SRF), which induces MRTF-A/SRF complex-dependent transcription pathways (Olson and Nordheim, 2010). Intriguingly, both YAP and MRTF-A control, in an autoregulatory manner, the expression of many target genes that are members of mechanotransductive machinery (Sun et al., 2006, Nardone et al., 2017, Foster et al., 2017, Kim et al., 2017) and ECM protein deposition, which, in turn, provide again a feedback on cell behavior (Humphrey et al., 2014, Loebel et al., 2019).

Another way in which the mechanotransductive sequence modulates the cellular state is through a bridge between the cytoskeleton and the nucleus, mediated by the LINC (linker of nuclear and cytoskeleton) complex. The F-actin of the stress fibers binds to the LINC complex via nesprin 1 or 2, which itself is embedded in the outer nuclear membrane. Also, the other components of the cytoskeleton can be connected to the nucleus, microtubules via dynein and/or kinesin to nesprin 1 and 2, and intermediate filaments through plectin to nesprin 3. The nesprins are connected to SUN dimers (in case of nesprin 1/2 together with emerin) in the inner nuclear membrane, which are linked to the nuclear lamina envelope. Many details are still elusive, but it has been hypothesized that this ECM/IAC/cytoskeleton/LINC/nuclear lamina bridge allows mechanoregulation of the 3D nuclear morphology and chromatin organization, influencing, eventually, gene expression (Osmanagic-Myers et al., 2015, Uhler and Shivashankar, 2017b, Martino et al., 2018). The connection enables force transmission from the cell/microenvironment interface to the nucleus, leading either to nuclear relaxation and motility in the case of low forces, or deformation and positional stability of the nucleus at high forces. Depending on the state of their condensation, chromatin regions can be associated to the nuclear lamina through lamin-binding proteins (e.g., Lap2α). Changes in the nuclear shape due to mechanotransductive processes can, therefore, affect the spatial organization of chromatin regions and lead to differential chromosome intermingling and gene clustering, as a consequence. This also causes the enrichment of different RNA polymerase II complexes and alterations in gene expression patterns (Isermann and Lammerding, 2013, Uhler and Shivashankar, 2017a, Uhler and Shivashankar, 2017b). Furthermore, the nuclear lamina is built up by lamin A/C and B and determines the structural integrity and stiffness of the nucleus. It has been shown that lamin A/C expression and phosphorylation levels are regulated by the mechanical state of tissues and cells. Lamin A levels increase with tissue stiffness, and in cells on rigid substrates with high actomyosin contractility lamin A/C is dephosphorylated at Ser22. Dephosphorylated lamin A/C is integrated in nuclear lamina, whereas Ser22-phosporylated lamin A/C resides in the nucleoplasm. Nuclear envelope-associated lamin A/C influences the YAP and SRF activity too, and thus the expression of their target genes, for example, myosin II (Wang et al., 2009, Swift et al., 2013, Ho et al., 2013, Buxboim et al., 2014, Uhler and Shivashankar, 2017b).

In these different ways, mechanotransductive processes converge into the nucleus and can influence the nuclear morphology and mechanics, as well as the spatial organization/positioning of chromosomes and gene clusters, which eventually modulates gene expression patterns (Figure 4.6.2A). The overall complexity of the mechanotransductive machinery is far from being understood in-depth. However, the insights obtained in recent years provide a fascinating picture of highly intricate and interlaced processes and signaling pathways with sophisticated autoregulatory feedback mechanisms that control many aspects of the cellular state and tissue homeostasis (Figure 4.6.2).

4.6.4 Aberrations of Mechanotransductive Structures and Processes in Diseases

Pathophysiological aberrations in components of the mechanotransductive machinery are often at the base of the modulations in cell and tissue mechanics related to diseases and described throughout this book (see Volume 2). In fact, numerous alterations along the mechanotransductive sequence have been reported that are linked to diseases and that are, despite still existing challenges, interesting in regard to diagnostic and therapeutic approaches (Ingber, 2003a, Jaalouk and Lammerding, 2009, Vogel, 2018, Guck, 2019, Sheridan, 2019, Tschumperlin and Lagares, 2020, Slack et al., 2021). In the following, some general examples of irregularities in mechanotransduction-related structures and processes at different levels that are relevant for diseases will be introduced.

In many pathologies, abnormal rearrangements in the cell/microenvironment interface affecting compositional, structural, and mechanical features of the ECM (Humphrey et al., 2014, Bonnans et al., 2014, Cox, 2021) and the glycocalyx (Tarbell and Cancel, 2016, Kuo et al., 2018) have been observed. This is, in particular, true for cancers (Montagner and Dupont, 2020, Cox, 2021) (see also Chapter 6.3). The ECM of solid tumors has been found to be denser and stiffer than its healthy counterparts as a consequence of locally increased cross-linking and overexpression of ECM proteins, which impacts tumor growth and metastatic potential through focal adhesion signaling (Erler et al., 2006, Provenzano et al., 2008, Levental et al., 2009, Pickup et al., 2014, Nebuloni et al., 2016). In cancer cells, the composition and physical properties (e.g., the bulkiness) of the glycocalyx are also aberrant (Pinho and Reis, 2015, Kuo et al., 2018), which can influence, for example, integrin clustering (Paszek et al., 2014). The potential clinical relevance of the combined ECM and glycocalyx effect on integrin mechanosignaling has recently been shown for glioblastoma multiforme (Barnes et al., 2018). Mechanosensing, integrins, and their signaling are indeed involved in basically all stages of tumor progression and the metastatic cascade (Hamidi and Ivaska, 2018, Papalazarou et al., 2018, Montagner and Dupont, 2020). Congruently, in literature-based bioinformatics analyses, it was found that dysregulations in many genes that encode for IAC proteins are associated with various human diseases: in particular, in cancer/metastasis, but also in many others diseases, such as musculoskeletal, cardiovascular, neurological, hematological, and blistering disorders (Winograd-Katz et al., 2014). Also, bacteria and viruses are known to exploit or hijack components of the IAC and/or the mechanotransductive machinery (Winograd-Katz et al., 2014, Case and Waterman, 2015), for example, during viral entry (Hussein et al., 2015) or bacterial infection (Hamiaux et al., 2006, Izard et al., 2006, Hoffmann et al., 2011) of host cells.

Due to their crucial role in the determination of cell shape and tension, as well as tissue morphogenesis and development, the drastic and manifold consequences of deregulations affecting the cytoskeleton and RhoGTPase signaling are long-known, again, in particular, in many cancers, but also, for example, in immunodeficiency syndromes (e.g., the Wiskott-Aldrich syndrome, which actually gave WAVE and WASP their names) or neurodegenerative disorders (Boettner and Van Aelst, 2002, Mammoto and Ingber, 2009, Newell-Litwa et al., 2015, Olson, 2018). Moreover, mechanotransductive pathways (e.g., related to ROCK/Myosin-II) that remodel IAC and the actin cytoskeleton might be involved in drug sensitivity of cancer cells against chemotherapeutic drugs, or even the development of drug resistance(Orgaz et al., 2020, Young et al., 2020, Kubiak et al., 2021). Also, the best characterized mechanosensitive transcription factor, that is, YAP/TAZ, has been shown to be implicated in various diseases, such as atherosclerosis and cardiovascular diseases, tissue fibrosis, inflammatory responses, muscular dystrophy, and different cancers (Panciera et al., 2017).

Diseases that are associated with mutations in genes encoding for proteins in the nuclear envelope have been grouped under the term laminopathies, and linked to faulty nuclear mechanotransduction. In line with the broad spectrum of functions of the nuclear lamina, these diseases encompass quite different clinical phenotypes. Some prominent examples are muscular dystrophies such as Emery-Dreifuss muscular dystrophy, neuropathies such as the Charcot-Marie-Tooth disease, lipoatrophic diseases, or progeroid (accelerated aging) disorders, such as the Hutchinson-Gilford progeria syndrome (Schreiber and Kennedy, 2013, Osmanagic-Myers et al., 2015).

The versatile and broad range of impact that aberrations in mechanotransductive processes and structures can have on cells and tissues, point out the importance of approaches aiming at obtaining precise mechanical fingerprints of diseases; further details can be found in Volume 2. These increasing insights leverage furthermore emergent therapeutic approaches targeting essential mechanotransductive components (in particular integrins) by so-called mechano-therapeutics in various diseases (Sheridan, 2019, Tschumperlin and Lagares, 2020, Slack et al., 2021).

References

Arnold, M., E. A. Cavalcanti-Adam, R. Glass, J. Blümmel, W. Eck, M. Kantlehner, H. Kessler and J. P. Spatz (2004). "Activation of integrin function by nanopatterned adhesive interfaces." ChemPhysChem **5**: 383–388.

Arthur, W. T. and K. Burridge (2001). "RhoA inactivation by p190RhoGAP regulates cell spreading and migration by promoting membrane protrusion and polarity." Molecular Biology of the Cell **12**: 2711–2720.

Arthur, W. T., L. A. Petch and K. Burridge (2000). "Integrin engagement suppresses RhoA activity via a c-Src-dependent mechanism." Current Biology **10**: 719–722.

Azoitei, M. L., J. Noh, D. J. Marston, P. Roudot, C. B. Marshall, T. A. Daugird, S. L. Lisanza, M.-J. Sandí, M. Ikura and J. Sondek (2019). "Spatiotemporal dynamics of GEF-H1 activation controlled by microtubule-and Src-mediated pathways." Journal of Cell Biology **218**: 3077–3097.

Baek, J., S.-Y. Cho, H. Kang, H. Ahn, W.-B. Jung, Y. Cho, E. Lee, S.-W. Cho, H.-T. Jung and S. G. Im (2018). "Distinct mechanosensing of human neural stem cells on extremely limited anisotropic cellular contact." ACS Applied Materials & Interfaces **10**: 33891–33900.

Barnes, J. M., S. Kaushik, R. O. Bainer, J. K. Sa, E. C. Woods, F. Kai, L. Przybyla, M. Lee, H. W. Lee and J. C. Tung (2018). "A tension-mediated glycocalyx–integrin feedback loop promotes mesenchymal-like glioblastoma." Nature Cell Biology **20**: 1203–1214.

Bass, M. D., K. A. Roach, M. R. Morgan, Z. Mostafavi-Pour, T. Schoen, T. Muramatsu, U. Mayer, C. Ballestrem, J. P. Spatz and M. J. Humphries (2007). "Syndecan-4–dependent Rac1 regulation determines directional migration in response to the extracellular matrix." The Journal of Cell Biology **177**: 527–538.

Boettner, B. and L. Van Aelst (2002). "The role of Rho GTPases in disease development." Gene **286**: 155–174.

Bonnans, C., J. Chou and Z. Werb (2014). "Remodeling the extracellular matrix in development and disease." Nature Reviews. Molecular Cell Biology **15**: 786–801.

Bouchet, B. P., R. E. Gough, Y.-C. Ammon, D. Van De Willige, H. Post, G. Jacquemet, A. M. Altelaar, A. J. Heck, B. T. Goult and A. Akhmanova (2016). "Talin-KANK1 interaction controls the recruitment of cortical microtubule stabilizing complexes to focal adhesions." Elife **5**: e18124.

Brugnera, E., L. Haney, C. Grimsley, M. Lu, S. F. Walk, A.-C. Tosello-Trampont, I. G. Macara, H. Madhani, G. R. Fink and K. S. Ravichandran (2002). "Unconventional Rac-GEF activity is mediated through the Dock180–ELMO complex." Nature Cell Biology **4**: 574–582.

Buxboim, A., J. Swift, J. Irianto, K. R. Spinler, P. D. P. Dingal, A. Athirasala, Y.-R. C. Kao, S. Cho, T. Harada and J.-W. Shin (2014). "Matrix elasticity regulates lamin-A, C phosphorylation and turnover with feedback to actomyosin." Current Biology **24**: 1909–1917.

Canales, J., D. Morales, C. Blanco, J. Rivas, N. Díaz, I. Angelopoulos and O. Cerda (2019). "A TR (i) P to cell migration: New roles of TRP channels in mechanotransduction and cancer." Frontiers in Physiology **10**.

Carisey, A., R. Tsang, A. M. Greiner, N. Nijenhuis, N. Heath, A. Nazgiewicz, R. Kemkemer, B. Derby, J. Spatz and C. Ballestrem (2013). "Vinculin regulates the recruitment and release of core focal adhesion proteins in a force-dependent manner." Current Biology **23**: 271–281.

Carragher, N. O., M. A. Westhoff, V. J. Fincham, M. D. Schaller and M. C. Frame (2003). "A novel role for FAK as a protease-targeting adaptor protein: Regulation by p42 ERK and Src." Current Biology **13**: 1442–1450.

Case, L., M. Baird, G. Shtengel, S. Campbell, H. Hess, M. Davidson and C. Waterman (2015a). "Molecular mechanism of vinculin activation and nanoscale spatial organization in focal adhesions." Nature Cell Biology **17**: 880–892.

Case, L. B., M. A. Baird, G. Shtengel, S. L. Campbell, H. F. Hess, M. W. Davidson and C. M. Waterman (2015b). "Molecular mechanism of vinculin activation and nanoscale spatial organization in focal adhesions." Nature Cell Biology **17**: 880–892.

Case, L. B. and C. M. Waterman (2015). "Integration of actin dynamics and cell adhesion by a three-dimensional, mechanosensitive molecular clutch." Nature Cell Biology **17**: 955–963.

Changede, R., H. Cai, S. J. Wind and M. P. Sheetz (2019). "Integrin nanoclusters can bridge thin matrix fibers to form cell–matrix adhesions." Nature Materials **18**: 1–10.

Changede, R. and M. Sheetz (2017). "Integrin and cadherin clusters: A robust way to organize adhesions for cell mechanics." BioEssays **39**: 1–12.

Changede, R., X. Xu, F. Margadant and M. P. Sheetz (2015). "Nascent integrin adhesions form on all matrix rigidities after integrin activation." Developmental Cell **35**: 614–621.

Chen, S., M. J. Hourwitz, L. Campanello, J. T. Fourkas, W. Losert and C. A. Parent (2019). "Actin cytoskeleton and focal adhesions regulate the biased migration of breast cancer cells on nanoscale asymmetric sawteeth." ACS Nano **13**: 1454–1468.

Chen, Y., H. Lee, H. Tong, M. Schwartz and C. Zhu (2017). "Force regulated conformational change of integrin αVβ3." Matrix Biology **60**: 70–85.

Chighizola, M., T. Dini, C. Lenardi, P. Milani, A. Podestà and C. Schulte (2019). "Mechanotransduction in neuronal cell development and functioning." Biophysical Reviews **11**: 701–720.

Chighizola, M., T. Dini, S. Marcotti, M. D'urso, C. Piazzoni, F. Borghi, A. Previdi, L. Ceriani, C. Folliero, B. Stramer, C. Lenardi, P. Milani, A. Podestà and C. Schulte (2021). "The glycocalyx affects force loading-dependent mechanotransductive topography sensing at the nanoscale." https://jnanobiotechnology.biomedcentral.com/articles/10.1186/s12951-022-01585-5.

Chighizola, M., A. Previdi, T. Dini, C. Piazzoni, C. Lenardi, P. Milani, C. Schulte and A. Podesta (2020). "Adhesion force spectroscopy with nanostructured colloidal probes reveals nanotopography-dependent early mechanotransductive interactions at the cell membrane level." Nanoscale **12**: 14708–14723.

Chodniewicz, D. and R. L. Klemke (2004). "Regulation of integrin-mediated cellular responses through assembly of a CAS/Crk scaffold." Biochimica et Biophysica Acta (BBA)-Molecular Cell Research **1692**: 63–76.

Choi, C. K., M. Vicente-Manzanares, J. Zareno, L. A. Whitmore, A. Mogilner and A. R. Horwitz (2008). "Actin and α-actinin orchestrate the assembly and maturation of nascent adhesions in a myosin II motor-independent manner." Nature Cell Biology **10**: 1039–1050.

Chrzanowska-Wodnicka, M. and K. Burridge (1996). "Rho-stimulated contractility drives the formation of stress fibers and focal adhesions." The Journal of Cell Biology **133**: 1403–1415.

Ciobanasu, C., B. Faivre and C. Le Clainche (2014). "Actomyosin-dependent formation of the mechanosensitive talin–vinculin complex reinforces actin anchoring." Nature Communications **5**: 3095.

Cortesio, C. L., L. R. Boateng, T. M. Piazza, D. A. Bennin and A. Huttenlocher (2011). "Calpain-mediated proteolysis of paxillin negatively regulates focal adhesion dynamics and cell migration." Journal of Biological Chemistry **286**: 9998–10006.

Cox, T. (2021). "The matrix in cancer." Nature Reviews. Cancer **21**: 217–238.

Dalby, M. J., N. Gadegaard and R. O. Oreffo (2014). "Harnessing nanotopography and integrin–matrix interactions to influence stem cell fate." Nature Materials **13**: 558–569.

Del Rio, A., R. Perez-Jimenez, R. Liu, P. Roca-Cusachs, J. M. Fernandez and M. P. Sheetz (2009). "Stretching single talin rod molecules activates vinculin binding." Science **323**: 638–641.

Dubash, A. D., K. Wennerberg, R. García-Mata, M. M. Menold, W. T. Arthur and K. Burridge (2007). "A novel role for Lsc/p115 RhoGEF and LARG in regulating RhoA activity downstream of adhesion to fibronectin." Journal of Cell Science **120**: 3989–3998.

Elosegui-Artola, A., I. Andreu, A. E. Beedle, A. Lezamiz, M. Uroz, A. J. Kosmalska, R. Oria, J. Z. Kechagia, P. Rico-Lastres and A.-L. Le Roux (2017). "Force triggers YAP nuclear entry by regulating transport across nuclear pores." Cell **171**: 1397–1410. e14.

Elosegui-Artola, A., R. Oria, Y. Chen, A. Kosmalska, C. Pérez-González, N. Castro, C. Zhu, X. Trepat and P. Roca-Cusachs (2016). "Mechanical regulation of a molecular clutch defines force transmission and transduction in response to matrix rigidity." Nature Cell Biology **18**: 540–548.

Erler, J. T., K. L. Bennewith, M. Nicolau, N. Dornhöfer, C. Kong, Q.-T. Le, J.-T. A. Chi, S. S. Jeffrey and A. J. Giaccia (2006). "Lysyl oxidase is essential for hypoxia-induced metastasis." Nature **440**: 1222–1226.

Ezratty, E. J., C. Bertaux, E. E. Marcantonio and G. G. Gundersen (2009). "Clathrin mediates integrin endocytosis for focal adhesion disassembly in migrating cells." Journal of Cell Biology **187**: 733–747.

Ezratty, E. J., M. A. Partridge and G. G. Gundersen (2005). "Microtubule-induced focal adhesion disassembly is mediated by dynamin and focal adhesion kinase." Nature Cell Biology **7**: 581–590.

Ferrari, A., M. Cecchini, M. Serresi, P. Faraci, D. Pisignano and F. Beltram (2010). "Neuronal polarity selection by topography-induced focal adhesion control." Biomaterials **31**: 4682–4694.

Ferraris, G. M. S., C. Schulte, V. Buttiglione, V. De Lorenzi, A. Piontini, M. Galluzzi, A. Podestà, C. D. Madsen and N. Sidenius (2014). "The interaction between uPAR and vitronectin triggers ligand-independent adhesion signaling by integrins." The EMBO Journal **33**: 2458–2472.

Foster, C. T., F. Gualdrini and R. Treisman (2017). "Mutual dependence of the MRTF–SRF and YAP–TEAD pathways in cancer-associated fibroblasts is indirect and mediated by cytoskeletal dynamics." Genes & Development **31**: 2361–2375.

Franco, S. J. and A. Huttenlocher (2005). "Regulating cell migration: Calpains make the cut." Journal of Cell Science **118**: 3829–3838.

Franco, S. J., M. A. Rodgers, B. J. Perrin, J. Han, D. A. Bennin, D. R. Critchley and A. Huttenlocher (2004). "Calpain-mediated proteolysis of talin regulates adhesion dynamics." Nature Cell Biology **6**: 977–983.

Garcin, C. and A. Straube (2019). "Microtubules in cell migration." Essays in Biochemistry **63**: 509–520.

Gasiorowski, J. Z., C. J. Murphy and P. F. Nealey (2013). "Biophysical cues and cell behavior: The big impact of little things." Annual Review of Biomedical Engineering **15**: 155–176.

Gauthier, N. C. and P. Roca-Cusachs (2018). "Mechanosensing at integrin-mediated cell–matrix adhesions: From molecular to integrated mechanisms." Current Opinion in Cell Biology **50**: 20–26.

Grashoff, C., B. D. Hoffman, M. D. Brenner, R. Zhou, M. Parsons, M. T. Yang, M. A. Mclean, S. G. Sligar, C. S. Chen and T. Ha (2010). "Measuring mechanical tension across vinculin reveals regulation of focal adhesion dynamics." Nature **466**: 263–266.

Green, H. J. and N. H. Brown (2019). "Integrin intracellular machinery in action." Experimental Cell Research **378**: 226–231.

Guck, J. (2019). "Some thoughts on the future of cell mechanics." Biophysical Reviews **11**: 667–670.

Guilluy, C., V. Swaminathan, R. Garcia-Mata, E. T. O'Brien, R. Superfine and K. Burridge (2011). "The Rho GEFs LARG and GEF-H1 regulate the mechanical response to force on integrins." Nature Cell Biology **13**: 722–727.

Hamadi, A., M. Bouali, M. Dontenwill, H. Stoeckel, K. Takeda and P. Rondé (2005). "Regulation of focal adhesion dynamics and disassembly by phosphorylation of FAK at tyrosine 397." Journal of Cell Science **118**: 4415–4425.

Hamiaux, C., A. Van Eerde, C. Parsot, J. Broos and B. W. Dijkstra (2006). "Structural mimicry for vinculin activation by IpaA, a virulence factor of Shigella flexneri." EMBO Reports **7**: 794–799.

Hamidi, H. and J. Ivaska (2018). "Every step of the way: Integrins in cancer progression and metastasis." Nature Reviews. Drug Discovery **17**: 31–46.

Ho, C. Y., D. E. Jaalouk, M. K. Vartiainen and J. Lammerding (2013). "Lamin A/C and emerin regulate MKL1–SRF activity by modulating actin dynamics." Nature **497**: 507–511.

Hodge, R. and A. Ridley (2016). "Regulating Rho GTPases and their regulators." Nature Reviews. Molecular Cell Biology **17**: 496–510. https://doi.org/10.1038/nrm.2016.67

Hoffmann, C., K. Ohlsen and C. R. Hauck (2011). "Integrin-mediated uptake of fibronectin-binding bacteria." European Journal of Cell Biology **90**: 891–896.

Huang, D. L., N. A. Bax, C. D. Buckley, W. I. Weis and A. R. Dunn (2017). "Vinculin forms a directionally asymmetric catch bond with F-actin." Science **357**: 703–706.

Huang, J., S. V. Grater, F. Corbellini, S. Rinck, E. Bock, R. Kemkemer, H. Kessler, J. Ding and J. P. Spatz (2009). "Impact of order and disorder in RGD nanopatterns on cell adhesion." Nano Letters **9**: 1111–1116.

Humphrey, J. D., E. R. Dufresne and M. A. Schwartz (2014). "Mechanotransduction and extracellular matrix homeostasis." Nature Reviews. Molecular Cell Biology **15**: 802–812.

Humphries, J. D., M. R. Chastney, J. A. Askari and M. J. Humphries (2019). "Signal transduction via integrin adhesion complexes." Current Opinion in Cell Biology **56**: 14–21.

Hussein, H. A., L. R. Walker, U. M. Abdel-Raouf, S. A. Desouky, A. K. M. Montasser and S. M. Akula (2015). "Beyond RGD: Virus interactions with integrins." Archives of Virology **160**: 2669–2681.

Huveneers, S. and E. H. Danen (2009). "Adhesion signaling–crosstalk between integrins, Src and Rho." Journal of Cell Science **122**: 1059–1069.

Ingber, D. (2003a). "Mechanobiology and diseases of mechanotransduction." Annals of Medicine **35**: 564–577.

Ingber, D. E. (2003b). "Tensegrity I. Cell structure and hierarchical systems biology." Journal of Cell Science **116**: 1157–1173.

Ingber, D. E. (2003c). "Tensegrity II. How structural networks influence cellular information processing networks." Journal of Cell Science **116**: 1397–1408.

Isermann, P. and J. Lammerding (2013). "Nuclear mechanics and mechanotransduction in health and disease." Current Biology **23**: R1113–R1121.

Izard, T., G. Tran Van Nhieu and P. R. Bois (2006). "Shigella applies molecular mimicry to subvert vinculin and invade host cells." The Journal of Cell Biology **175**: 465–475.

Jaalouk, D. E. and J. Lammerding (2009). "Mechanotransduction gone awry." Nature Reviews. Molecular Cell Biology **10**: 63–73.

Jiang, G., G. Giannone, D. R. Critchley, E. Fukumoto and M. P. Sheetz (2003). "Two-piconewton slip bond between fibronectin and the cytoskeleton depends on talin." Nature **424**: 334–337.

Juanes, M. A., D. Isnardon, A. Badache, S. Brasselet, M. Mavrakis and B. L. Goode (2019). "The role of APC-mediated actin assembly in microtubule capture and focal adhesion turnover." Journal of Cell Biology **218**: 3415–3435.

Kalappurakkal, J. M., A. A. Anilkumar, C. Patra, T. S. Van Zanten, M. P. Sheetz and S. Mayor (2019). "Integrin mechano-chemical signaling generates plasma membrane nanodomains that promote cell spreading." Cell **177**: 1738–1756. e23.

Kaverina, I., O. Krylyshkina and J. V. Small (1999). "Microtubule targeting of substrate contacts promotes their relaxation and dissociation." The Journal of Cell Biology **146**: 1033–1044.

Kechagia, J. Z., J. Ivaska and P. Roca-Cusachs (2019). "Integrins as biomechanical sensors of the microenvironment." Nature Reviews. Molecular Cell Biology **1**.

Kim, M.-K., J.-W. Jang and S.-C. Bae (2018). "DNA binding partners of YAP/TAZ." BMB Reports **51**: 126.

Kim, N.-G. and B. M. Gumbiner (2015). "Adhesion to fibronectin regulates Hippo signaling via the FAK–Src–PI3K pathway." Journal of Cell Biology **210**: 503–515.

Kim, T., D. Hwang, D. Lee, J. H. Kim, S. Y. Kim and D. S. Lim (2017). "MRTF potentiates TEAD-YAP transcriptional activity causing metastasis." The EMBO Journal **36**: 520–535.

Kim, Y. and S. Kumar (2014). "CD44-mediated adhesion to hyaluronic acid contributes to mechanosensing and invasive motility." Molecular Cancer Research **12**: 1416–1429.

Kong, F., A. J. García, A. P. Mould, M. J. Humphries and C. Zhu (2009). "Demonstration of catch bonds between an integrin and its ligand." Journal of Cell Biology **185**: 1275–1284.

Kubiak, A., M. Chighizola, C. Schulte, N. Bryniarska, J. Wesolowska, M. Pudelek, M. Lasota, D. Ryszawy, A. Basta-Kaim, P. Laidler, A. Podesta and M. Lekka (2021). "Stiffening of DU145 prostate cancer cells driven by actin filaments – microtubule crosstalk conferring resistance to microtubule-targeting drugs." Nanoscale **13**: 6212–6226.

Kuo, J.-C., X. Han, C.-T. Hsiao, J. R. Yates III and C. M. Waterman (2011). "Analysis of the myosin-II-responsive focal adhesion proteome reveals a role for β-Pix in negative regulation of focal adhesion maturation." Nature Cell Biology **13**: 383–393.

Kuo, J. C.-H., J. G. Gandhi, R. N. Zia and M. J. Paszek (2018). "Physical biology of the cancer cell glycocalyx." Nature Physics **14**: 658–669.

Laflamme, S. E., S. Mathew-Steiner, N. Singh, D. Colello-Borges and B. Nieves (2018). "Integrin and microtubule crosstalk in the regulation of cellular processes." Cellular and Molecular Life Sciences **75**: 4177–4185.

Last, J. A., P. Russell, P. F. Nealey and C. J. Murphy (2010). "The applications of atomic force microscopy to vision science." Investigative Ophthalmology & Visual Science **51**: 6083–6094.

Lawson, C., S.-T. Lim, S. Uryu, X. L. Chen, D. A. Calderwood and D. D. Schlaepfer (2012). "FAK promotes recruitment of talin to nascent adhesions to control cell motility." Journal of Cell Biology **196**: 223–232.

Lawson, C. D. and A. J. Ridley (2018). "Rho GTPase signaling complexes in cell migration and invasion." Journal of Cell Biology **217**: 447–457.

Leclech, C., C. Natale and A. Barakat (2020). "The basement membrane as a structured surface – role in vascular health and disease." Journal of Cell Science **133**.

Levental, K. R., H. Yu, L. Kass, J. N. Lakins, M. Egeblad, J. T. Erler, S. F. Fong, K. Csiszar, A. Giaccia and W. Weninger (2009). "Matrix crosslinking forces tumor progression by enhancing integrin signaling." Cell **139**: 891–906.

Li, J. and T. A. Springer (2017). "Integrin extension enables ultrasensitive regulation by cytoskeletal force." Proceedings of the National Academy of Sciences **114**: 4685–4690.

Lim, Y., S.-T. Lim, A. Tomar, M. Gardel, J. A. Bernard-Trifilo, X. L. Chen, S. A. Uryu, R. Canete-Soler, J. Zhai and H. Lin (2008). "PyK2 and FAK connections to p190Rho guanine nucleotide exchange factor regulate RhoA activity, focal adhesion formation, and cell motility." The Journal of Cell Biology **180**: 187–203.

Liu, Y., R. Medda, Z. Liu, K. Galior, K. Yehl, J. P. Spatz, E. A. Cavalcanti-Adam and K. Salaita (2014). "Nanoparticle tension probes patterned at the nanoscale: Impact of integrin clustering on force transmission." Nano Letters **14**: 5539–5546.

Loebel, C., R. L. Mauck and J. A. Burdick (2019). "Local nascent protein deposition and remodeling guide mesenchymal stromal cell mechanosensing and fate in three-dimensional hydrogels." Nature Materials **18**: 883–891.

Maffioli, E., C. Schulte, S. Nonnis, F. G. Scalvini, C. Piazzoni, C. Lenardi, A. Negri, P. Milani and G. Tedeschi (2017). "Proteomic Dissection of Nanotopography-Sensitive Mechanotransductive Signaling Hubs that Foster Neuronal Differentiation in PC12 Cells." Frontiers in Cellular Neuroscience **11**.

Mammoto, A. and D. E. Ingber (2009). "Cytoskeletal control of growth and cell fate switching." Current Opinion in Cell Biology **21**: 864–870.

Martino, F., A. R. Perestrelo, V. Vinarský, S. Pagliari and G. Forte (2018). "Cellular mechanotransduction: From tension to function." Frontiers in Physiology **9**: 824.

Mendoza, P., R. Ortiz, J. Díaz, A. F. Quest, L. Leyton, D. Stupack and V. A. Torres (2013). "Rab5 activation promotes focal adhesion disassembly, migration and invasiveness in tumor cells." Journal of Cell Science **126**: 3835–3847.

Miller, N. L., E. G. Kleinschmidt and D. D. Schlaepfer (2014). "RhoGEFs in cell motility: Novel links between Rgnef and focal adhesion kinase." Current Molecular Medicine **14**: 221–234.

Miller, N. L., C. Lawson, E. G. Kleinschmidt, I. Tancioni, S. Uryu and D. D. Schlaepfer (2013). "A non-canonical role for Rgnef in promoting integrin-stimulated focal adhesion kinase activation." Journal of Cell Science **126**: 5074–5085.

Mitchison, T. and M. Kirschner (1988). "Cytoskeletal dynamics and nerve growth." Neuron **1**: 761–772.

Mitra, S. K. and D. D. Schlaepfer (2006). "Integrin-regulated FAK–Src signaling in normal and cancer cells." Current Opinion in Cell Biology **18**: 516–523.

Montagner, M. and S. Dupont (2020). "Mechanical forces as determinants of disseminated metastatic cell fate." Cells **9**.

Morgan, M. R., H. Hamidi, M. D. Bass, S. Warwood, C. Ballestrem and M. J. Humphries (2013). "Syndecan-4 phosphorylation is a control point for integrin recycling." Developmental Cell **24**: 472–485.

Mui, K. L., C. S. Chen and R. K. Assoian (2016). "The mechanical regulation of integrin–cadherin crosstalk organizes cells, signaling and forces." Journal of Cell Science **129**: 1093–1100.

Nardone, G., J. Oliver-De La Cruz, J. Vrbsky, C. Martini, J. Pribyl, P. Skládal, M. Pešl, G. Caluori, S. Pagliari and F. Martino (2017). "YAP regulates cell mechanics by controlling focal adhesion assembly." Nature Communications **8**: 1–13.

Nebuloni, M., L. Albarello, A. Andolfo, C. Magagnotti, L. Genovese, I. Locatelli, G. Tonon, E. Longhi, P. Zerbi and R. Allevi (2016). "Insight on colorectal carcinoma infiltration by studying perilesional extracellular matrix." Scientific Reports **6**: 22522.

Newell-Litwa, K. A., R. Horwitz and M. L. Lamers (2015). "Non-muscle myosin II in disease: Mechanisms and therapeutic opportunities." Disease Models & Mechanisms **8**: 1495–1515.

Nobes, C. D. and A. Hall (1995). "Rho, rac, and cdc42 GTPases regulate the assembly of multimolecular focal complexes associated with actin stress fibers, lamellipodia, and filopodia." Cell **81**: 53–62.

Noordstra, I. and A. Akhmanova (2017). "Linking cortical microtubule attachment and exocytosis." F1000Res **6**: 469.

Nourse, J. L. and M. M. Pathak (2017). "How cells channel their stress: Interplay between Piezo1 and the cytoskeleton." In Seminars in cell & developmental biology. Elsevier, vol **71**: 3–12.

Olson, E. N. and A. Nordheim (2010). "Linking actin dynamics and gene transcription to drive cellular motile functions." Nature Reviews. Molecular Cell Biology **11**: 353–365.

Olson, M. F. (2018). "Rho GTPases, their post-translational modifications, disease-associated mutations and pharmacological inhibitors." Small GTPases **9**: 203–215.

Orgaz, J., E. Crosas-Molist, A. Sadok, A. Perdrix-Rosell, O. Maiques, I. Rodriguez-Hernandez, J. Monger, S. Mele, M. Georgouli, V. Bridgeman, P. Karagiannis, R. Lee, P. Pandya, L. Boehme, F. Wallberg, C. Tape, S. Karagiannis, I. Malanchi and V. Sanz-Moreno (2020). "Myosin II reactivation and cytoskeletal remodeling as a hallmark and a vulnerability in melanoma therapy resistance." Cancer Cell **37**: 85-+.

Oria, R., T. Wiegand, J. Escribano, A. Elosegui-Artola, J. J. Uriarte, C. Moreno-Pulido, I. Platzman, P. Delcanale, L. Albertazzi and D. Navajas (2017). "Force loading explains spatial sensing of ligands by cells." Nature **552**: 219–224.

Osmanagic-Myers, S., T. Dechat and R. Foisner (2015). "Lamins at the crossroads of mechanosignaling." Genes & Development **29**: 225–237.

Panciera, T., L. Azzolin, M. Cordenonsi and S. Piccolo (2017). "Mechanobiology of YAP and TAZ in physiology and disease." Nature Reviews. Molecular Cell Biology **18**: 758.

Papalazarou, V., M. Salmeron-Sanchez and L. M. Machesky (2018). "Tissue engineering the cancer microenvironment – challenges and opportunities." Biophysical Reviews **10**: 1695–1711.

Park, J., D.-H. Kim, H.-N. Kim, C. J. Wang, M. K. Kwak, E. Hur, K.-Y. Suh, S. S. An and A. Levchenko (2016). "Directed migration of cancer cells guided by the graded texture of the underlying matrix." Nature Materials **15**: 792–801.

Park, J., D.-H. Kim and A. Levchenko (2018). "Topotaxis: A new mechanism of directed cell migration in topographic ECM gradients." Biophysical Journal **114**: 1257–1263.

Paszek, M. J., C. C. Dufort, O. Rossier, R. Bainer, J. K. Mouw, K. Godula, J. E. Hudak, J. N. Lakins, A. C. Wijekoon and L. Cassereau (2014). "The cancer glycocalyx mechanically primes integrin-mediated growth and survival." Nature **511**: 319–325.

Pathak, M. M., J. L. Nourse, T. Tran, J. Hwe, J. Arulmoli, T. L. Dai Trang, E. Bernardis, L. A. Flanagan and F. Tombola (2014). "Stretch-activated ion channel Piezo1 directs lineage choice in human neural stem cells." Proceedings of the National Academy of Sciences **111**: 16148–16153.

Pellinen, T., A. Arjonen, K. Vuoriluoto, K. Kallio, J. A. Fransen and J. Ivaska (2006). "Small GTPase Rab21 regulates cell adhesion and controls endosomal traffic of β1-integrins." The Journal of Cell Biology **173**: 767–780.

Pickup, M. W., J. K. Mouw and V. M. Weaver (2014). "The extracellular matrix modulates the hallmarks of cancer." EMBO Reports **15**: 1243–1253.

Pinho, S. S. and C. A. Reis (2015). "Glycosylation in cancer: Mechanisms and clinical implications." Nature Reviews. Cancer **15**: 540–555.

Premont, R. T., S. J. Perry, R. Schmalzigaug, J. T. Roseman, Y. Xing and A. Claing (2004). "The GIT/PIX complex: An oligomeric assembly of GIT family ARF GTPase-activating proteins and PIX family Rac1/Cdc42 guanine nucleotide exchange factors." Cellular Signaling **16**: 1001–1011.

Provenzano, P. P., D. R. Inman, K. W. Eliceiri, J. G. Knittel, L. Yan, C. T. Rueden, J. G. White and P. J. Keely (2008). "Collagen density promotes mammary tumor initiation and progression." BMC Medicine **6**: 11.

Rafiq, N. B. M., Y. Nishimura, S. V. Plotnikov, V. Thiagarajan, Z. Zhang, S. Shi, M. Natarajan, V. Viasnoff, P. Kanchanawong and G. E. Jones (2019). "A mechano-signaling network linking microtubules, myosin IIA filaments and integrin-based adhesions." Nature Materials **18**: 638.

Ray, A., O. Lee, Z. Win, R. M. Edwards, P. W. Alford, D.-H. Kim and P. P. Provenzano (2017). "Anisotropic forces from spatially constrained focal adhesions mediate contact guidance directed cell migration." Nature Communications **8**: 1–17.

Ren, X. D., W. B. Kiosses and M. A. Schwartz (1999). "Regulation of the small GTP-binding protein Rho by cell adhesion and the cytoskeleton." The EMBO Journal **18**: 578–585.

Ridone, P., M. Vassalli and B. Martinac (2019). "Piezo1 mechanosensitive channels: What are they and why are they important." Biophysical Reviews **11**(5): 795–805.

Sawada, Y., M. Tamada, B. J. Dubin-Thaler, O. Cherniavskaya, R. Sakai, S. Tanaka and M. P. Sheetz (2006). "Force sensing by mechanical extension of the Src family kinase substrate p130Cas." Cell **127**: 1015–1026.

Scheswohl, D. M., J. R. Harrell, Z. Rajfur, G. Gao, S. L. Campbell and M. D. Schaller (2008). "Multiple paxillin binding sites regulate FAK function." Journal of Molecular Signaling **3**: 1.

Schreiber, K. H. and B. K. Kennedy (2013). "When lamins go bad: Nuclear structure and disease." Cell **152**: 1365–1375.

Schulte, C., G. M. S. Ferraris, A. Oldani, M. Galluzzi, A. Podestà, L. Puricelli, V. De Lorenzi, C. Lenardi, P. Milani and N. Sidenius (2016a). "Lamellipodial tension, not integrin/ligand binding, is the crucial factor to realize integrin activation and cell migration." European Journal of Cell Biology **95**: 1–14.

Schulte, C., S. Rodighiero, M. A. Cappelluti, L. Puricelli, E. Maffioli, F. Borghi, A. Negri, E. Sogne, M. Galluzzi and C. Piazzoni (2016b). "Conversion of nanoscale topographical information of cluster-assembled zirconia surfaces into mechanotransductive events promotes neuronal differentiation." Journal of Nanobiotechnology **14**: 18.

Schvartzman, M., M. Palma, J. Sable, J. Abramson, X. Hu, M. P. Sheetz and S. J. Wind (2011). "Nanolithographic control of the spatial organization of cellular adhesion receptors at the single-molecule level." Nano Letters **11**: 1306–1312.

Seidlits, S. K., Z. Z. Khaing, R. R. Petersen, J. D. Nickels, J. E. Vanscoy, J. B. Shear and C. E. Schmidt (2010). "The effects of hyaluronic acid hydrogels with tunable mechanical properties on neural progenitor cell differentiation." Biomaterials **31**: 3930–3940.

Sharifi, M. N., E. E. Mowers, L. E. Drake, C. Collier, H. Chen, M. Zamora, S. Mui and K. F. Macleod (2016). "Autophagy promotes focal adhesion disassembly and cell motility of metastatic tumor cells through the direct interaction of paxillin with LC3." Cell Reports **15**: 1660–1672.

Sheridan, C. (2019). "Pancreatic cancer provides testbed for first mechanotherapeutics." Nature Biotechnology **37**: 829–831.

Shi, F. and J. Sottile (2008). "Caveolin-1-dependent β1 integrin endocytosis is a critical regulator of fibronectin turnover." Journal of Cell Science **121**: 2360–2371.

Slack, R., S. Macdonald, J. Roper, R. Jenkins and R. Hatley (2021). "Emerging therapeutic opportunities for integrin inhibitors." Nature Reviews. Drug Discovery **21**: 60–78.

Stehbens, S. and T. Wittmann (2012). "Targeting and transport: How microtubules control focal adhesion dynamics." Journal of Cell Biology **198**: 481–489.

Stehbens, S. J., M. Paszek, H. Pemble, A. Ettinger, S. Gierke and T. Wittmann (2014). "CLASPs link focal-adhesion-associated microtubule capture to localized exocytosis and adhesion site turnover." Nature Cell Biology **16**: 558–570.

Strohmeyer, N., M. Bharadwaj, M. Costell, R. Fässler and D. J. Müller (2017). "Fibronectin-bound α5β1 integrins sense load and signal to reinforce adhesion in less than a second." Nature Materials **16**: 1262–1270.

Sun, Q., G. Chen, J. W. Streb, X. Long, Y. Yang, C. J. Stoeckert and J. M. Miano (2006). "Defining the mammalian CArGome." Genome Research **16**: 197–207.

Sun, Z., M. Costell and R. Fässler (2019). "Integrin activation by talin, kindlin and mechanical forces." Nature Cell Biology **21**: 25–31.

Swift, J., I. L. Ivanovska, A. Buxboim, T. Harada, P. D. P. Dingal, J. Pinter, J. D. Pajerowski, K. R. Spinler, J.-W. Shin and M. Tewari (2013). "Nuclear lamin-A scales with tissue stiffness and enhances matrix-directed differentiation." Science **341**: 1240104.

Tamkun, J. W., D. W. Desimone, D. Fonda, R. S. Patel, C. Buck, A. F. Horwitz and R. O. Hynes (1986). "Structure of integrin, a glycoprotein involved in the transmembrane linkage between fibronectin and actin." Cell **46**: 271–282.

Tarbell, J. and L. Cancel (2016). "The glycocalyx and its significance in human medicine." Journal of Internal Medicine **280**: 97–113.

Theodosiou, M., M. Widmaier, R. T. Böttcher, E. Rognoni, M. Veelders, M. Bharadwaj, A. Lambacher, K. Austen, D. J. Müller and R. Zent (2016). "Kindlin-2 cooperates with talin to activate integrins and induces cell spreading by directly binding paxillin." Elife **5**: e10130.

Tomar, A., S.-T. Lim, Y. Lim and D. D. Schlaepfer (2009). "A FAK-p120RasGAP-p190RhoGAP complex regulates polarity in migrating cells." Journal of Cell Science **122**: 1852–1862.

Totaro, A., T. Panciera and S. Piccolo (2018). "YAP/TAZ upstream signals and downstream responses." Nature Cell Biology **20**: 888–899.

Tschumperlin, D. and D. Lagares (2020). "Mechano-therapeutics: Targeting Mechanical Signaling in Fibrosis and Tumor Stroma." Pharmacology & Therapeutics **212**: 107575.

Uhler, C. and G. Shivashankar (2017a). "Chromosome intermingling: Mechanical hotspots for genome regulation." Trends in Cell Biology **27**: 810–819.

Uhler, C. and G. Shivashankar (2017b). "Regulation of genome organization and gene expression by nuclear mechanotransduction." Nature Reviews. Molecular Cell Biology **18**: 717–727.

Vicente-Manzanares, M. and A. R. Horwitz (2011). "Adhesion dynamics at a glance." Journal of Cell Science **124**: 3923–3927.

Vogel, V. (2018). "Unraveling the mechanobiology of extracellular matrix." Annual Review of Physiology **80**: 353–387.

Wang, N., J. P. Butler and D. E. Ingber (1993). "Mechanotransduction across the cell surface and through the cytoskeleton." Science **260**: 1124–1127.

Wang, N., J. D. Tytell and D. E. Ingber (2009). "Mechanotransduction at a distance: Mechanically coupling the extracellular matrix with the nucleus." Nature Reviews. Molecular Cell Biology **10**: 75–82.

Webb, D. J., K. Donais, L. A. Whitmore, S. M. Thomas, C. E. Turner, J. T. Parsons and A. F. Horwitz (2004). "FAK–Src signaling through paxillin, ERK and MLCK regulates adhesion disassembly." Nature Cell Biology **6**: 154–161.

Winograd-Katz, S. E., R. Fässler, B. Geiger and K. R. Legate (2014). "The integrin adhesome: From genes and proteins to human disease." Nature Reviews. Molecular Cell Biology **15**: 273–288.

Yang, B., Z. Z. Lieu, H. Wolfenson, F. M. Hameed, A. D. Bershadsky and M. P. Sheetz (2016). "Mechanosensing controlled directly by tyrosine kinases." Nano Letters **16**: 5951–5961.

Yao, M., B. T. Goult, H. Chen, P. Cong, M. P. Sheetz and J. Yan (2014). "Mechanical activation of vinculin binding to talin locks talin in an unfolded conformation." Scientific Reports **4**: 1–7.

Young, J., X. Hua, H. Somsel, F. Reichart, H. Kessler and J. Spatz (2020). "Integrin subtypes and nanoscale ligand presentation influence drug sensitivity in cancer cells." Nano Letters **20**: 1183–1191.

Young, J. L., A. W. Holle and J. P. Spatz (2016). "Nanoscale and mechanical properties of the physiological cell–ECM microenvironment." Experimental Cell Research **343**: 3–6.

Yue, J., M. Xie, X. Gou, P. Lee, M. D. Schneider and X. Wu (2014). "Microtubules regulate focal adhesion dynamics through MAP4K4." Developmental Cell **31**: 572–585.

Zaidel-Bar, R., R. Milo, Z. Kam and B. Geiger (2007). "A paxillin tyrosine phosphorylation switch regulates the assembly and form of cell-matrix adhesions." Journal of Cell Science **120**: 137–148.

Sample Preparation

Carmela Rianna, Carsten Schulte

5.1 Cell Culture

5.1.1 Introduction

Cell culture is one of the major techniques used in life science to study biological and molecular cellular processes, in vitro. In the 1950s, the first cell line, HeLa, was cultured from human cervical cancer (Gey, 1952), but large-scale use of cell culture was only achieved from the mid-1980s. Since then, the development of cell culture, in vitro, has led to seminal findings on tissue physiology and pathophysiology outside the organism. Decades of cell culture experiments have provided the base for our interpretation of complex biological phenomena, such as stem cell differentiation (Jaiswal et al., 1997) or tissue morphogenesis (Schnaper et al., 1993). More recently, being of pivotal significance in the framework of this book, by employing cell culture substrates with different structural and biophysical properties (particularly in regards to rigidity and topography) the fundamental importance of the interplay between cells and microenvironment has been disclosed (Engler et al., 2006, Dalby et al., 2007, Engler et al., 2007, Dalby et al., 2014, Crowder et al., 2016, Young et al., 2016). It is now evident that almost all cells are indeed not solitary entities but that they strongly interact and cross-talk with their native extracellular matrix (ECM). A dynamic interplay of biochemical and biophysical signals deriving from the microenvironment orchestrates intricate intracellular signaling cascades that influence the phenotypic and functional fate by altering gene and protein expression (Birgersdotter et al., 2005).

After decades of performing cell culture on conventional cell culture systems, based mainly on rigid and flat 2D glass/plastic flasks or dishes, a growing interest has been shown towards novel methods that reproduce the structural and biophysical features of the in vivo environment, such as micro/nanostructured substrates or 3D systems with different elastic properties and mesh sizes (Place et al., 2009, Kollmannsberger et al., 2011, Mendes, 2013, Chen et al., 2014, Dalby et al., 2014, Crowder et al., 2016, Duval et al., 2017). This chapter aims to give a short introduction into common terminologies, concepts, necessities, and practices in cell culture that need to be considered. It is intended to provide a general impression of cell culture handling and necessities for interested readers of this book. With this goal, we will highlight the major differences between cell culture types (i.e., cell lines and primary cell culture), as well as traditional cell culture systems and new systems that take into account biophysical features.

Carmela Rianna, Institute of Biophysics, University of Bremen, Bremen, Germany
Carsten Schulte, Interdisciplinary Centre for Nanostructured Materials and Interfaces (C.I.Ma.I.Na.), Università degli Studi di Milano, Milan, Italy

https://doi.org/10.1515/9783110989380-007

5.1.1.1 Cell Lines and Primary Culture

There are different types of cell culture, mainly divided into "culture of established cell lines" (or simply "cell lines") and "primary culture" (Wang, 2006, Kaur and Dufour, 2012, Uysal et al., 2018). Cell lines are permanently established cell cultures that will proliferate indefinitely, given appropriate conditions, such as specific medium, space, and controlled environment. They are usually derived from cancerous tissues, or have been immortalized in a specific way (e.g., by overriding of the cell cycle by viral gene, such as human papilloma virus or SV-40; or by expression of essential proteins in control of cell senescence, like the telomerase hTERT). However, this tumorous origin and/or the necessary transformations are the main disadvantage of cell lines. Although they retain characteristics of their non-immortal counterparts, they have incurred significant alterations (Wang, 2006, Pan et al., 2009, Pastor et al., 2010, Kaur and Dufour, 2012, Geraghty et al., 2014b, Uysal et al., 2018).

Primary cells, on the contrary, are cells isolated directly from human or animal tissues, by using enzymatic or mechanical methods. The advantage of primary cell culture is that they have been directly removed from the in vivo condition and, therefore, they more accurately resemble functions and properties of the in vivo cells and tissues. On the other side, setting up culture conditions for primary cells is more complicated and their lifespan is limited (Wang, 2006, Uysal et al., 2018).

In both cases, to reproduce the physiological conditions of in vivo environment and to ensure cell viability, proliferation, and growth, cell culture needs to be performed in a sterile environment at optimal pH, humidity, and temperature, within a culture medium. The choice of the appropriate cell culture medium or growth medium is crucial. Different cell types might need different media; however, the typical composition of culture media is a mixture of components that enhance and facilitate cell viability and growth. Most of these components include amino acids, vitamins, lipids, glucose, carbohydrates, inorganic salts, and growth factors (Arora, 2013). Additionally, serum is often added to culture media, as a further source of growth factors to stimulate cell division and growth, and proteins (e.g., fibronectin and vitronectin) to improve cell adhesion. The most common source of serum is bovine blood. The use of serum however comes with a relatively high cost and the disadvantage that it is not a standardized component, with batch-to-batch variability. For the latter reason, in cell biological experiments, most of the time, the amount of serum in the medium is reduced to the bare minimum necessary for the functioning and well-being of the cells. For some cell types, the exigencies to guarantee cellular proliferation and well-being have been determined, so that defined serum-free media exist that provide all necessary nutrients and growth factors (Wang, 2006, Uysal et al., 2018).

What type of cells to choose between cell lines and primary culture depends pretty much on the type of experiment planned, as well as on the available infrastructure and expertise available (particularly in the case of primary cells).

5.1.1.2 Conventional 2D Cell Culture Versus ECM-Mimicking Culture Systems

When performing cell culture in vitro, it is crucial to reproduce the conditions of in vivo tissues, as much as possible. For decades, the conventional cell culture system employed was based on a 2D approach in which cells are grown on a plate, which is rigid and flat. These cell culture systems, such as flasks or Petri dishes with plastic or glass bottoms, have been routinely used in thousands of laboratories worldwide, providing interesting findings that helped to understand better cell physiology and behavior. However, these conventional 2D systems with basically featureless surfaces at the nanoscale level are not able to recapitulate important aspects of the in vivo context and lack the geometrical, structural, and mechanical properties of native ECM (Place et al., 2009, Kollmannsberger et al., 2011, Gasiorowski et al., 2013, Mendes, 2013, Chen et al., 2014, Dalby et al., 2014, Crowder et al., 2016).

Furthermore, in 2D systems, the cell membrane can only partially face the substrates and the neighboring cells. This can lead to unnatural interaction with the environment and soluble factors (Gieni and Hendzel, 2008). Thanks to the progress in the field of mechanobiology, it became evident that these biophysical properties are essentially involved in virtually all cell biological processes (Place et al., 2009, Kollmannsberger et al., 2011, Gasiorowski et al., 2013, Mendes, 2013, Chen et al., 2014, Dalby et al., 2014, Crowder et al., 2016, Young et al., 2016). In order to overcome the limitation of conventional cell culture, in the last few decades, many techniques and methods have been applied to further the development of biomaterial systems that enable the culturing of cells on more biomimetic supports. Examples are micro/nanostructured surfaces or 3D ECM-like networks, which allow the cells to interact with a substrate that provide 3D biophysical cues (such as elasticity and/or nanotopography), mimicking those found in natural ECM (Place et al., 2009, Haycock, 2011, Mendes, 2013, Chen et al., 2014, Dalby et al., 2014, Andersen et al., 2015, Ravi et al., 2015, Crowder et al., 2016, Young et al., 2016). Differences in cellular behavior induced by the interaction with biophysical substrate characteristics depend on the way cells experience their microenvironment and the induced signaling cascades. These processes have been described under the terms mechanosensing and mechanotransduction, and can influence numerous and versatile cell biological key processes, such as survival, proliferation, adhesion, migration, and differentiation, or also on a broader scale, tissue development and homeostasis (Gauthier and Roca-Cusachs, 2018, Kechagia et al., 2019, Sun et al., 2019). Accordingly, it has been shown that cell behavior and fate are strongly influenced by the type of cell culture system used. For example, adhesion of cells to flat and rigid 2D surfaces can alter the cell metabolism and functionality, leading to results that may not reproduce the in vivo context (Lee et al., 2008). The use of 2D or 3D systems influences processes such as immune system activation, defense response, cell adhesion, apoptosis, and tissue formation (Chen et al., 1997, Birgersdotter et al., 2005, Kenny et al., 2007, Weigelt et al., 2010, Duval et al., 2017, Kapałczyńska et al., 2018). Further details on mechanotransduction can be found in chapter 4.6.

In nature, many cells are fully embedded in the ECM; 3D systems can therefore be biologically more relevant than 2D systems (Gevaert, 2012, Duval et al., 2017, Kapałczyńska et al., 2018), because they are able to recapitulate better critical cues present in the native ECM in three dimensions and facilitate biological processes such as tissue organization (Lee et al., 2008). 3D cell culture can grow within a scaffold or with a scaffold-free technique. Most used scaffolds are based on natural and polymer hydrogels (Tibbitt and Anseth, 2009) or matrices (Grinnell, 2003, Lee et al., 2008). Examples of polymer hydrogels suited to mimic natural ECM are polyethylene glycol (PEG), poly(hydroxyethyl methacrylate) (polyHEMA), polyvinyl alcohol (PVA), and polycaprolactone (PCL). Examples of natural polymers (and proteins) that are able to form hydrogels for 3D cell culture are alginate, chitosan, hyaluronan, dextran, collagen, and fibrin (Place et al., 2009). Furthermore, protocols have been established to decellularize natural ECM, which can then serve as scaffold for cell culture (Hoshiba, 2017, Taylor et al., 2018).

Examples of scaffold-free methods include cellular spheroids. They are created with different techniques, such as hanging drop, rotating culture, and concave plate methods (Kelm et al., 2003, Timmins et al., 2005, Ivascu and Kubbies, 2006, Pampaloni et al., 2007). 3D spheroid cultures have been largely employed to study cancer processes, especially because they are able to capture the complexity of solid tumors. Moreover, their arrangement and migratory features closely mimic tumor microregions (Desoize, 2000).

5.1.1.3 Cocultures

Apart from the ECM, many cells also interact with other cells (including other cell types, for example, neurons and astrocytes in the brain). Therefore, it can be of major interest to study these interactions of cell populations in coculture conditions. Replicating these kinds of interactions in a controlled manner in cell culture is challenging because, often, it requires achieving a physical separation of the two (or more) cell types and/or populations that still enable the typical type of communication and/or interaction of these cells. Various approaches of coculture systems have been developed, for example, transwell systems (i.e., inserts for cell culture wells that can be populated by one cell population and introduced into the well of cell culture plates that host the other cell population; pores in the transwell insert permit the diffusion of signaling proteins secreted by the cells), microfluidic systems with separated compartments that are connected by fluid channels, or micropatterning techniques that allow to control the adhesion of cells in specific areas in the same device (Théry, 2010, Goers et al., 2014, D'Arcangelo and McGuigan, 2015).

5.1.2 Cell Culture Equipment

Here, we will give a short introduction of standard cell culture equipment and requirements, focusing on the exigencies for culturing cells derived from the mammalian body (Wang, 2006, Geraghty et al., 2014b, Uysal et al., 2018). Equipment and requirements for the culture and handling of other cell types from common non-mammalian animal models such as fish cells (e.g., zebrafish), amphibian cells (e.g., xenopus), insect cells (e.g., drosophila), plant cells (e.g., *Arabidopsis*), or essentials for work on pathogens or BSL-4 materials (which require particular training), can differ substantially from the ones mentioned here and will not be covered in this chapter.

5.1.2.1 Laboratory Equipment

The laboratory equipment must ensure the in vitro reproduction of the in vivo physiological conditions in which cells live and grow (e.g., with respect to temperature, pH, or CO_2), as well as the avoidance of stress (chemical or physical) for the cell culture.

Basic equipment and supplies:

- Incubator (incl. CO_2 supply and settings).
- Biosafety cabinets (designed for work comprising biosafety levels BSL-1, -2, -3 materials).
  Some remarks:
  The working area within the biosafety cabinets has to be large enough to permit easy and comfortable cell culture handling of one person at a time, as also cleaning procedures (inside and outside). Also, appropriate illumination has to be provided. Disinfection with UV light and 70% alcohol before and after every use is mandatory.
- Phase contrast microscope (e.g., to control the well-being of cells or the proceedings of subculturing procedures).

Furthermore, various other equipment and supplies are necessary to be able to perform cell culture:

Centrifuges, vortex, fridge, freezer, cryostorage (cryoboxes, liquid nitrogen, tanks, etc.), cell culture plasticware (flasks +/− vented caps, plates, pipette tips, centrifuge tubes, caps, filters, etc.), syringes, needles, pH meter, autoclave, laboratory scale, waste containers, media (plus supplements depending on cell types), and cell counting devices (counting chambers, or automated cell counting apparatus).

5.1.2.2 Physical Environment

The physical environment should, if possible, be dedicated only to cell/tissue culture and should be as aseptic as possible, for example, with air lock. If that is not possible, there should, at least, be a designated working area for cell/tissue culture with minimized entry and exit (especially of personnel without cell culture expertise). The cell/tissue culture environment should be under positive pressure and equipped with HEPA (high-efficiency particulate air) filtered air flowing through.

There should be specific and adequate storage areas for liquids, chemicals, and cell culture supplies, particularly, if they are sterile.

Always read instructions that come with the products carefully to store them in the right manner; some reagents are sensitive to light and have to be stored in the dark; many reagents for cell culture have to be stored in cool places or frozen.

5.1.3 Cell Culture Procedures

Many different continuous cell lines share the same protocols for cell culture. However, some cells require special treatments, for example, coated flasks, special media, or growth factors (Meleady and O'Connor, 2006, Geraghty et al., 2014a). In the following paragraphs, we describe some of the most commonly and routinely used cell culture protocols in biology, such as subculture, cell cryopreservation, and thawing.

5.1.3.1 Subculture of Adherent Cells

Subculture is a technique routinely performed during cell culture. In this procedure, also known as passaging, cells from a previous culture are transferred to fresh growth medium in a new culture. Adherent cell lines, in fact, will grow in vitro until they are "confluent," that is, until they have covered all the surface area available. At this point, to prevent the culture from dying, cells need to be subcultured. First of all, cells have to be suspended; therefore, adherent cells are first washed with PBS and then incubated with a proteolytic enzyme to detach them from the substrate. The enzyme most frequently used for this purpose is trypsin, a serine protease acting on the C-terminal side of lysine or arginine (Rick, 1974, Evnin et al., 1990). The trypsin is usually of mammalian origin, which means that it has an optimal operating temperature at ~ 37 °C. Therefore, the trypsinization is performed in the incubator, and after a few minutes (depending on the cell type and their adhesion strength), the cells will be detached from the substrate (due to the cleavage of adhesion-mediating surface receptors) and suspended. Often, trypsin is used in combination with EDTA (ethylenediamine tetraacetic acid), which is a metal ion chelator and, as such, it can chelate divalent cations that are important for the activation and functioning of integrins. For some cell lines, the exposure to proteases is harmful; in this case, EDTA alone or cell scrapers can be used as alternative to trypsin, to detach the cells.

Once in suspension, the cells can be collected and centrifuged (usually 5 min) to obtain a cell pellet. The supernatant medium (still containing the trypsin, which, at longer exposure, could damage the cells) can be discarded by pipetting, so that the remaining cell pellet can be finally re-suspended in fresh medium. In some cases, before starting the new culture, a cell counting process will be necessary, usually performed with a hemocytometer, that is, a counting-chamber device (Tolnai, 1975, Phelan and Lawler, 1997), or a cell counting apparatus, in order to control the amount of cells to be subcultured. Once cells have been subcultured in fresh medium in tissue culture flasks or Petri dishes, they can be stored in the incubator and allowed to adhere to the surface overnight. Afterwards, the medium is changed to remove cell debris or dead/floating cells. The process can be repeated when cells are confluent again, recording the passage time, every time. The passage time is the number of times the culture has been subcultured and will help keeping track of the "age" of a cell culture. This number is particularly important in controlling experiment reproducibility, because cells might change characteristics if they have been passaged too often (Geraghty et al., 2014b).

5.1.3.2 Cryopreservation and Thawing

Cryopreservation allows storage of cells at low temperatures and avoids having to keep the cell lines in culture for a long time, if not needed. As with many other cell culture procedures, every cell line will require a specific protocol for the best results of cryopreservation. However, some of the criteria that should always be respected are:
- Cells should be 90% viable, ensuring absence of contamination.
- High concentration of serum should be used.
- A cryoprotectant (such as dimethyl sulphoxide, DMSO) should be used to protect the cells from the formation of ice crystals.
- Cultured cells should be frozen at a high concentration and low passage number.

Overall, the basic principle for successful cryopreservation is based on "slow freezing and quick thawing." In fact, cells should be frozen slowly, reducing the temperature by approximately 1 °C per minute. Sterile cryovials are used to store frozen cells at ultra-low temperature, for example, in liquid nitrogen or in the gas phase above liquid nitrogen (below −135 °C). During the cryopreservation process, special care should be taken when storing the cryovials in liquid nitrogen. In fact, liquid nitrogen is associated with potential hazard, and it is, therefore, important that all the users wear personal protective equipment and are trained to minimize the risk of incidents during this process.

In contrast to cryopreservation, the process of thawing cryopreserved cells is used when cells need to be brought in culture after freezing them. In this case, the procedure needs to be fast to avoid the cells coming into long contact with DMSO, which is cytotoxic above 4 °C. The cryovials containing frozen cells are therefore removed from liquid nitrogen and allowed to thaw at 37 °C (usually in a heated water bath). Then, cells are rapidly transferred to previously aliquoted medium (to dilute the DMSO) and centrifuged for 3–5 min, which leads to a formation of a cell pellet. In this way, the supernatant can be removed (and with it, the DMSO) and the remaining cell pellet can be resuspended in fresh medium and plated into appropriate cell culture flasks or dishes. If needed, the cells can also be counted at this stage and added to a culture flask or dish with an appropriate volume of medium. The cells are allowed to adhere to the surface of the flasks overnight and, afterwards, the medium can be replaced to remove any dead cells or debris.

5.1.4 Safety Requirements During Cell Culture

A cell culture laboratory has many specific hazards to consider in order to avoid harming individuals and the environment. Here, we can only give a very short overview, but every cell culture facility has specific procedures and training for people that are working in cell culture. The main hazard is the risk of contact with potentially infective, toxic, or corrosive solvents, which are often required when manipulating human or animal cells or to maintain cell culture as well as during various experimental procedures. Every time a specific substance is handled, it is important to read carefully the Safety Data Sheet (SDS) that contains all information regarding the properties of the substance, such as melting and boiling points, infection and toxicity information, disposal, and procedure to handle spills.

Safety equipment is also fundamental to protect users; for example, personal protective equipment (PPE) and primary barriers like biosafety cabinet (or cell culture hood). PPE represents a barrier between the user and hazardous agents, and includes items for personal protection such as gloves, coats, safety glasses, goggles, and face shields. Biosafety cabinet prevents contamination in both users and cell culture.

Simple safe laboratory practices can also help protect persons and cell culture, and include the following:
- The work surfaces need to be decontaminated and cleaned before and after any experiments.

- Appropriate PPE needs to be used.
- Gloves need to be changed if contaminated.
- All the materials (i.e., needles, broken glasses, pipettes, and so on) need to be disposed of according to the institutional policies.
- Hands need to be washed before leaving the laboratory.

5.1.5 Conclusions

Since its introduction, cell culture has proved to be a valuable tool in studying biological and molecular processes and functions, in vitro. Advances in technology have also provided tools to overcome the limitations of conventional cell culture systems, which insufficiently recapitulate biophysical and biochemical properties of the natural ECM. Overall, despite the fact that in vitro culture represents only an approximation of the in vivo context, if consciously practiced, it can provide a powerful tool to answer fundamental biological questions.

In this chapter, we proposed a brief overview on different cell types, the equipment required to perform cell culture, some of the most commonly used procedures, and safety requirements. The procedures described in this chapter are meant to be of general introduction and can largely differ for some cell types and often need to be adapted according to the planned experiments. We, therefore, suggest to always seek information on the specific cell type intended to be used and following guidance from experts in biology, before starting a cell culture, in vitro.

References

Andersen, T., P. Auk-Emblem and M. Dornish (2015). "3D cell culture in alginate hydrogels." Microarrays **4**(2): 133–161.

Arora, M. (2013). "Cell culture media: A review." Mater Methods **3**(175): 24.

Birgersdotter, A., R. Sandberg and I. Ernberg (2005). Gene expression perturbation in vitro – a growing case for three-dimensional (3D) culture systems. Seminars in cancer biology, Elsevier.

Chen, C. S., M. Mrksich, S. Huang, G. M. Whitesides and D. E. Ingber (1997). "Geometric control of cell life and death.." Science **276**(5317): 1425–1428.

Chen, W., Y. Shao, X. Li, G. Zhao and J. Fu (2014). "Nanotopographical surfaces for stem cell fate control: Engineering mechanobiology from the bottom." Nano Today **9**(6): 759–784.

Crowder, S. W., V. Leonardo, T. Whittaker, P. Papathanasiou and M. M. Stevens (2016). "Material cues as potent regulators of epigenetics and stem cell function." Cell Stem Cell **18**(1): 39–52.

D'Arcangelo, E. and A. P. McGuigan (2015). "Micropatterning strategies to engineer controlled cell and tissue architecture in vitro." BioTechniques **58**(1): 13–23.

Dalby, M. J., N. Gadegaard and R. O. Oreffo (2014). "Harnessing nanotopography and integrin–matrix interactions to influence stem cell fate." Nature Materials **13**(6): 558–569.

Dalby, M. J., N. Gadegaard, R. Tare, A. Andar, M. O. Riehle, P. Herzyk, C. D. Wilkinson and R. O. Oreffo (2007). "The control of human mesenchymal cell differentiation using nanoscale symmetry and disorder." Nature Materials **6**(12): 997–1003.

Desoize, B. (2000). "Contribution of three-dimensional culture to cancer research." Critical Reviews in Oncology/hematology **36**(2–3): 59.

Duval, K., H. Grover, L.-H. Han, Y. Mou, A. F. Pegoraro, J. Fredberg and Z. Chen (2017). "Modeling physiological events in 2D vs. 3D cell culture." Physiology **32**(4): 266–277.

Engler, A. J., F. Rehfeldt, S. Sen and D. E. Discher (2007). "Microtissue elasticity: Measurements by atomic force microscopy and its influence on cell differentiation." Methods in Cell Biology **83**: 521–545.

Engler, A. J., S. Sen, H. L. Sweeney and D. E. Discher (2006). "Matrix elasticity directs stem cell lineage specification." Cell **126**(4): 677–689.

Evnin, L. B., J. R. Vásquez and C. S. Craik (1990). "Substrate specificity of trypsin investigated by using a genetic selection." Proceedings of the National Academy of Sciences **87**(17): 6659–6663.

Gasiorowski, J. Z., C. J. Murphy and P. F. Nealey (2013). "Biophysical cues and cell behavior: The big impact of little things." Annual Review of Biomedical Engineering **15**: 155–176.

Gauthier, N. C. and P. Roca-Cusachs (2018). "Mechanosensing at integrin-mediated cell–matrix adhesions: From molecular to integrated mechanisms." Current Opinion in Cell Biology **50**: 20–26.

Geraghty, R., A. Capes-Davis, J. Davis, J. Downward, R. Freshney, I. Knezevic, R. Lovell-Badge, J. Masters, J. Meredith and G. Stacey (2014a). "Guidelines for the use of cell lines in biomedical research." British Journal of Cancer **111**(6): 1021.

Geraghty, R., A. Capes-Davis, J. Davis, J. Downward, R. Freshney, I. Knezevic, R. Lovell-Badge, J. Masters, J. Meredith and G. Stacey (2014b). "Guidelines for the use of cell lines in biomedical research." British Journal of Cancer **111**(6): 1021–1046.

Gevaert, M. (2012). "Engineering 3D tissue systems to better mimic human biology." Bridge **42**: 48–55.

Gey, G. (1952). "Tissue culture studies of the proliferative capacity of cervical carcinoma and normal epithelium." Cancer Research **12**: 264–265.

Gieni, R. S. and M. J. Hendzel (2008). "Mechanotransduction from the ECM to the genome: Are the pieces now in place?." Journal of Cellular Biochemistry **104**(6): 1964–1987.

Goers, L., P. Freemont and K. M. Polizzi (2014). "Co-culture systems and technologies: Taking synthetic biology to the next level." Journal of the Royal Society Interface **11**(96): 20140065.

Grinnell, F. (2003). "Fibroblast biology in three-dimensional collagen matrices." Trends in Cell Biology **13**(5): 264–269.

Haycock, J. W. (2011). "3D cell culture: A review of current approaches and techniques." Methods in Molecular Biology **695**: 1–15.

Hoshiba, T. (2017). "Cultured cell-derived decellularized matrices: A review towards the next decade." Journal of Materials Chemistry B **5**(23): 4322–4331.

Ivascu, A. and M. Kubbies (2006). "Rapid generation of single-tumor spheroids for high-throughput cell function and toxicity analysis." Journal of Biomolecular Screening **11**(8): 922–932.

Jaiswal, N., S. E. Haynesworth, A. I. Caplan and S. P. Bruder (1997). "Osteogenic differentiation of purified, culture-expanded human mesenchymal stem cells in vitro." Journal of Cellular Biochemistry **64**(2): 295–312.

Kapałczyńska, M., T. Kolenda, W. Przybyła, M. Zajączkowska, A. Teresiak, V. Filas, M. Ibbs, R. Bliźniak, Ł. Łuczewski and K. Lamperska (2018). "2D and 3D cell cultures–a comparison of different types of cancer cell cultures." Archives of Medical Science: AMS **14**(4): 910.

Kaur, G. and J. M. Dufour (2012). "Cell lines: Valuable tools or useless artifacts". Spermatogenesis **2**(1): 1–5.

Kechagia, J. Z., J. Ivaska and P. Roca-Cusachs (2019). Integrins as biomechanical sensors of the microenvironment. Nature reviews molecular cell biology. 1.

Kelm, J. M., N. E. Timmins, C. J. Brown, M. Fussenegger and L. K. Nielsen (2003). "Method for generation of homogeneous multicellular tumor spheroids applicable to a wide variety of cell types." Biotechnology and Bioengineering **83**(2): 173–180.

Kenny, P. A., G. Y. Lee, C. A. Myers, R. M. Neve, J. R. Semeiks, P. T. Spellman, K. Lorenz, E. H. Lee, M. H. Barcellos-Hoff and O. W. Petersen (2007). "The morphologies of breast cancer cell lines in three-dimensional assays correlate with their profiles of gene expression." Molecular Oncology **1**(1): 84–96.

Kollmannsberger, P., C. Bidan, J. Dunlop and P. Fratzl (2011). "The physics of tissue patterning and extracellular matrix organization: How cells join forces." Soft Matter **7**(20): 9549–9560.

Lee, J., M. J. Cuddihy and N. A. Kotov (2008). "Three-dimensional cell culture matrices: State of the art." Tissue Engineering. Part B, Reviews **14**(1): 61–86.

Meleady, P. and R. O'Connor (2006). General Procedures for Cell Culture. In: Celis J.E., ed. Cell biology: a laboratory handbook. 3rd ed. New York: Elsevier, 13–20.

Mendes, P. M. (2013). "Cellular nanotechnology: Making biological interfaces smarter." Chemical Society Reviews **42**(24): 9207–9218.

Pampaloni, F., E. G. Reynaud and E. H. Stelzer (2007). "The third dimension bridges the gap between cell culture and live tissue." Nature Reviews. Molecular Cell Biology **8**(10): 839–845.

Pan, C., C. Kumar, S. Bohl, U. Klingmueller and M. Mann (2009). "Comparative proteomic phenotyping of cell lines and primary cells to assess preservation of cell type-specific functions." Molecular & Cellular Proteomics **8**(3): 443–450.

Pastor, D. M., L. S. Poritz, T. L. Olson, C. L. Kline, L. R. Harris III, W. A. Koltun, V. M. Chinchilli and R. B. Irby (2010). "Primary cell lines: False representation or model system? a comparison of four human colorectal tumors and their coordinately established cell lines." International Journal of Clinical and Experimental Medicine **3**(1): 69.

Phelan, M. C. and G. Lawler (1997). "Cell counting." Current Protocols in Cytometry (1): A. 3A 1-A. 3A. 4.

Place, E. S., J. H. George, C. K. Williams and M. M. Stevens (2009). "Synthetic polymer scaffolds for tissue engineering." Chemical Society Reviews **38**(4): 1139–1151.

Ravi, M., V. Paramesh, S. Kaviya, E. Anuradha and F. P. Solomon (2015). "3D cell culture systems: Advantages and applications." Journal of Cellular Physiology **230**(1): 16–26.

Rick, W. (1974). Trypsin. Methods of enzymatic analysis, Elsevier: 1013–1024.

Schnaper, H. W., D. S. Grant, W. G. Stetler-Stevenson, R. Fridman, G. D'Orazi, A. N. Murphy, R. E. Bird, M. Hoythya, T. R. Fuerst and D. L. French (1993). "Type IV collagenase (s) and TIMPs modulate endothelial cell morphogenesis in vitro." Journal of Cellular Physiology **156**(2): 235–246.

Sun, Z., M. Costell and R. Fässler (2019). "Integrin activation by talin, kindlin and mechanical forces." Nature Cell Biology **21**(1): 25–31.

Taylor, D. A., L. C. Sampaio, Z. Ferdous, A. S. Gobin and L. J. Taite (2018). "Decellularized matrices in regenerative medicine." Acta biomaterialia **74**: 74–89.

Théry, M. (2010). "Micropatterning as a tool to decipher cell morphogenesis and functions." Journal of Cell Science **123**(24): 4201–4213.

Tibbitt, M. W. and K. S. Anseth (2009). "Hydrogels as extracellular matrix mimics for 3D cell culture." Biotechnology and Bioengineering **103**(4): 655–663.

Timmins, N., F. Harding, C. Smart, M. Brown and L. Nielsen (2005). "Method for the generation and cultivation of functional three-dimensional mammary constructs without exogenous extracellular matrix." Cell and Tissue Research **320**(1): 207–210.

Tolnai, S. (1975). "A method for viable cell count." Methods in Cell Science **1**(1): 37–38.

Uysal, O., T. Sevimli, M. Sevimli, S. Gunes and A. E. Sariboyaci (2018). Cell and Tissue Culture: The Base of Biotechnology. Omics Technologies and Bio-Engineering, Elsevier: 391–429.

Wang, F. (2006). "Culture of animal cells: A manual of basic technique." In Vitro Cellular & Developmental Biology-Animal **42**(5): 169–169.

Weigelt, B., A. T. Lo, C. C. Park, J. W. Gray and M. J. Bissell (2010). "HER2 signaling pathway activation and response of breast cancer cells to HER2-targeting agents is dependent strongly on the 3D microenvironment." Breast Cancer Research and Treatment **122**(1): 35–43.

Young, J. L., A. W. Holle and J. P. Spatz (2016). "Nanoscale and mechanical properties of the physiological cell–ECM microenvironment." Experimental Cell Research **343**(1): 3–6.

Manuela Brás, Joanna Zemła, Ramon Farré

5.2 Tissue Preparation

5.2.1 Fresh Native Tissue

The authors refer "fresh tissue samples" as those that come directly from the patient or animal after surgery without being submitted to any freezing and chemical process.

Fresh tissue samples are always a challenge for AFM analysis since they have to be adherent and completely immobilized onto the sample plate over long periods of time, sometimes hours, as in mechanical properties assessment tests. If the immobilization is poor, the AFM measurements cannot be performed or the results may present some artifacts. An example can be the measurement of mechanical properties on fibers; if there are even the slightest movements (vertical or horizontal) during the acquisition of force-distance curves, they may lead to an incorrect interpretation of Young's modulus, and the elasticity maps will be unusable (Kammoun et al., 2019). Fresh tissue samples can be very different with regard to roughness, thickness, and consistence (stiffer or "fluffier"). These parameters depend on the type of tissue (epithelium, muscle, connective, or nervous tissue). Different tissues may be present in the same organ, such as the brain, heart, or colon. Taking into consideration the organ functionalities, some tissues may produce substances, such as the mucosa epithelia of uterus, esophagus, gastric, intestinal, nasal, and vaginal. The preparation of these samples can be even more demanding due to the eventual contamination of the AFM tip.

After excision, fresh tissue samples are immersed in a liquid, such as feeding medium or cell culture medium, and transferred from the surgery room to the AFM room. If human cancer samples are the object of further AFM examination, it is important to have the tumor sample investigated by a pathomorphologist at first, who identifies the cancerous tissue sections and healthy margins. It also has to be kept in mind that every human tissue is always potentially biohazardous; thus, adequate safety protocols must be followed (Deptuła et al., 2020).

The first step for a successful AFM analysis is an efficient but gentle tissue cleaning. Some tissues produce mucous, requiring a very careful cleaning process for its observation by AFM; since the mucous is always a source of contamination for the AFM cantilever tips.

Manuela Brás, i3S – Instituto de Investigação e Inovação em Saúde, Universidade do Porto, Portugal, INEB – Instituto de Engenharia Biomédica, Porto, Portugal, FEUP – Faculdade de Engenharia da Universidade do Porto, Portugal
Joanna Zemła, Institute of Nuclear Physics, Polish Academy of Sciences, Krakow, Poland
Ramon Farré, Unitat de Biofísica i Bioenginyeria, Facultat de Medicina i Ciències de la Salut, Universitat de Barcelona, Barcelona, Spain, CIBER de Enfermedades Respiratorias, Madrid, Spain, Institut d'Investigacions Biomediques August Pi Sunyer, Barcelona, Spain

https://doi.org/10.1515/9783110989380-008

The operator must use polymer tweezers without teeth to avoid tissue damage and use PBS or other biological liquids to wash the tissues carefully, preserving all the structures and layers as much as possible. There are some tissues that require measurement as they come from the surgery room, due to their size and/ or low stiffness cannot be cut for the AFM analysis. Other tissues, due to their size and/ or stiffness, can be easily cut before being analyzed.

Use of vibratomes is a common practice to obtain fresh tissue slices that are cut with a very high precision. Another advantage of using this equipment is that the sample can be hydrated, when needed, during the cutting process. This equipment does not require tissue fixation, meaning that cells viability and native tissue structures are preserved (Mattei et al., 2015). Tissues just need to be embedded in an agarose gel, with a melting point up to 37 °C, to maintain their structure and viability. Vibratomes present some disadvantages as a consequence of the lack of standard protocols, specifically regarding cutting parameters, such as thickness, speed, oscillation amplitude, and blade angle, for certain applications. The main difficulty is to optimize the cutting speed, partly because it depends too much on the tissue type (Zimmermann et al., 2009). In spite of the above mentioned reasons, vibratomes continue to be an option, still highly used in this biological field.

Several works have been published regarding the study of mechanical properties of fresh animal and human tissues. Jorba et al., used the AFM to measure the stiffness of mice brains to assess if the obstructive sleep apnea could contribute to increased risk of Alzheimer's disease, inducing alterations in brain stiffness (Jorba et al., 2017). After animal euthanasia, mice brains were placed in ice-cold Krebs-Henseleit buffer and 200 μm slices were obtained with a vibratome. The fixing of the brain piece is assured by placing it inside a Petri dish under a special compliant mesh (2 mm spacing) of silicone thread (0.25 mm in diameter) that does not compress the sample (Jorba et al., 2017). Antonovaite et al., also used the vibratome to study the correlation between viscoelasticity of the hippocampus of the mouse brain slices in the development of neurological diseases (Antonovaite et al., 2018). Murine embryonic of murine forelimbs or bovine cartilage plugs from bovine femoral condyles were sectioned with the vibratome and AFM force mapping was used to assess the compressive modulus (Xin et al., 2016).

Cui and colleagues studied the nanomechanical properties of cervical cancer and intraepithelial neoplasia tissues by AFM. The samples were collected from the patients and preserved in a precooled lactated Ringer's solution containing a protease inhibitor cocktail and stored at 4 °C for no more than 72 h until the AFM measurements (Cui et al., 2017). The samples were analyzed as if they were without any reduction in size, and immobilized on a 35-mm dish using a 1-minute™ biocompatible epoxy gel. To avoid interfering with the sample's mechanical properties, the manipulation was performed as gently as possible immediately after the specimens were glued to the dishes (1–1.5 min). Lekka and colleagues also studied the mechanical properties of the endometrium. Tissues were harvested and transported in DMEM till the AFM room. Thick slices of normal and tumor tissues were cut using two razors in a liquid (6 mm × 2 mm × 2 mm). Sections were kept in DMEM at 4 °C for AFM measurements. The slice was removed from the medium and glued onto a glass coverslip using two 0.5 µL droplets of cyanoacrylate adhesive (AXIA, Kitta, Korea) placed at both extremities of the sample. Special care was taken to preserve the media on a surface devoted to AFM measurements. After gluing (2–3 min), the slice was immediately immersed in DMEM and AFM measurements were carried out within the next 2–3 h at room temperature (Lekka et al., 2012).

Di Mundo et al., studied the mechanical properties of human corneal Descemet's membrane (DM) by AFM nanoindentation. After the patient's death, the tissues were preserved in a storage/ transport medium for 20 days on average (Di Mundo et al., 2017). The corneas were centered on the base of a trephine punch using the peripheral holes of the suction area as the reference. The corneal tissue, with endothelium facing the air, was secured on the base using vacuum. A 9.5 mm diameter punch (Moria, Antony, France) was used to create a superficial cut. The endothelium was covered with the medium to create a thin film of fluid. The membrane was lifted using a cleavage hook throughout the circumference to limit the peripheral tearing of this delicate tissue. A critical issue for the AFM analysis in liquid was to keep the membrane well opened and firmly adhered to the ground of the AFM liquid cell as the DM folds spontaneously in the liquid environment (Di mundo et al., 2017).

Lambordo et al., obtained corneas from patients (6 and 10 h after death) and immediately preserved them at 4 °C in a corneal storage medium (Eusol-C; Alchimia Srl, Padova, Italy) enriched with 15% dextran. Each cornea was trephined to a size of 8.0 mm diameter with a hand trephine (MicroKeratron, Geuder AG, Heidelberg, Germany) after gently removing the epithelium with a sponge (Merocel; Medtronic, Minneapolis, MN) soaked in deionized water. The whole cornea, devoid of epithelium, was preserved in a corneal storage medium enriched with 15% dextran and used for experiments within 12 h (Lombardo et al., 2012)

Lui et al., studied lung tissue. This tissue is highly compliant and elastic, being difficult to cut into strips for AFM characterization. To stabilize the lung structure for cutting, isolated mouse lungs must be inflated intra tracheal with 50 mL/kg body weight of 2% low gel point agarose (prepared in PBS), warmed to 37 °C. The agarose will gelify and stiffen in the airspaces to gently stabilize the lung structure (Lui et al., 2011).

Some tissues can have a cadaveric source. Gouveia and coworkers assessed the biomechanics by AFM of corneal substrates and their effects on stem cell maintenance and differentiation. Human corneas were obtained from cadaveric donors using a scleral ring. In order to maintain the hydration, thickness, and transparency, the samples were kept at dextran containing Carry-C preservative medium in refrigeration conditions during transportation, and analyzed not later than 2 weeks (Gouveia et al., 2019).

Fresh sample tissues could be difficult to obtain since they do not always have sufficient size (mainly regarding the tumor/not healthy tissue) for diagnostic and research purposes. In terms of handling, preparation, and analysis, they are more demanding (mainly if they are soft) when compared to frozen and/or stiff tissues. The analysis of fresh tissues has several advantages once they are closer to the native tissue conditions.

5.2.2 Preserved Native Tissue Samples

In any type of tissue samples examination, the time between the sample excision and measurement/analysis, as well as sample preservation conditions are crucial since, quite often, the samples have to be transported from one institution to another as it is in the case of AFM studies. The availability of equipment and qualified human operators at the exact time may be limited; thus, tissue samples preservation protocols must be established.

The common biospecimens processing methods involve snap freezing, paraffin embedding, or formalin/paraformaldehyde (PFA)/glutaraldehyde (GA) fixation (Mendy et al., 2018, Taylor et al., 2019). All three approaches allow for a long-term storage of biological material. Formalin, PFA, and GA fixation methods preserve the morphological structure of the sample well, which is efficient for histological observations. However, they change the sample's molecular structure via protein crosslinking, which results in changes of physical-chemical properties of the specimen. Paraffin fixation method keeps the morphological structure of the tissue untouched, and additionally, it does not compromise proteins or the nucleic acids structure. The fixed biospecimens are kept at room temperature (25 °C), which is an advantage, making these techniques commonly used. The third way for biological samples storage is freezing. The biobanking protocols state that human tissues should be cut into 0.5 cm^3 pieces, prior quick freezing (Mendy et al., 2018). In addition, an optimal cutting temperature (OCT) compound can be used to allow for future cryosections (Sicard et al., 2017, Sicard et al., 2018). The freezing procedure itself may be performed either by snap freezing, that is, by immersing the sample in liquid nitrogen (LN2) (van Zwieten et al., 2014), followed by transferring the sample in cold ice to a −80 °C freezer or directly to an LN2 dewar (Lindner et al., 2019). The above listed protocols are commonly practiced in biobanks or biological resource centers, where tissue samples (e.g., blood, cancer tissues, and urine) are stored. These institutions play an important role in the study of infectious and noncommunicable disease etiology and identification of new potential diagnostic markers. They are crucial to the development of personalized drug treatment and translational research as well (Mendy et al., 2018).

Biospecimens of higher volume require a slow freezing procedure. For the slow freezing process, a sample is transferred to the freezing medium, which consists of cell culture medium supplemented with 10% FBS and n% of a cryoprotectant agent. The most common cryoprotectant agents are dimethyl sulfoxide* (DMSO), Propane-1,2-diol* (PrOH), ethane-1,2-diol* (EtOH), or Propane-1,2,3-triol*. The amount of the cryoprotectant agent depends on the tissue type, although it usually does not exceed 15% v/v. Hence, tissue samples are placed in cryovials and embedded in a freezing medium. After 15–30 min, they are placed in a CoolCell box (Cell Freezing Container) and stored for 8–12 h in a −80 °C freezer for slow freezing. Subsequently, the cryovials are transferred to a −150 °C mechanical freezer or liquid nitrogen freezer (submersion or vapor at −96 or −132 °C, respectively) for a long-term storage. The efficiency of freezing protocols is usually evaluated by specimen morphology observation (Isildar et al., 2019) or histological examination (Castro et al., 2011, Mendy et al., 2018, Porter et al., 2019).

The thawing process is conducted at room temperature just before the sample examination. Tissue sectioning, prior freezing, is performed in order to reduce freezing and thawing cycles, which might negatively influence the biospecimen.

Recently, the influence of GA fixing and slow freezing (DMEM with 10% DMSO) on the mechanical properties of bronchial tissue samples has been studied (Zemla et al., 2018b) Fresh tissue samples were investigated as well. AFM-force spectroscopy was used to investigate the influence of sample preservation techniques on healthy and asthmatic tissue specimens. Pulmonary diseases often cause tissue remodeling, which involves fibrosis, changes in ECM, hyperplasia and so on; thus, variation of tissue elasticity can be expected. It was found that fresh healthy tissues are more rigid than asthmatic ones. The same behavior was observed for thawed samples, and additionally, there was no statistically significant difference between the E values obtained for fresh and frozen tissues (Figure 5.2.1, left). GA fixing method resulted in the reversed dependency of elasticity parameter value E. Additionally, relative tissue stiffness index (RTSI) has been defined as the ratio of E_{asthma} to $E_{healthy}$, and compared for different indentation depths as well as sample preparation protocols (Figure 5.2.1, right). It was shown that freezing did not affect the mechanical properties of tissues, and in both cases, the values of RTSI did not depend on the indentation depths. Contrary to these, GA fixation changed the mechanical properties of the tissue samples. Additionally, the RTSI was found to vary depending on the indentation depth (Figure 5.2.1 right). At 200 nm indentation depth, the RTSI of GA fixed samples is close to 1, while at 600 nm indentation, it shows that asthmatic tissue is almost twice stiffer than the healthy fixed samples. This inconsistency discriminates GA-fixing as an efficient method of tissue sample storage.

A few years earlier, Van Zwieten compared the elasticity parameters obtained for fresh human gluteus maximus muscles and also snap-freezed in LN2 or LN2-cooled isopentane. DMD muscles have also been studied. It has been shown that AFM measurements allow distinguishing healthy and DMD tissues, regardless of the tested sample preparation protocols (van Zwieten et al., 2014). These results as well as control experiments (DNA and biochemical characteristics, histology tests) performed in biobanks

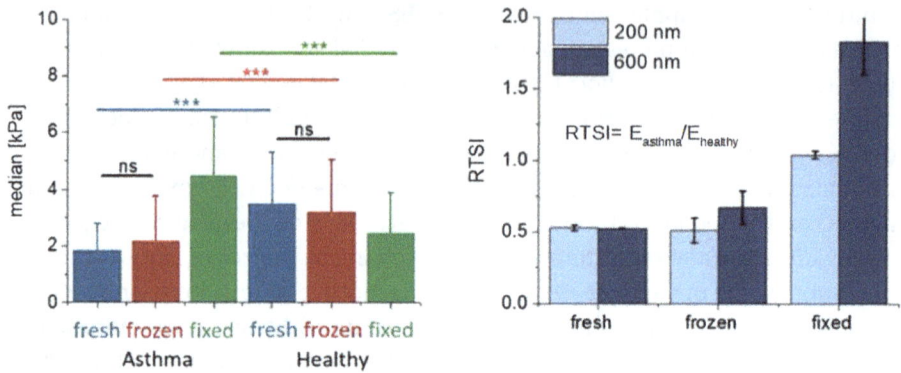

Figure 5.2.1: Medians of the Young's modulus determined for airways tissue samples at indentation depth 600 nm. Tissues were compared depending on the applied preparation protocol (i.e., fresh, frozen, and GA fixed samples). Data are presented as a median ± median deviation (left). Relative tissue stiffness index at different indentation depths as a function of sample preservation protocol (right) (reprinted with permission from Zemła et al. (2018b)).

confirm that slow freezing as well as snap freezing are harmless tissue preservation methods (Groelz et al., 2018, Lindner et al., 2019).

5.2.3 Decellularized Tissues

Mechanical assessment of native tissue samples, either fresh or preserved, provides information on the viscoelastic properties of the 3D structure present on the different types of cells in the tissue and extracellular matrix (ECM). Whereas this information on native tissue is of high interest to know the mechanical properties of tissues in their natural configuration, to determine the specific contribution of ECM in tissue viscoelasticity is also of considerable importance. This is particularly relevant considering that in prevalent diseases, ECM mechanics plays a substantial role in characterizing the severity of the disease (e.g., collagen accumulation in different organ fibrosis) or determining the risk of disease progression (e.g., tumor infiltration and cancer metastasis).

Preparation of ECM tissue samples for mechanical assessment requires performing a decellularization process to separate the ECM scaffold and the tissue cells. Ideally, decellularization must completely eliminate all cells in the tissue and leave the ECM scaffold absolutely intact. Hence, the process should be considerably aggressive to extract all cells and cell debris (e.g., small membrane fragments attached to the ECM by focal adhesions) and at the same the time should be smooth enough to avoid any alteration of the ECM. Obviously, such an ideal process does not exist in real life, and thus selection of the optimal decellularization protocol requires a good balance to achieve a satisfactory trade-off using the different decellularization methods, which should be combined into carefully selected and tested protocols. For instance, regarding the effectiveness of decellularization, it is usually considered acceptable to reduce the concentration of DNA to less than 50 ng double-stranded DNA (dsDNA) per mg ECM dry weight, to have less than 200 bp DNA fragment length, and observing no visible nuclear material by 4',6-diamidino-2-phenylindole (DAPI) staining (Crapo et al., 2011).

The specific decellularization technique must be chosen depending of the intended ECM final scope, for instance to preserve the original mechanical properties as much as possible. As the ECM features vary depending on the tissues and organs, the optimal decellularization protocol should also be tissue-dependent. There are three main different mechanisms/agents to decellularize an organ or tissue: physical, biological (enzymatic), and chemical (Badylak et al., 2011). Physical procedures (freezing, mechanical forces, and sonication) can enhance tissue decellularization through intracellular ice crystals formation, tissue layer delamination, and ultrasonic cell disruption, respectively. However, these procedures are not sufficient to perform total decellularization and an efficient removal of cell residues. Biological agents, such as enzymes, can be used as decellularization agents in order to degrade specific biological components. Proteases, lipase, glycosidases, and nucleases can be used to help chemical cell lysis and to foster target molecules removal. However, these enzymes can damage the ECM to different extents; so a titration of their activity must be tested for each tissue and organ. Chemical agents (detergents and acid/base buffers) are also used to decellularize tissue and organs. Different compounds and their combinations promote cellular lysis through membrane permeabilization, protein denaturalization, and nucleic acid disruption. Chemical agents also affect the ECM structure and biochemical properties; so their use must be well characterized. Furthermore, after their use, extensive washes are required in order to clean the ECM sample.

The several potential decellularization processes that are available have different effects on the ECM properties (Gilbert et al., 2006), which can be critical for assessing the preserved mechanical properties in the natural scaffold. Thus, some issues should be taken into account before choosing the decellularization method. If the 3D structure and its stiffness are important, collagens must be preserved. In this case, ionic detergents would be the optimal choice, and enzymatic or alkaline-acid methods should be avoided. This consideration is valid if any protein preservation is important. Osmotic buffers are a more conservative solution to obtain a decellularized ECM, but are slower and cannot penetrate in thicker compact organs; for example, the heart. In dense tissues or organs, detergents can help buffers to penetrate but they affect the proteic ultra-structure due to the disruption of the protein-protein interactions. Alcohols and other solvents are very efficient to remove lipids from tissues, but they can cross-link proteins and modify the ECM ultrastructure, and hence the mechanics. Enzymatic treatments guarantee a specific removal of the determined proteins with a certain grade of unintended targets, and they are not sufficient for a complete decellularization of the whole tissue. Therefore, they must be used in combination with other methods. Physical methods, such as temperature, force, pressure, and electricity, can be applied but they have limited efficacy and a high probability to damage the ECM components and structure, and thereby the mechanical properties.

There are different techniques to apply the decellularization agents, each one with its advantages and drawbacks. The effectiveness of tissue and organ decellularization depends on the intrinsic tissue properties as the specific cell density, thickness and compaction, lipid content, as well on the selected decellularization agents (Crapo et al., 2011). However, the existence of an inverse correlation between the decellularization efficiency and the ECM quality preservation must be considered. Whole organ perfusion, which is achieved by the administration of decellularization agents through the organ or tissue vasculature, has shown high efficiency. Antegrade or retrograde perfusion preserves organ structures, reaches densely packed cell areas, and simultaneously allows the removal of cell debris. Whereas, the vasculature is the only ubiquitous circuit for applying decellularization, in some organs, there are additional routes. For instance, the lung can also be decellularized through the trachea (Melo et al., 2014). Application of pressure gradient during decellularization increases the contact and forces decellularization agents to pass across the tissues. This method, designated as convective flow, can also help cell residues' removal, with less impact to ECM architecture. Supercritical fluids application of an inert gas for decellularization promotes cell debris removal and minimal alteration of ECM mechanical properties. Organs and tissues can be decellularized by immersion in chemical and biological agents, while being subjected to mechanical agitation. Cellular density of the tissue and compaction, together with the selected decellularization buffer, determines the protocol duration. It is remarkable that this procedure is particularly useful to directly decellularize thin pieces of native tissue, which is of relevance, for instance, when trying to compare the mechanical properties of a native tissue and ECM in the very same sample (Farre et al., 2018). As a general rule, a specific combination of mild physical, biological, and chemical methods, together with the type of administration should be tested for each tissue/organ and the type of application, to obtain the best result (Keane et al., 2015). However, it is advisable to specifically test how different variants of decellularization procedures affect the mechanical properties of the obtained ECM (Nonaka et al., 2014, da Palma et al., 2015, da Palma et al., 2016).

Given that the ECM obtained by decellularizing the different organs/tissues may be used for tissue engineering or regenerative medicine applications, they can be subjected to sterilization processes to comply with the regulations in force for each application and country. It is therefore important to assess the potential effect of such a sample preparation procedure on the mechanical properties of ECM (Keane et al., 2015). Classical sterilization methods, such as ethylene oxide exposure, gamma irradiation, and electron beam irradiation, are known to potentially alter ECM ultrastructure and thus the mechanical properties. For instance, it has been reported that conventional gamma sterilization significantly modifies the lung ECM mechanics

(Nonaka et al., 2014). Tissue ECM can be sterilized by simple treatment with acids or solvents, but such methods could not provide sufficient penetration, depending on the size of the organ or tissue, and may damage key ECM components. An important issue is whether it is possible to couple decellularization and sterilization in one process, to ensure a clinically safe ECM for the recipient, without affecting its ultrastructure and mechanics. To this end, peracetic acid has so far demonstrated to be a reasonable approach to minimize bacterial, fungal, and spore content. Tributyl phosphate organic solvent has shown viricidal properties. Supercritical carbon dioxide is under investigation as an alternative method for sterilizing natural ECM. This agent should reduce the bacterial and viral loads, with minor changes in mechanical properties relative to other sterilization methods. However, finding a sterilization procedure that is optimal for both biological safety and preservation of mechanical properties is still an open issue. Moreover, it is to be noted that in view of potential routine use for clinical applications, ECM may require frozen storage from obtention to final use. Therefore, it is also important to characterize to what extent freezing-thawing modifies the ECM mechanics (Nonaka et al., 2014), which is expected to depend on the type of the decellularized tissue/organ.

5.2.4 Tissue Sample Immobilization

5.2.4.1 Native Tissue

AFM-based studies of nanomechanical properties of tissue samples require sample immobilization, in a way that prevents it from mechanical damaging or chemical contamination. The common way of tissue sample immobilization is gluing it with epoxy gel (Cui et al., 2017) and cyanoacrylate adhesive (Lekka et al., 2012, Zemla et al., 2018b). A novel interesting approach of brain tissue sample immobilization was presented by Jorba and coworkers (2017) who have designed a ring-shaped sample holder with a mesh, which was gently pressing the sample from the top (Jorba et al., 2017). Di Mundo et al., used the following procedure for the cornea membrane adherence to a solid support under liquid: polycarbonate (PC) slabs (1 mm thick) were roughened with sandpaper (P1000 grade) to improve adhesion and covered it with a cyanoacrilic glue drop (2–4 μL). The tissues were lifted and spread out on an aluminum (Al) foil, with the endothelial side facing the top. The tissue was stained with Trypan Blue to make it visible. The tissue was kept open on the foil and was placed in contact (inverted) with the PC slab point, where the glue drop (2–4 μL) had been positioned (Di Mundo et al., 2017). After 2 min (adherence time), the Al foil was detached. The tissue was then firmly adhered with cyanoacrilic glue on the PC slabs, with the endothelial side touching the base and the DM facing the air; hence the AFM probe (Di Mundo et al., 2017). Morgan et al., immobilized sample tissues to be analyzed in a liquid for AFM, using a diagram of soft-clamping immobilizing retainer of tissue (SCIRT) usage. A 13 mm Thermanox coverslip was modified using a biopsy punch to allow for a tissue access window. Small (~0.5 mm) droplets of cyanoacrylate glue were added to the perimeter of the SCIRT using a 22 gauge needle. The SCIRT was placed over the tissue, clamping it into place. The dish was then filled with a buffer to cure the glue and keep the tissue hydrated (Morgan et al., 2014). Babu et al., also used the same methodology to study the tissue brain samples, immobilized with punched Thermanox coverslips, glued to the Petri dish at their borders; thus avoiding direct contact of the tissue with glue (Babu et al., 2019). To guarantee sample immobilization, Kreplak et al., used a nitrocellulose membrane, which has a huge affinity for proteins; thus it is a reliable way of soft tissue immobilization (Kreplak et al., 2007). None of these methods can be named as the universal one. Some tissues are known to be quite adherent (small human pulmonary arteries); so poly-L-lysine is often a sufficient adhesive, but for nonadherent specimens such as muscle, cyanoacrylate adhesive is used (Sicard et al., 2017). Lui et al., also used poly-L-lysine, immediately before AFM characterization. The tissue strip was attached to a poly-L-lysine–coated 15-mm coverslip by lifting the coverslip from below the floating tissue, making sure the tissue strip spreads evenly on the coverslip surface; if necessary, it is possible to sandwich the strip with a second clean, uncoated coverslip, and apply a mild pressure to assist with tissue attachment to the poly-L-lysine–coated coverslip (Liu et al., 2011).

5.2.4.2 Decellularized Tissue

Preparation of decellularized tissue samples for mechanical assessment depends on the specific measurement to be applied to the ECM. For instance, micromechanical assessment of ECM stiffness at specific local sites in the sample can be measured by AFM. To this end, the decellularized tissue can be embedded in optimal cutting temperature compound (OCT), frozen and cryosectioned into thin slices (10–50 μm). These ECM slices are placed on top of the positively charged glass slides, and OCT is removed by thawing at room temperature and washing the samples in PBS (Luque et al., 2013, Jorba et al., 2017, Jorba et al., 2019). In that way, the ECM is prepared for stiffness measurement by AFM, which provides data on local mechanics at the microscale; thus allowing detection of local inhomogeneities as in the case of collagen deposition in fibrosis (Melo et al., 2014). In addition to locally measuring ECM stiffness by AFM, bulk stiffness of the decellularized tissues can be measured by application of tensile stretch to macroscopic ECM samples. To this end, the decellularized tissue is cut into a strip ($\approx$7 × 2 × 2 mm) with a scalpel and gently dried with a tissue paper. Then, one end of the strip is fixed with cyanoacrylate glue to a hook attached to the lever arm of a servo-controlled displacement system, allowing to stretch the strip, thereby measuring its force–deformation, and hence stiffness. Interestingly, multiscale assessment of ECM mechanics allows assessing how the local fiber mechanics and 3D mesh structure contribute to tissue mechanics (Uriarte et al., 2016, Farre et al., 2018, Perea-Gil et al., 2018). Zhao and colleagues used the decellularized tongue tissue for studying tongue cancer and tongue regeneration. In this study, paraffin-embedded tongue squamous cell carcinoma tissue was prepared after surgical excision. Tongue extracellular matrix was harvested and flatten-fixed on glass slides (Long et al., 2017).

As it seems clear, quite a number of efficient sample preservation protocols have already been proposed. However, there is still much research ahead in the field of tissue sample immobilization techniques (Uriarte et al., 2016, Farre et al., 2018, Perea-Gil et al., 2018), which is a highly challenge step for AFM analysis.

This book chapter was funded in part by the Spanish Ministry of Sciences, Innovation and Universities (SAF2017-85574-R and DPI2017-83721-P), and by the Marie Sklodowska-Curie Action, Innovative Training Networks 2018, EU grant agreement no. 812772.

References

Antonovaite, N., S. V. Beekmans, E. M. Hol, W. J. Wadman and D. Iannuzzi (2018). "Regional variations in stiffness in live mouse brain tissue determined by depth-controlled indentation mapping". Scientific Reports **8**: 12517.

Babu, P. K. V. and M. Radmacher (2019). "Mechanics of brain tissues studied by atomic force microscopy: A perspective". Frontiers in Neuroscience **13**: 600.

Badylak, S. F., D. Taylor and K. Uygun (2011). "Whole-organ tissue engineering: Decellularization and recellularization of three-dimensional matrix scaffolds". Annual Review of Biomedical Engineering **13**: 27–53.

Castro, S. V., A. A. De Carvalho, C. M. Da Silva, L. R. Faustino, C. C. Campello, C. M. Lucci, S. N. Bao, J. R. De Figueiredo and A. P. Rodrigues (2011). "Freezing solution containing dimethylsulfoxide and fetal calf serum maintains survival and ultrastructure of goat preantral follicles after cryopreservation and in vitro culture of ovarian tissue". Cell and Tissue Research **346**: 283–292.

Crapo, P. M., T. W. Gilbert and S. F. Badylak (2011). "An overview of tissue and whole organ decellularization processes". Biomaterials **32**: 3233–3243.

Cui, Y., X. Zhang, K. You, Y. Guo, C. Liu, X. Fang and L. Geng (2017). "Nanomechanical characteristics of cervical cancer and cervical intraepithelial neoplasia revealed by atomic force microscopy". Medical Science Monitor: International Medical Journal of Experimental and Clinical Research **23**: 4205–4213.

Da Palma, R. K., N. Campillo, J. J. Uriarte, L. V. Oliveira, D. Navajas and R. Farre (2015). "Pressure- and flow-controlled media perfusion differently modify vascular mechanics in lung decellularization". Journal of the Mechanical Behavior of Biomedical Materials **49**: 69–79.

Da Palma, R. K., P. N. Nonaka, N. Campillo, J. J. Uriarte, J. J. Urbano, D. Navajas, R. Farre and L. V. F. Oliveira (2016). "Behavior of vascular resistance undergoing various pressure insufflation and perfusion on decellularized lungs". Journal of Biomechanics **49**: 1230–1232.

Deptula, P., D. Lysik, K. Pogoda, M. Ciesluk, A. Namiot, J. Mystkowska, G. Krol, S. Gluszek, P. A. Janmey and R. Bucki. (2020). "Tissue Rheology as a Possible Complementary Procedure to

Advance Histological Diagnosis of Colon Cancer." Acs Biomaterials Science & Engineering 6: 5620–5631.

Di Mundo, R., G. Recchia, M. Parekh, A. Ruzza, S. Ferrari and G. Carbone (2017). "Sensing inhomogeneous mechanical properties of human corneal Descemet's membrane with AFM nano-indentation". Journal of the Mechanical Behavior of Biomedical Materials 74: 21–27.

Farre, N., J. Otero, B. Falcones, M. Torres, I. Jorba, D. Gozal, I. Almendros, R. Farre and D. Navajas (2018). "Intermittent hypoxia mimicking sleep apnea increases passive stiffness of myocardial extracellular matrixa multiscale study". Front Physiology 9: 1143.

Gilbert, T. W., T. L. Sellaro and S. F. Badylak (2006). "Decellularization of tissues and organs". Biomaterials 27: 3675–3683.

Gouveia, R. M., G. Lepert, S. Gupta, R. R. Mohan, C. Paterson and C. J. Connon (2019). "Assessment of corneal substrate biomechanics and its effect on epithelial stem cell maintenance and differentiation". Nature Communications 10: 1496.

Groelz, D., C. Viertler, D. Pabst, N. Dettmann and K. Zatloukal (2018). "Impact of storage conditions on the quality of nucleic acids in paraffin embedded tissues". PLoS One 13: e0203608–e.

Isildar, B., S. Ozkan, M. Oncul, Z. Baslar, S. Kaleli, M. Tasyurekli and M. Koyuturk (2019). "Comparison of different cryopreservation protocols for human umbilical cord tissue as source of mesenchymal stem cells". Acta histochemica 121: 361–367.

Iyer, P. S., L. O. Mavoungou, F. Ronzoni, J. Zemla, E. Schmid-Siegert, S. Antonini, L. A. Neff, O. M. Dorchies, M. Jaconi, M. Lekka, G. Messina and N. Mermod (2018). "Autologous cell therapy approach for duchenne muscular dystrophy using piggybac transposons and mesoangioblasts". Molecular Therapy: The Journal of the American Society of Gene Therapy 26: 1093–1108.

Jorba, I., M. J. Menal, M. Torres, D. Gozal, G. Piñol-Ripoll, A. Colell, J. M. Montserrat, D. Navajas, R. Farré and I. Almendros (2017a). "Ageing and chronic intermittent hypoxia mimicking sleep apnea do not modify local brain tissue stiffness in healthy mice". Journal of the Mechanical Behavior of Biomedical Materials 71: 106–113.

Jorba, I., J. J. Uriarte, N. Campillo, R. Farre and D. Navajas (2017b). "Probing micromechanical properties of the extracellular matrix of soft tissues by atomic force microscopy". Journal of Cellular Physiology 232: 19–26.

Jorba, I., G. Beltran, B. Falcones, B. Suki, R. Farre, J. M. Garcia-Aznar and D. Navajas (2019). "Nonlinear elasticity of the lung extracellular microenvironment is regulated by macroscale tissue strain". Acta Biomater 92: 265–276.

Kammoun, M., R. Ternifi, V. Dupres, P. Pouletaut, S. Même, F. Szeremeta, J. Landoulsi, J.-M. Constans, F. Lafont, M. Subramaniam, J. R. Hawse and S. F. Bensamoun (2019). "Development of a novel multiphysical approach for the characterization of mechanical properties of musculotendinous tissues". Scientific Reports 9: 7733.

Keane, T. J., I. T. Swinehart and S. F. Badylak (2015). "Methods of tissue decellularization used for preparation of biologic scaffolds and in vivo relevance". Methods 84: 25–34.

Kreplak, L., H. Wang, U. Aebi and X.-P. Kong (2007). "Atomic force microscopy of mammalian urothelial surface". Journal of Molecular Biology 374: 8.

Lekka, M., D. Gil, K. Pogoda, J. Dulinska-Litewka, R. Jach, J. Gostek, O. Klymenko, S. Prauzner-Bechcicki, Z. Stachura, J. Wiltowska-Zuber, K. Okon and P. Laidler (2012). "Cancer cell detection in tissue sections using AFM". Archives of Biochemistry and Biophysics 518: 151–156.

Lekka, M., K. Pogoda, J. Gostek, O. Klymenko, S. Prauzner-Bechcicki, J. Wiltowska-Zuber, J. Jaczewska, J. Lekki and Z. Stachura (2012). "Cancer cell recognition–mechanical phenotype". Micron 43: 1259–1266.

Lindner, M., A. Morresi-Hauf, A. Stowasser, A. Hapfelmeier, R. A. Hatz and I. Koch (2019). "Quality assessment of tissue samples stored in a specialized human lung biobank". PLoS One **14**: e0203977.

Liu, F. and D. J. Tschumperlin (2011). "Micro-mechanical characterization of lung tissue using atomic force microscopy". Journal of Visualized Experiments: JoVE **54**: e2911.

Lombardo, M., G. Lombardo, G. Carbone, M. P. De Santo, R. Barberi and S. Serrao (2012). "Biomechanics of the Anterior Human Corneal Tissue Investigated with Atomic Force Microscopy". Investigative Ophthalmology & Visual Science **53**: 7.

Long, Z., H. Linxuan, Y. Shuyi, Z. Junheng, W. Hua and Z. Yan (2017). "Decellularized tongue tissue as an in vitro model for studying tongue cancer and tongue regeneration". Acta Biomater **58**: 122–135.

Luque, T., E. Melo, E. Garreta, J. Cortiella, J. Nichols, R. Farre and D. Navajas (2013). "Local micromechanical properties of decellularized lung scaffolds measured with atomic force microscopy". Acta Biomater **9**: 6852–6859.

Mattei, G., I. Cristiani, C. Magliaro and A. Ahluwalia (2015). "Profile analysis of hepatic porcine and murine brain tissue slices obtained with a vibratome". PeerJ **3**: e932.

Melo, E., E. Garreta, T. Luque, J. Cortiella, J. Nichols, D. Navajas and R. Farre (2014). "Effects of the decellularization method on the local stiffness of acellular lungs". Tissue Engineering. Part C, Methods **20**: 412–422.

Mendy, M., R. T. Lawlor, A. L. Van Kappel, P. H. J. Riegman, F. Betsou, O. D. Cohen and M. K. Henderson (2018). "Biospecimens and Biobanking in Global Health". Clinics in Laboratory Medicine **38**: 183–207.

Morgan, J. T., V. K. Raghunathan, S. M. Thomasy, C. J. Murphy and P. Russell (2014). "Robust and artifact-free mounting of tissue samples for atomic force microscopy". Biotechniques **56**: 40–42.

Nonaka, P. N., J. J. Uriarte, N. Campillo, E. Melo, D. Navajas, R. Farre and L. V. Oliveira (2014). "Mechanical properties of mouse lungs along organ decellularization by sodium dodecyl sulfate". Respiratory Physiology & Neurobiology **200**: 1–5.

Perea-Gil, I., C. Galvez-Monton, C. Prat-Vidal, I. Jorba, C. Segu-Verges, S. Roura, C. Soler-Botija, O. Iborra-Egea, E. Revuelta-Lopez, M. A. Fernandez, R. Farre, D. Navajas and A. Bayes-Genis (2018). "Head-to-head comparison of two engineered cardiac grafts for myocardial repair: From scaffold characterization to pre-clinical testing". Scientific Reports **8**: 6708.

Porter, L. H., M. G. Lawrence, H. Wang, A. K. Clark, A. Bakshi, D. Obinata, D. Goode, M. Papargiris, D. Mural, Clouston, A. Ryan, S. Norden, E. Corey, P. S. Nelson, J. T. Isaacs, J. Grummet, J. Kourambas, S. Sandhu, D. G. Murphy, D. Pook, M. Frydenberg, R. A. Taylor and G. P. Risbridger (2019). "Establishing a cryopreservation protocol for patient-derived xenografts of prostate cancer". Prostate **79**: 1326–1337.

Sicard, D., L. E. Fredenburgh and D. J. Tschumperlin (2017). "Measured pulmonary arterial tissue stiffness is highly sensitive to AFM indenter dimensions". Journal of the Mechanical Behavior of Biomedical Materials **74**: 118–127.

Sicard, D., A. J. Haak, K. M. Choi, A. R. Craig, L. E. Fredenburgh and D. J. Tschumperlin (2018). "Aging and anatomical variations in lung tissue stiffness". American Journal of Physiology. Lung Cellular and Molecular Physiology **314**: L946–l55.

Taylor, M. J., B. P. Weegman, S. C. Baicu and S. E. Giwa (2019). "New approaches to cryopreservation of cells, tissues, and organs". Transfusion Medicine and Hemotherapy: Offizielles Organ Der Deutschen Gesellschaft Fur Transfusionsmedizin Und Immunhamatologie **46**: 197–215.

Uriarte, J. J., T. Meirelles, D. Gorbenko Del Blanco, P. N. Nonaka, N. Campillo, E. Sarri, D. Navajas, G. Egea and R. Farre (2016). "Early impairment of lung mechanics in a murine model of marfan syndrome". PLoS One **11**: e0152124.

Van Zwieten, R. W., S. Puttini, M. Lekka, G. Witz, E. Gicquel-Zouida, I. Richard, J. A. Lobrinus, F. Chevalley, H. Brune, G. Dietler, A. Kulik, T. Kuntzer and N. Mermod (2014). "Assessing dystrophies and other muscle diseases at the nanometer scale by atomic force microscopy". Nanomedicine (Lond) **9**: 393–406.

Xin, X., L. Zhiyu, C. Luyao, C. Sarah and P. N. Corey (2016). "Mapping the nonreciprocal micromechanics of individual cells and the surrounding matrix within living tissues". Scientific Reports | 6:24272 | 101038/srep24272. **6**. 24272: 1–9.

Zemla, J., J. Danilkiewicz, B. Orzechowska, J. Pabijan, S. Seweryn and M. Lekka (2018a). "Atomic force microscopy as a tool for assessing the cellular elasticity and adhesiveness to identify cancer cells and tissues". Seminars in Cell & Developmental Biology **73**: 115–124.

Zemla, J., T. Stachura, I. Gross-Sondej, K. Górka, K. Okoń, G. Pyka-Fościak, J. Soja, K. Sładek and M. Lekka (2018b). "AFM-based nanomechanical characterization of bronchoscopic samples in asthma patients". Journal of Molecular Recognition: JMR **12**: e2752.

Zimmermann, M., J. Lampe, S. Lange, I. Smirnow, A. Königsrainer, C. Hann-Von-Weyhern, F. Fend, M. Gregor, M. Bitzer and U. M. Lauer (2009). "Improved reproducibility in preparing precision-cut liver tissue slices". Cytotechnology **61**: 145–152.

Zhao, L., L. Huang, S. Yu, J. Zheng, H. Wang and Y. Zhang. (2017). "Decellularized tongue tissue as an in vitro model for studying tongue cancer and tongue regeneration." Acta Biomater **58**: 122–135. doi: 10.1016/j.actbio.2017.05.062. Epub 2017 Jun 7.

Arnaud Millet, Carmela Rianna

5.3 Soft Hydrogels Mimicking the Extracellular Matrix

5.3.1 Historical and Conceptual Background

Since the beginning of cell culture method's development, the need to find a relevant cellular environment appeared as obvious. This quest progressively transformed to a motto, which could be summarized by: "we need to think in 3D." As soon as 1906, Harrison developed the so-called hanging drop method that authorizes cells to grow in a 3D liquid environment without adhesion to a stiff surface (Harrison et al., 1907). This technique was soon completed by the tube-culture method of Strangeways and Fell (1926). Despite these techniques, 2D cell cultures invaded cell labs during the 1950s and 1960s. It needed the breakthrough of Ehrmann and Gey (1956) who succeeded to dissolve collagen from rats' tails using an acidic solution to open the field of collagen-based 3D culture. The discovery of fibronectin in 1973 (Gahmberg and Hakomori, 1973, Hynes, 1973, Ruoslahti et al., 1973) and the characterization of laminin in 1979 (Timpl et al., 1979), associated with the production of matrigel from chondrosarcoma cells in 1977 (Orkin et al., 1977) closed the era of hydrogels pioneers (Simian and Bissell, 2017). The 1980s begun with a theoretical paper by Bissell et al. (1982), which hypothesized that the extracellular matrix (ECM) is able to regulate gene expression. The field of 3D cell culture was born, and the need to design hydrogels mimicking ECM was prevalent in the community of cell biologists. One of the questions that is at the center of the 3D cell culture field is: How complex a hydrogel mimicking the ECM should be?

The first degree of complexity that a hydrogel could give us is the third dimension. 2D and 3D environment do have adhesive, topographical, mechanical, and soluble properties that differ (Sachlos and Czernuszka, 2003) and real tissues definitely present 3D environments (Figure 5.3.1). Adhesion, spreading, and migration have been described as different in 2D and 3D, but it should be remembered that these differences are not only quantitative. For example, the 3D migration of fibrosarcoma cells has velocity profiles displaying different speed and self-correlation processes in different directions, rendering the classical persistent random walk model of cell migration inadequate in 3D (Wu et al., 2014). The second degree of

Acknowledgments: AM is supported by the ATIP/Avenir Young group leader program (Inserm/ CNRS) and by La Ligue nationale contre le cancer.

Arnaud Millet, Institute for Advanced Biosciences, Grenoble-Alpes University, Inserm and Research Department University Hospital of Grenoble Alpes, Grenoble, France
Carmela Rianna, Institute of Biophysics, University of Bremen, Bremen, Germany

https://doi.org/10.1515/9783110989380-009

complexity is the spatial architecture of the scaffold. Native ECM displays a wide variety of fibrillar structures that can change dynamically (Levental et al., 2009). The third element of complexity is the scaffold composition, and ECM is mainly composed of collagen type I in mammals, but many other molecules could compose the ECM (cf. Chapter 4.5). Of course, the desired complexity of a hydrogel is directly related to the biological question addressed. Hydrogels are unable to reproduce the heterogeneity of real tissues, and in vitro organoids are the main tool to address the multicellular and the complex chemical composition of a tissue. Despite this fact, hydrogels are able to help to decipher how cells integrate relevant physical and chemical signals. In this chapter, we will describe the various types of hydrogels used in cell biology and the various strategies to control their properties. We will then illustrate the usefulness of hydrogels with some biomedical applications.

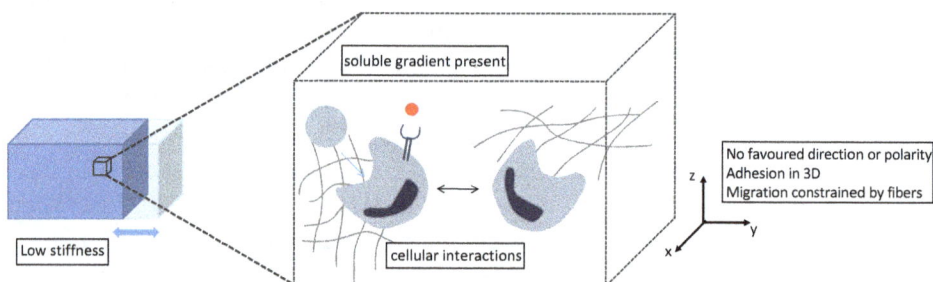

Figure 5.3.1: Numerous cues are different in 3D when compared to 2D. We have highlighted on this schematic diagram the main differences that cell encounter in a 3D ECM environment.

5.3.2 Various Hydrogels for Various Applications

Hydrogels are water-swollen polymeric networks composed of cross-linked hydrophilic polymers (Zhu and Marchant, 2011). The cross-linking mechanism could be physical or chemical. The physical cross-linking is noncovalent but usually sufficient to maintain the hydrogel structure and avoid its dissolving in the aqueous media. Chemical cross-linking is based on covalent links through the polymerization process. The huge diversity of hydrogels is related to the possibility to modify the cross-linking that will impact the porosity and elastic properties of the gel, its ionic charge, structure (crystalline or amorphous), and composition (homo or multipolymer). The polymers that could be used to create hydrogels could have a natural origin or being synthetic. The hybridization is also possible.

5.3.2.1 Natural Polymers

Natural polymers offer many important properties such as their biocompatibility, biochemical relevant composition, and biodegradability. These polymers could be:
- Proteins: collagen, gelatin (obtained from collagen), fibrin, complex mix of protein such as Matrigel™. Protein-based hydrogels are generally formed by thermal gelation, and covalent bonds to increase their stiffness could be done using a formaldehyde.
- Polysaccharides: hyaluronic acid, agarose, dextran, alginate, and chitosan. The chemistry of polysaccharides offers a large panel of possible functionalization.
- DNA (desoxyribonucleic acid).

Various combinations could be performed to obtain hybrid protein/polysaccharide polymers. Despite their biocompatibility, these natural polymers could drive immunogenic reactions when used in biomedical applications.

5.3.2.2 Collagen Hydrogels as an Example

Collagen is the most abundant component of the ECM in vertebrates. Many types of collagen are known, and collagen of type I is the most abundant of them and widespread in noncartilaginous connective tissues. Collagens are composed of three helical polypeptide chains. Type I collagen could be obtained from many sources, but the most classical one is the acidic dissolving of collagen from rat tails. The relevance of biological functions assessed by using rat collagen when physiological or pathophysiological process is at stake could be addressed from a structural point of view. Indeed, the comparison of protein sequences (triple-helical region) between rat collagen I (*Rattus norvegicus*) and human collagen I (*Homo sapiens*) is as follows: collagen alpha-1 (I) chain (coded by COL1A1 gene) sequences between the two species have a 96% amino acid identity and 97% positive matches (taking into account shared biochemical properties), collagen alpha-2 (I) chain (coded by COL1A2 gene) sequences between the two species have a 92% amino acid identity and 95% positive matches (Uniprot database https://www.uniprot.org/, alignment performed the 2019-03-05). Another source of collagen type I is from Porcine. Porcine collagen I (*Sus scrofa*) is unreviewed in the Uniprot database, which means that we do not have precise information concerning signal peptide, N-terminal propeptide, and C-terminal propeptide, but the total sequence of collagen alpha-1 chain is 91.9% identical to the rat sequence, and sequence alignment shows that propeptides contain the main mismatches of the two sequences. As phylogenetically *Sus scrofa* is closer to human compared to rats, the above analysis demonstrates that the high homology of protein sequences reasonably advocates for similar biological results with the different types of collagen type I. Apart from the structural point of view, collagen hydrogels could be obtained easily in the laboratory as the polymerization is thermally controlled. The main trouble when cells are directly embedded in the gel prior to polymerization is the need to adjust the pH in order to protect cells as well as to authorize the polymerization.

5.3.2.3 Synthetic Polymers

Synthetic polymers present several advantages when compared to natural ones. First, their physicochemical properties such as network parameters, mechanical strength, and diffusion profile are more reproducible and can be tailored by molecular design. Second, they can be designed as biodegradable or not, which could be of interest when ECM remodeling is taken into account. Third, the controlled chemistry authorizes to obtain bioactives hydrogels by modifying their adhesion properties or by controlling drug release. It is also possible to incorporate natural polymers associated with these synthetic hydrogels.

When mechanical stability is chosen, nonbiodegradable hydrogels could be prepared using various vinylated monomers or macromers such as (nonexhaustive list): acrylamide, acrylic acid, 2-hydroxyethyl methacrylate, 2-hydroxypropyl methacrylate, methoxyl polyethyleneglycol monoacrylate, and polyvinyl alcohol. Cross-linkers that could be used are: N,N'-methylenebisacrylamide, ethyleneglycoldiacrylate (EGDA), and poly(ethylene glycol) diacrylate (PEGDA). PEG is one of the most used polymers due to its high solubility in various solvent, nontoxicity, low protein adhesion, and low immunogenicity. PEG-based hydrogels chemistry offers many possibilities to functionalize these scaffolds (Zhu, 2010).

Biodegradable scaffolds are usually based on polyesters, including poly(lactic acid) (PLA), poly(glycolic acid), and poly (ε-caprolactone) as well as various combinations of copolymers. These esters could be used to obtain PEG-based biodegradable hydrogels by forming amphiphilic polymers. Examples of triblock polymers are PLA-PEG-PLA or PEG-PLA-PEG. The photopolymerization of these polyester-containing macromers gives hydrolytically degradable hydrogels (Clapper et al., 2007).

5.3.2.4 Construction of a 3D Scaffold and Bioprinting

Conventional scaffold fabrications techniques are gas foaming, fiber networking, solution casting, melt molding, and so on (Lee et al., 2007b, Mooney et al., 1996, Pitarresi et al., 2008, Salerno et al., 2009). However, a more recent technology offering fascinating perspectives in the field of 3D scaffold realization and tissue engineering is the so-called 3D bioprinting or organ printing.

3D bioprinting is a novel approach with the potential to revolutionize the field of tissue engineering and medical treatment. This technique benefits from the same high precision and resolution of a conventional 3D printer; however, it employs biocompatible ink instead of the usual synthetic material. The principle of bioprinting has been proposed for the first time in 1988 by Klebe (1988), naming the procedure as cytoscribing process, based on micropositioning of cells to build 2D and 3D synthetic tissues. Since then, 3D bioprinting techniques continue to expand offering a straightforward method to fabricate 3D scaffolds for engineering complex tissue structures. One of the crucial aspects of bioprinting is the selection of materials to use as bioinks (Malda and Groll, 2016).

Due to their biocompatibility and high water content, hydrogels are among the most suitable classes of bioink materials for the formation of 3D scaffolds able to mimic or replace the native tissue microenvironment for biomedical applications (Fisher et al., 2010, Gaharwar et al., 2014, Slaughter et al., 2009, Stanton et al., 2015, Tibbitt and Anseth, 2009). Both natural (such as alginate, chitosan, and fibrin) and synthetic hydrogels (such as poly(hydroxyethyl methacrylate), PVA, and PEG) have been proved in fact to be good candidate for suitable bioinks in bioprinting processes. When designing hydrogels for bioprinting, several properties need to be considered in order to find a compromise between cell viability and printing accuracy (Murphy et al., 2013, Stanton et al., 2015). For example, increasing polymer cross-link density will facilitate the printing process; but will reduce gel porosity, hence prevent cell spreading and migration (Bertassoni et al., 2014, Topuz et al., 2018).

5.3.2.5 Biophysical Opportunities: Imprinting Topographical Signals

The versatility of hydrogels gives rise to biophysical and biochemical opportunities to mimic the properties of natural microenvironment, like topography, stiffness, and protein functionalization. Combining patterning techniques with hydrogel chemistry, it is in fact possible to create hybrid hydrogel structures, which recapitulate multiple properties of natural tissues. The effects of topography and "contact guidance" have been extensively studied in terms of cell–material interaction (Kim and Kim, 2018). Many cell properties are in fact influenced by the surrounding topography, such as adhesion, elongation, migration, and differentiation (Dalby et al., 2014, Kim et al., 2012). Comparing cell behavior on flat and patterned surfaces several trends have been found, like the findings that surface topography generally increases cell adhesion, elongation, and migration speed. Moreover, also feature sizes affect those properties, for example in the case of nanogrooved topographies consisting of alternating grooves and ridge, several studies found that contact guidance is not initiated on groove depths below 35 nm or ridge widths < 100 nm (Biggs, Richards et al, 2010).

One of the most used process of topographic imprinting on hydrogels is based on photolithography and soft lithography techniques, involving first the realization of a patterned silicon wafer and then the replica molding of a soft hydrogel polymerized in contact with the silicon wafer (Yu and Ober, 2003). Other techniques used to create topographies on soft hydrogels have been for example two-photon polymerization (Lee et al., 2008, Torgersen et al., 2013), the combination of photoresist lithography and dry etching (Lei et al., 2004), or soft embossing based on PDMS stamps (Kobel et al., 2009). Moreover, using UV light exposure through a photomask, photopolymerizable hydrogels can also be directly patterned via photolithography. With all these techniques, the fabrication of micropatterned hydrogels has been shown to be a versatile and helpful tool for several biological applications, such as cell culture substrates, scaffolds for tissue engineering, and high-throughput analytical platforms (Khademhosseini and Langer, 2007).

5.3.2.6 Biophysical Opportunities: Stiffness of Hydrogels

Together with ECM topography, mechanical properties or stiffness of the surrounding microenvironment also has a critical impact on many cell processes, including cell proliferation, migration, or differentiation. Human tissues are very heterogeneous and present disparate features and mechanical properties. Going from brain to liver to muscles or bone, we will find stiffness values ranging from 1 to 3 kPa to 15 MPa, respectively. Therefore, when approaching in vitro studies to investigate cell behavior or cell–material interaction, it is crucial to mimic the in vivo microenvironment trying to replicate the features and properties of natural cell environment. In this perspective, hydrogels have proved to be very helpful and versatile substrates, taking advantage of the possibility to easily modulate their mechanical properties.

The use of hydrogels as cell culture supports allows indeed to mimic the mechanical properties of natural tissues and ECM, overcoming the limitation of standard cell culture system, like Petri-dishes, which are way stiffer (>GPa) than natural tissues.

Stiffness-tunable hydrogels helped to discern many fundamental phenomena of cell biology, gaining deep insights in many phenomena, such as differentiation, migration, proliferation, apoptosis, and so on. Pelham and Wang (1997) first reported the effect on the motility and spread of fibroblasts and kidney epithelial cells to substrate stiffness. Since then, many other studies have been focused on the effect of substrate stiffness and cell interactions.

For example, cells spread more on stiff substrates (Lo et al., 2000, Yeung et al., 2005) and in case of gradient-stiffness gel, they migrate in the direction of the stiffer substrate (Smith et al., 2006). Stem cells' differentiation also depends on substrate stiffness, and Engler et al. (2006) reported that stem cells differentiate into different specific lineages when seeded on matrices mimicking the stiffness of different tissues like brain (0.1–1 kPa), muscle (8–17 kPa), and bone (25–40 kPa). Different stiffness gels also helped to reveal some differences in biophysical properties of cancer cells when compared with normal cells. Metastatic cancer cells in fact increase their mechanical properties when seeded on soft hydrogels compared to their normal counterparts, also evidencing a stronger attitude to invade the hydrogel resembling phenomena of tissue penetration during the first step of metastasis formation (Rianna and Radmacher, 2017).

5.3.2.7 Biochemical Opportunities: Hydrogel Functionalization

Many properties of hydrogels could be modified, benefiting from chemical functionalization of synthetic hydrogels. One property is the ability of hydrogel to engage the adhesion of a cell to it. This behavior is fundamental, as in natural tissues, cells harbour various receptors such as integrins, selectins, CD44, or syndecan that could interact with specific peptide sequence of the ECM. These interactions have been implicated in many processes during development, organization, and maintenance of tissues. From the molecular point of view, specific peptides have been identified from ECM proteins, including fibronectin, vitronectin, bone sialoprotein laminin, and collagen. These peptides could be used as tools to modify hydrogel ability to induce cell adhesion. In Table 5.3.1, we have summarized the various cell adhesive peptides that have been identified.

Table 5.3.1: Sequences of specific cell adhesive peptides and their cell receptors (adapted from Zhu and Marchant, 2011).

Origin	Cell adhesive peptides	Cell receptor
Fibronectin	RGD	Integrin
	PHSRN	Integrin $\alpha_5\beta_1$
	EILDV	Integrin $\alpha_4\beta_1$
	KQAGDV	Integrin
	REDV	Integrin $\alpha_4\beta_1$
	LIGRKK	Heparin
	SPPRRARV	Heparin
	WQPPRARI	Heparin
Vitronectin	GKKQRFRHRNRKG	Heparin
Bone sialoprotein	FHRRIKA	Heparin
Laminin	RGD	110 kDa protein
	IKVAV	67 kDa protein
	YIGSR	Integrin
	PDGSR	Integrin
	LRGDN	Integrin
	LRE	Integrin
	IKLLI	Heparin
Collagen	RGD	Integrin
	DGEA	Integrin $\alpha_2\beta_1$
	GFOGER	Integrin
	GDR, GRD	Integrin $\alpha_2\beta_1$
Elastin	VAPG	67 kDa protein

Synthetic hydrogels offer the possibility to control the spatial peptide expression. One example is the insertion of a RGD sequence in a PEGDA chain, which maintain a similar chemical structure between RGD-PEGDA and PEGDA with the two acrylate groups at both ends (Zhu et al., 2009).

A second property of hydrogel that should be taken into account is the natural turn-over of the ECM engaged by specific enzymes that degrade collagen, laminin, or fibrin. Similarly, to the cell adhesive peptides, the specific known cleavage sites of ECM proteins are summarized in Table 5.3.2.

Table 5.3.2: Sequences of cleavage site for enzyme degradation on ECM proteins and peptide library (adapted from Zhu and Marchant, 2011).

Origin	Enzyme-sensitive peptide	Specific enzyme
Collagen-I	GPQGIAGQ	MMP-1
Laminin	QLLADTPV	MMP
	YSGDENP	MMP
	DENPDIE	MMP-12
Fibrinogen	YKNR, YKNRD	Plasmin
	YKNS, YKND	Plasmin
	NRV, NRD	Plasmin
Aggrecan	PENFF	MMP-13
Peptide library	GPQGIWGQ	MMP-1, MMP-12
	GPQGILGQ	MMP-1
	GPQGLA	MMP-13
	LGPA	MMP-1
	APGL	MMP-1
	AAAAAAAAA	Elastase
	AAPV	Elastase
	AAPVRGMG	Elastase
	GGYRG	Chymotrypsin
	GL, GFL, GFGL	Papain

These various enzyme-sensitive peptides could be incorporated in PEGDA polymer in the same way as RGD sequences (Lee et al., 2007a). Or a Michael addition of enzyme-sensitive peptides with two thiols using multi-branched PEG with vinyl sulfone groups (Jo et al., 2010). Increasing the relevance of the functionalization, it has been reported that enzyme-sensitive peptides could be coupled with adhesive-specific peptides. One study has used the CPENFFRGD peptide incorporated in a PEG hydrogel. This peptide was sensitive to MMP-13, and the adhesion motif was the RGD peptide. This type of hydrogels could be used as a dynamical platform to study the interaction between cells and the ECM (Casadio et al., 2010).

5.3.2.8 Challenges for Hydrogels

Hydrogel scaffolds have demonstrated their ability to help biologist to think in 3D. They offer the possibility to mimic ECM by controlling their chemistry, their physical properties like elasticity, and also to address biological relevant questions through bioactive gels. Despite the huge amount of published work in this area, there are still real challenges in the hydrogel field:
- Scaffolding methods have, during the last years, opened new way to realize complex architectures. Even if these architectures could mimic structural aspects of various tissues, from the experimental point of view, cell seeding in these scaffolds is notoriously difficult. It is usually found that cell penetration, in these structures, is poor.
- Native tissues are vascularized and hydrogels are not. This simple observation opens a new field of research in order to control nutrients, oxygen, pH, and signaling in hydrogels in a way similar than what is performed by the vascular transport.
- Tissues are crowded with cells. In human, there is around 10^8 cells/cm^3 in a normal liver (Wilson et al., 2003). Obtaining this kind of density in in vitro hydrogels is a real challenge. A recent work using a photopatterning and photorelease sortase-tag enhanced protein ligation-modified proteins in hydrogels is an elegant new approach, even if its technicality forbids yet its transfer to biological laboratories (Shadish et al., 2019).
- Biological processes are taking place in 4D. The spatiotemporal control of hydrogels is notoriously difficult, especially when it envisioned the manipulation of full-length proteins.

5.3.3 Biomedical Applications

In order to illustrate the use of hydrogels as tools to mimic the ECM, we will present a recent work on the influence of a 3D collagen environment on the innate immune response driven by macrophages and then we will present a brief overview of the use of hydrogels to model macrophage–ECM interactions in a cancerous context.

5.3.3.1 3D Hydrogels as an Opportunity to Understand the Innate Immune Response

Macrophages are innate immune cells present in every tissue playing a critical role in homeostasis. As first-line defenders, these cells are prone to modify their phenotype in response to their surrounding environment, sensing various signals, and displaying a large panel of activation states in order to cope with various pathogens (Okabe and Medzhitov, 2016). Even if it is now accepted that the spectrum of activation states of these cells is better viewed as a continuum, it is still interesting to understand how the environment is able to favor a pro-inflammatory (M1) or an anti-inflammatory (M2) phenotype which could represent the extremes of this spectrum (Sica and Mantovani, 2012). M1 macrophages are specialized in the removal of pathogens and are classically obtained in vitro using a combination of IFN-γ and LPS (lipopolysaccharide). M1 macrophages are associated with the production of reactive oxygen species and pro-inflammatory cytokines secretion like TNF-α, IL-6, or chemokines like CCL-20 (Martinez et al., 2006, Murray et al., 2014). M2 macrophages are obtained using a stimulation with IL-4 (±IL-13) and are also called alternatively activated macrophages (Martinez and Gordon, 2014). These cells are described as anti-inflammatory and seem to participate to wound healing. This polarization is notably associated with the membrane expression of the mannose receptor also named CD206. Modifying the polarization of macrophages has emerged as a new therapeutical approach in inflammatory diseases and in cancer (Sica and Mantovani, 2012). This goal, in order to be attained, needs that macrophage polarizations are properly defined and the influence of their cellular environment clarified. Even if it is possible to modify the activation state of macrophages using various chemical signals, it has been recently recognized that macrophages are also sensitive to their physical environment (McWhorter et al., 2015). The question of how a three-dimensional environment impacts immune cell functions has been recently recognized as a key element in our understanding of the tumor microenvironment. In this context, hydrogels offer a clear opportunity to decipher molecular and cellular processes involved in the macrophage–ECM–cancerous cell interactions (Springer and Fischbach, 2016).

In order to address that question, 3D fibrillary matrices from naturally derived collagen-based networks appear as naturally interesting as they reconstitute the microstructure of the in vivo ECM (Green and Elisseeff, 2016). We have recently reported a study using 3D collagen gels that give us the opportunity to reassess the classification of human macrophage polarization (Court et al., 2019). We first studied how transcription factors associated with polarizations of macrophages were impacted and find that their response was modulated by the 3D environment. For IFNγ/LPS-stimulated macrophages (M1), we found an integrin β2 induction of pSTAT1^{ser727} in 3D collagen gel. Previous works have reported that macrophages seem to interact with collagen through integrin β1 but denatured collagen was used as adhesion substratum (Pacifici et al., 1994). THP1 monocytic cell line is also believed to interact with fibronectin (Lin et al., 1995) or gelatin methacryloyl (Cha et al., 2017) through integrin β1. Despite these results, when native collagen is used, human monocytes interact with collagen mainly using integrin β2 (Garnotel et al., 2000). STAT6 transcription factor is phosphorylated under IL-4 or IL-13 stimulation. We found that 3D macrophages present an increased response to IL-4/IL-13 as demonstrated by the phosphorylation of tyrosine 641 (Y641). This response was not related to an integrin signaling. This 3D environment could also impact other immune functions of macrophages, and we found that the expression of class II HLA molecules presents a different expression pattern in 2D and 3D. These molecules were found to be differentially expressed in 2D notably in response to LPS, and we confirm that IL4/IL13 macrophages present a stronger expression of these molecules. This differential expression was not maintained in 3D where M1 and M2 macrophages were indistinguishable according to HLA class II expression levels. That finding was not similar for class I HLA molecules that conserve in 3D the differential expression of these molecules in M1 macrophages expressing a higher level than their M2 counterparts.

In our study, we found that the response of macrophages to IFNγ/LPS is more complex than previously thought. We notably found that no general inflammatory or anti-inflammatory pattern could be drawn when 2D and 3D are compared. We also found that TNF-α is not modulated contrary to IL-6, and we demonstrated that integrins β_1, β_2 but also β_3 are involved in this increased expression in 3D (Figure 5.3.2). The implication of integrin β_3 is particularly interesting as it is known to be an IL-1β

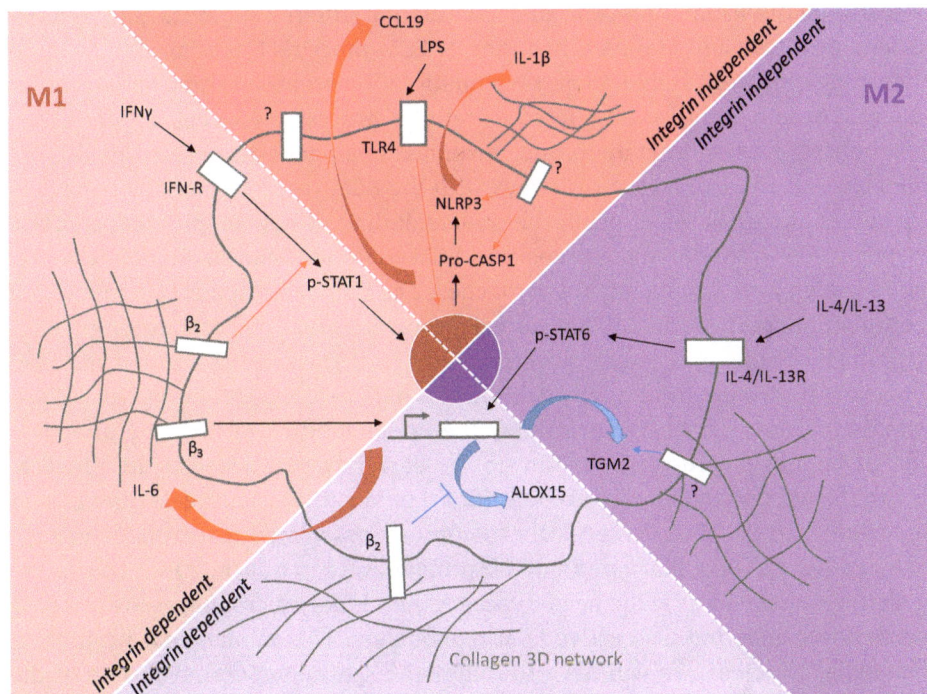

Figure 5.3.2: Schematic overview of the results obtained by the team on the various integrin-dependent and integrin-independent mechanisms implied in the control of the STAT1 and STAT6 signaling pathways. We have also demonstrated that the interaction with a 3D environment could enhance the NLRP3 inflammasome activation in an integrin-independent mechanism.

receptor (Takada et al., 2017). Accordingly, we find an IL-1β signature in 3D macrophages in response to IFNγ/LPS. We also demonstrated that this IL-1β signature is under the control of the NLRP3 inflammasome in an integrin independent mechanism. As the growing field of biomaterial research is progressively moving from biocompatible "immunoevasive" materials to "immune-modulating" materials, the implication of inflammasome in macrophage response to biomaterials is of outstanding interest (Vasconcelos et al., 2019). This study, in a 3D collagen type I context, gave us the opportunity to reassess the classification of human macrophage polarizations. These results are of a particular interest in the field of immune-oncology where the macrophage involvement and targeting need a thoughtful understanding of cellular environmental clues leading to various activation states.

5.3.3.2 Hydrogels as Tools to Understand the Macrophage–ECM Interactions in the Tumor Microenvironment

The ECM regulation of development and tissue homeostasis is deeply annoyed in cancer. Tumors usually present an accumulation, altered organization, and enhanced post-translational modifications of ECM proteins (Lu et al., 2012). In an authoritative review, Hanahan and Weinberg (2011) proposed some hallmarks for cancer. These hallmarks are related to cellular processes like cell proliferation, metabolism, evasion from apoptosis, sustained angiogenesis, or immune recruitment that all share common pathways with the ECM regulation of cell behaviors (Pickup et al., 2014). As a result, the understanding of how biophysical and biochemical cues from the tumor-associated ECM could influence malignancy is capital to design new therapeutical approaches. The immune response component of the tumor microenvironment has gained strong attention recently. Macrophages, in particular, appear to be key players and are by themselves associated with bad prognosis (DeNardo and Ruffell, 2019, Ritter and Greten, 2019, Ruffell and Coussens, 2015). As we have demonstrated, the understanding of the state of activation of human macrophages could not be correctly performed if the physical environment is not taken into account (Court et al., 2019). Hydrogels offer a clear opportunity to decipher molecularly the fibrosis-associated macrophage mechanosignaling that take place in tumors (Springer and Fischbach, 2016). One of the challenges ahead of us is the need to keep in mind that macrophages are not only mechanosensitive but also active ECM re-modelers. This implies that when their density is sufficient, we will need to follow the spatiotemporal evolution of the ECM modeled by a hydrogel scaffold to characterize the feedback loop between ECM and macrophages.

5.3.4 Conclusion

Reconstructing the 3D tissular environment of cells is a key goal of cell culture. This is only possible if the chemistry and physical properties of the ECM could be reproduced in the laboratory. The field of soft hydrogels for cell culture is devoted to this task. In this chapter, we have described the variety of natural and synthetic compounds used to mimic various aspects of the ECM. We have illustrated the usefulness of these approaches by presenting important results on the modulation of the innate immune response induced by the physical microenvironment mimicked by a collagen network. This field of research is growing rapidly, and new important results are expected to emerge in the near future.

References

Bertassoni, L. E., M. Cecconi, V. Manoharan, M. Nikkhah, J. Hjortnaes, A. L. Cristino, G. Barabaschi, D. Demarchi, M. R. Dokmeci and Y. Yang (2014). "Hydrogel bioprinted microchannel networks for vascularization of tissue engineering constructs". Lab on a Chip **14**(13): 2202–2211.

Biggs, M. J. P., R. G. Richards and M. J. Dalby (2010). "Nanotopographical modification: A regulator of cellular function through focal adhesions". Nanomedicine: Nanotechnology, Biology and Medicine **6**(5): 619–633.

Bissell, M. J., H. G. Hall and G. Parry (1982). "How does the extracellular matrix direct gene expression?". Journal of Theoretical Biology **99**(1): 31–68.

Casadio, Y. S., D. H. Brown, T. V. Chirila, H.-B. Kraatz and M. V. Baker (2010). "Biodegradation of poly (2-hydroxyethyl methacrylate)(PHEMA) and poly {(2-hydroxyethyl methacrylate)-co-[poly (ethylene glycol) methyl ether methacrylate]} hydrogels containing peptide-based cross-linking agents". Biomacromolecules **11**(11): 2949–2959.

Cha, B. H., S. R. Shin, J. Leijten, Y. C. Li, S. Singh, J. C. Liu, N. Annabi, R. Abdi, M. R. Dokmeci and N. E. Vrana (2017). "Integrin-mediated interactions control macrophage polarization in 3D hydrogels". Advanced Healthcare Materials **6**(21): 1700289.

Clapper, J. D., J. M. Skeie, R. F. Mullins and C. A. Guymon (2007). "Development and characterization of photopolymerizable biodegradable materials from PEG–PLA–PEG block macromonomers". Polymer **48**(22): 6554–6564.

Court, M., M. Malier and A. Millet (2019). "3D type I collagen environment leads up to a reassessment of the classification of human macrophage polarizations". Biomaterials **208**: 98–109.

Dalby, M. J., N. Gadegaard and R. O. Oreffo (2014). "Harnessing nanotopography and integrin–matrix interactions to influence stem cell fate". Nature Materials **13**(6): 558–569.

DeNardo, D. G. and B. Ruffell (2019). "Macrophages as regulators of tumour immunity and immunotherapy". Nature Reviews Immunology **19**(6): 369–382.

Ehrmann, R. L. and G. O. Gey (1956). "The growth of cells on a transparent gel of reconstituted rat-tail collagen". Journal of the National Cancer Institute **16**(6): 1375–1403.

Engler, A. J., S. Sen, H. L. Sweeney and D. E. Discher (2006). "Matrix elasticity directs stem cell lineage specification". Cell **126**(4): 677–689.

Fisher, O. Z., A. Khademhosseini, R. Langer and N. A. Peppas (2010). "Bioinspired materials for controlling stem cell fate". Accounts of Chemical Research **43**(3): 419–428.

Gaharwar, A. K., N. A. Peppas and A. Khademhosseini (2014). "Nanocomposite hydrogels for biomedical applications". Biotechnology and Bioengineering **111**(3): 441–453.

Gahmberg, C. G. and S.-I. Hakomori (1973). "Altered growth behavior of malignant cells associated with changes in externally labeled glycoprotein and glycolipid". Proceedings of the National Academy of Sciences **70**(12): 3329–3333.

Garnotel, R., L. Rittié, S. Poitevin, J.-C. Monboisse, P. Nguyen, G. Potron, F.-X. Maquart, A. Randoux and P. Gillery (2000). "Human blood monocytes interact with type I collagen through αXβ2 integrin (CD11c-CD18, gp150-95)". The Journal of Immunology **164**(11): 5928–5934.

Green, J. J. and J. H. Elisseeff (2016). "Mimicking biological functionality with polymers for biomedical applications". Nature **540**(7633): 386–394.

Hanahan, D. and R. A. Weinberg (2011). "Hallmarks of cancer: The next generation". cell **144**(5): 646–674.

Harrison, R. G., M. Greenman, F. P. Mall and C. Jackson (1907). "Observations of the living developing nerve fiber". The Anatomical Record **1**(5): 116–128.

Hynes, R. O. (1973). "Alteration of cell-surface proteins by viral transformation and by proteolysis". Proceedings of the National Academy of Sciences **70**(11): 3170–3174.

Jo, Y. S., S. C. Rizzi, M. Ehrbar, F. E. Weber, J. A. Hubbell and M. P. Lutolf (2010). "Biomimetic PEG hydrogels crosslinked with minimal plasmin-sensitive tri-amino acid peptides". Journal of Biomedical Materials Research Part A: An Official Journal of the Society for Biomaterials, the Japanese Society for Biomaterials, and the Australian Society for Biomaterials and the Korean Society for Biomaterials **93**(3): 870–877.

Khademhosseini, A. and R. Langer (2007). "Microengineered hydrogels for tissue engineering". Biomaterials **28**(34): 5087–5092.

Kim, D.-H., P. P. Provenzano, C. L. Smith and A. Levchenko (2012). "Matrix nanotopography as a regulator of cell function". Journal of Cell Biology **197**(3): 351–360.

Kim, H. N. and J. Kim (2018). "Effect of Topographical Feature Size on the Trend of Cell Behaviors". IEEE Transactions on Nanotechnology **17**: 377–380.

Klebe, R. J. (1988). "Cytoscribing: A method for micropositioning cells and the construction of two- and three-dimensional synthetic tissues". Experimental Cell Research **179**(2): 362–373.

Kobel, S., M. Limacher, S. Gobaa, T. Laroche and M. P. Lutolf (2009). "Micropatterning of hydrogels by soft embossing". Langmuir **25**(15): 8774–8779.

Lee, S.-H., J. J. Moon, J. S. Miller and J. L. West (2007a). "Poly (ethylene glycol) hydrogels conjugated with a collagenase-sensitive fluorogenic substrate to visualize collagenase activity during three-dimensional cell migration". Biomaterials **28**(20): 3163–3170.

Lee, S.-H., J. J. Moon and J. L. West (2008). "Three-dimensional micropatterning of bioactive hydrogels via two-photon laser scanning photolithography for guided 3D cell migration". Biomaterials **29**(20): 2962–2968.

Lee, S. H., S. C. Seong, J. H. Lee, I. K. Han, S. H. Oh, K. J. Cho, H. S. Han and M. C. Lee (2007b). Porous polymer prosthesis for meniscal regeneration. Key Engineering Materials, Trans Tech Publ.

Lei, M., Y. Gu, A. Baldi, R. A. Siegel and B. Ziaie (2004). "High-resolution technique for fabricating environmentally sensitive hydrogel microstructures". Langmuir **20**(21): 8947–8951.

Leventhal, K. R., H. Yu, L. Kass, J. N. Lakins, M. Egeblad, J. T. Erler, S. F. Fong, K. Csiszar, A. Giaccia and W. Weninger (2009). "Matrix crosslinking forces tumor progression by enhancing integrin signaling". Cell **139**(5): 891–906.

Lin, T. H., C. Rosales, K. Mondal, J. B. Bolen, S. Haskill and R. L. Juliano (1995). "Integrin-mediated Tyrosine Phosphorylation and Cytokine Message Induction in Monocytic Cells A POSSIBLE SIGNALING ROLE FOR THE Syk TYROSINE KINASE". Journal of Biological Chemistry **270**(27): 16189–16197.

Lo, C.-M., H.-B. Wang, M. Dembo and Y.-L. Wang (2000). "Cell movement is guided by the rigidity of the substrate". Biophysical Journal **79**(1): 144–152.

Lu, P., V. M. Weaver and Z. Werb (2012). "The extracellular matrix: A dynamic niche in cancer progression". Journal of Cell Biology **196**(4): 395–406.

Malda, J. and J. Groll (2016). "A Step Towards Clinical Translation of Biofabrication". Trends in Biotechnology **34**(5): 356.

Martinez, F. O. and S. Gordon (2014). "The M1 and M2 paradigm of macrophage activation: Time for reassessment". F1000Prime Rep. 2014 Mar 3;6:13. doi: 10.12703/P6-13 **6**.

Martinez, F. O., S. Gordon, M. Locati and A. Mantovani (2006). "Transcriptional profiling of the human monocyte-to-macrophage differentiation and polarization: New molecules and patterns of gene expression". The Journal of Immunology **177**(10): 7303–7311.

McWhorter, F. Y., C. T. Davis and W. F. Liu (2015). "Physical and mechanical regulation of macrophage phenotype and function". Cellular and Molecular Life Sciences **72**(7): 1303–1316.

Mooney, D., C. Mazzoni, C. Breuer, K. McNamara, D. Hern, J. Vacanti and R. Langer (1996). Stabilized polyglycolic acid fibre-based tubes for tissue engineering. Biomaterials. 1996 Jan; **17**(2): 115–124. doi: 10.1016/0142-9612(96)85756-5.

Murphy, S. V., A. Skardal and A. Atala (2013). "Evaluation of hydrogels for bio-printing applications". Journal of Biomedical Materials Research. Part A **101**(1): 272–284.

Murray, P. J., J. E. Allen, S. K. Biswas, E. A. Fisher, D. W. Gilroy, S. Goerdt, S. Gordon, J. A. Hamilton, L. B. Ivashkiv and T. Lawrence (2014). "Macrophage activation and polarization: Nomenclature and experimental guidelines". Immunity **41**(1): 14–20.

Okabe, Y. and R. Medzhitov (2016). "Tissue biology perspective on macrophages". Nature Immunology **17**(1): 9.

Orkin, R., P. Gehron, E. B. Mcgoodwin, G. Martin, T. Valentine and R. Swarm (1977). "A murine tumor producing a matrix of basement membrane". The Journal of Experimental Medicine **145**(1): 204–220.

Pacifici, R., J. Roman, R. Kimble, R. Civitelli, C. M. Brownfield and C. Bizzarri (1994). "Ligand binding to monocyte alpha 5 beta 1 integrin activates the alpha 2 beta 1 receptor via the alpha 5 subunit cytoplasmic domain and protein kinase C". The Journal of Immunology **153**(5): 2222–2233.

Pelham, R. J. and Y.-L. Wang (1997). "Cell locomotion and focal adhesions are regulated by substrate flexibility". Proceedings of the National Academy of Sciences **94**(25): 13661–13665.

Pickup, M. W., J. K. Mouw and V. M. Weaver (2014). "The extracellular matrix modulates the hallmarks of cancer". EMBO Reports **15**(12): 1243–1253.

Pitarresi, G., F. Palumbo, R. Calabrese, E. Craparo and G. Giammona (2008). "Crosslinked hyaluronan with a protein-like polymer: Novel bioresorbable films for biomedical applications". Journal of Biomedical Materials Research. Part A **84**(2): 413–424.

Rianna, C. and M. Radmacher (2017). "Influence of microenvironment topography and stiffness on the mechanics and motility of normal and cancer renal cells". Nanoscale **9**(31): 11222–11230.

Ritter, B. and F. R. Greten (2019). "Modulating inflammation for cancer therapy". Journal of Experimental Medicine **216**(6): 1234–1243.

Ruffell, B. and L. M. Coussens (2015). "Macrophages and therapeutic resistance in cancer". Cancer Cell **27**(4): 462–472.

Ruoslahti, E., A. Vaheri, P. Kuusela and E. Linder (1973). "Fibroblast surface antigen: A new serum protein". Biochimica Et Biophysica Acta (Bba)-protein Structure **322**(2): 352–358.

Sachlos, E. and J. Czernuszka (2003). "Making tissue engineering scaffolds work. Review: The application of solid freeform fabrication technology to the production of tissue engineering scaffolds". European Cells & Materials **5**(29): 39–40.

Salerno, A., P. Netti, E. Di Maio and S. Iannace (2009). "Engineering of foamed structures for biomedical application". Journal of Cellular Plastics **45**(2): 103–117.

Shadish, J. A., G. M. Benuska and C. A. DeForest (2019). "Bioactive site-specifically modified proteins for 4D patterning of gel biomaterials". Nature Materials **18**(9): 1005–1014.

Sica, A. and A. Mantovani (2012). "Macrophage plasticity and polarization: In vivo veritas". The Journal of Clinical Investigation **122**(3): 787–795.

Simian, M. and M. J. Bissell (2017). "Organoids: A historical perspective of thinking in three dimensions". Journal of Cell Biology **216**(1): 31–40.

Slaughter, B. V., S. S. Khurshid, O. Z. Fisher, A. Khademhosseini and N. A. Peppas (2009). "Hydrogels in regenerative medicine". Advanced Materials **21**(32–33): 3307–3329.

Smith, J. T., J. T. Elkin and W. M. Reichert (2006). "Directed cell migration on fibronectin gradients: Effect of gradient slope". Experimental Cell Research **312**(13): 2424–2432.

Springer, N. L. and C. Fischbach (2016). "Biomaterials approaches to modeling macrophage–extracellular matrix interactions in the tumor microenvironment". Current Opinion in Biotechnology **40**: 16–23.

Stanton, M., J. Samitier and S. Sanchez (2015). "Bioprinting of 3D hydrogels". Lab on a Chip **15**(15): 3111–3115.

Strangeways, T. S. and H. B. Fell (1926). "Experimental studies on the differentiation of embryonic tissues growing in vivo and in vitro. – I. The development of the undifferentiated limb-bud (a) when subcutaneously grafted into the post-embryonic chick and (b) when cultivated in vitro". Proceedings of the Royal Society of London. Series B, Containing Papers of a Biological Character **99**(698): 340–366.

Takada, Y. K., J. Yu, M. Fujita, J. Saegusa, C.-Y. Wu and Y. Takada (2017). "Direct binding to integrins and loss of disulfide linkage in interleukin-1β (IL-1β) are involved in the agonistic action of IL-1β". Journal of Biological Chemistry **292**(49): 20067–20075.

Tibbitt, M. W. and K. S. Anseth (2009). "Hydrogels as extracellular matrix mimics for 3D cell culture". Biotechnology and Bioengineering **103**(4): 655–663.

Timpl, R., H. Rohde, P. G. Robey, S. I. Rennard, J.-M. Foidart and G. R. Martin (1979). "Laminin–a glycoprotein from basement membranes". Journal of Biological Chemistry **254**(19): 9933–9937.

Topuz, M., B. Dikici, M. Gavgali and H. Yilmazer (2018). "A review on the hydrogels used in 3D bio-printing". International Journal of 3D Printing Technologies and Digital Industry **2**(2): 68–75.

Torgersen, J., X. H. Qin, Z. Li, A. Ovsianikov, R. Liska and J. Stampfl (2013). "Hydrogels for two-photon polymerization: A toolbox for mimicking the extracellular matrix". Advanced Functional Materials **23**(36): 4542–4554.

Vasconcelos, D. P., A. P. Aguas, M. A. Barbosa, P. Pelegrin and J. N. Barbosa (2019). "The inflammasome in host response to biomaterials: Bridging inflammation and tissue regeneration". Acta biomaterialia **83**: 1–12.

Wilson, Z., A. Rostami-Hodjegan, J. Burn, A. Tooley, J. Boyle, S. Ellis and G. Tucker (2003). "Inter-individual variability in levels of human microsomal protein and hepatocellularity per gram of liver". British Journal of Clinical Pharmacology **56**(4): 433–440.

Wu, P.-H., A. Giri, S. X. Sun and D. Wirtz (2014). "Three-dimensional cell migration does not follow a random walk". Proceedings of the National Academy of Sciences **111**(11): 3949–3954.

Yeung, T., P. C. Georges, L. A. Flanagan, B. Marg, M. Ortiz, M. Funaki, N. Zahir, W. Ming, V. Weaver and P. A. Janmey (2005). "Effects of substrate stiffness on cell morphology, cytoskeletal structure, and adhesion". Cell Motility and the Cytoskeleton **60**(1): 24–34.

Yu, T. and C. K. Ober (2003). "Methods for the topographical patterning and patterned surface modification of hydrogels based on hydroxyethyl methacrylate". Biomacromolecules **4**(5): 1126–1131.

Zhu, J. (2010). "Bioactive modification of poly (ethylene glycol) hydrogels for tissue engineering".
    Biomaterials **31**(17): 4639–4656.
Zhu, J. and R. E. Marchant (2011). "Design properties of hydrogel tissue-engineering scaffolds".
    Expert Review of Medical Devices **8**(5): 607–626.
Zhu, J., C. Tang, K. Kottke-Marchant and R. E. Marchant (2009). "Design and synthesis of
    biomimetic hydrogel scaffolds with controlled organization of cyclic RGD peptides".
    Bioconjugate Chemistry **20**(2): 333–339.

Harinderbir Kaur, Christian Godon, Jean-Marie Teulon,
Thierry Desnos, Jean-Luc Pellequer

5.4 Preparation and Deposition of Plant Roots for AFM Nanomechanical Measurements

5.4.1 Introduction

In general, cell growth is a mechanical process that balances internal and external stresses allowing or limiting expansion. Plant cells are often compared to "hydraulic machines" due to the similar concept of balanced counterforces between the primary wall stresses and the turgor pressure. Knowledge of the mechanics of plant root cells is essential to understand how plant wall works and, therefore, how plants grow. One method for qualitative and quantitative analysis of mechanical properties is atomic force microscopy (AFM) that measures mechanical properties of living cells or tissues under conditions close to relevant physiological environments (Arnould et al., 2017, Kozlova et al., 2019, Milani et al., 2011, 2014, Peaucelle et al., 2012, 2011, Torode et al., 2018, Yakubov et al., 2016, Zdunek and Kurenda, 2013, Zhao et al., 2005). In our recent study, AFM has been used to investigate properties of root epidermal cells (Balzergue et al., 2017), which are the cells that form the outermost layer of the root. These epidermal cells were studied in situ on living seedlings and were therefore still influenced by the inherent multicellular properties of the root, which is an improvement over studies conducted on isolated living cells.

This chapter explains and illustrates the different steps required to prepare and deposit the seedling, and in particular the root tip, of the plant on a glass slide ready for nanomechanical measurements by AFM.

Acknowledgments: IBS acknowledges integration into the Interdisciplinary Research Institute of Grenoble (IRIG, CEA). This work acknowledges the AFM platform at the IBS. This work was partly funded by the Agence Nationale de la Recherche (ANR-18-CE20-0023-03; ANR-09-BLAN-0118; ANR-12-ADAP-0019), CEA (APTTOX021401, APTTOX021403), and Investissements d'avenir (DEMETERRES). We acknowledge Marjorie Chery for help in handling seedlings.

Harinderbir Kaur, Jean-Marie Teulon, Jean-Luc Pellequer, Univ. Grenoble Alpes, CEA, CNRS, IBS, Grenoble, France
Christian Godon, Aix Marseille Université, CNRS, CEA, Institut de Biosciences et Biotechnologies Aix-Marseille, Laboratoire de Signalisation pour l'adaptation des végétaux à leur environnement, CEA Cadarache, Saint-Paul-lez-Durance, France
Thierry Desnos, Aix Marseille Université, CNRS, CEA, Institut de Biosciences et Biotechnologies Aix-Marseille, Equipe Bioénergies et Microalgues, CEA Cadarache, Saint-Paul-lez-Durance, France

https://doi.org/10.1515/9783110989380-010

5.4.2 Materials

Seedlings are from the *Arabidopsis thaliana* Col^{er105} background hereafter named wild type (WT) (Balzergue et al., 2017). The arabidopsis WT (Col^{er105}), *stop1^{48}*, *almt1^{32}*, and *lpr1-2* lines were described previously (Balzergue et al., 2017). To homogenize the germination rate and seedling size, small seeds were screened out with a nylon mesh. During all the setup work, plant preparation is done in a sterile environment to avoid contaminations by bacteria or fungi spores; it is thus performed in a sterile laminar flow hood.

MES-(2-(*N*-morpholino)ethanesulfonic acid (Sigma, M8250) buffer (MES, pH 5.8, 50× concentration). pH is adjusted to 5.8 by adding 10 N of KOH.
SDS (sodium dodecyl sulfate) solution at 0.05% (Sigma, L4390).
Ethanol (#4145872, Carlo Erba, Val de Reuil, France).
Peltier-cooled incubator (IPP 110 +, Memmert, Schwabach, Germany).
$FeCl_2$: 15 M stock solution (Sigma, 44939).
Petri dish: 4-well plates 125 × 125 mm.
Microtape: Anapore, 9.14 m × 1.25 cm (Euromedis, Neuilly-sous-Clermont, France).
Artificial LED light box: Indoor Led, 45 W, 169 LEDs, 276 × 276 × 14 mm, full spectrum (www.cultureindoor.com).
Glass slides: StarFrost 3 × 1 in. (Knittel Glass, Braunschweig, Germany).
Cover slips: Glass 24 × 32 mm (Biosigma, Cona, Italy).
2 mL Standard Screw Thread Glass Vials, 12 × 32 mm (Thermo Scientific, C4013-2).
Superglue 3: Loctite (Henkel, France).
Silicone adhesive: NuSil MED1-1356 (NuSil Technology LLC, Carpinteria, CA, USA).
Cantilevers: The Pyrex-Nitride Probe (PNP-TR-50, NanoWorld AG, Neuchatel, Switzerland). (cant #2) has 200-μm-long silicon nitride cantilevers and integrated oxide sharpened, pyramidal tips with a height of 3.5 μm (and a 4-μm edge setback), a resonance frequency of 17 kHz, and a 70-nm-thick gold coating.
Growth solution:[1] 20 μL of 50× MES buffer (3.5 mM, pH 5.5–5.8) + 1 mL of MS/10 liquid medium (Balzergue et al., 2017).
MS/10 liquid medium: Sucrose (5 g/1 L) + 20 mL solution 1 (see Table 5.4.1).

---

1 To be kept for 1 day max.

Table 5.4.1: Composition of solution 1.

Components	Final concentration
NH_4NO_3	2.1 mM
KNO_3	1.9 mM
$CaCl_2$	0.3 mM
$MgSO_4$	0.15 mM
H_3BO_3	10 μM
$MnSO_4$	10 μM
KI	0.5 μM
$ZnSO_4$	3 μM
$Na_2 MoO_4$	0.1 μM
$CoCl_2$	0.01 μM
$CuSO_4$	0.01 μM

5.4.3 Sterilization of the Seeds

Seeds are surface-sterilized in a clean environment under a laminar flow hood using an Eppendorf tube (1.5 mL) and with a solution (SDS 0.05%, ethanol 70%) and then washed using ethanol 96%.

- Place a small quantity of seeds (an equivalent of around 50 μL or less) in an Eppendorf tube. Remember to mention the name of the seeds on it with date and cover the label with a transparent tape.
- Add 1 mL of the washing solution (SDS 0.05%, ethanol 70%) to the Eppendorf tube. Close the cap and shake for 1–5 min.
- With a pipette, push two times in the medium so that seeds are settled down. Then, pump out the solution carefully avoiding to take out seeds and discard the liquid.
- Rinse with 1 mL of pure ethanol. Shake the Eppendorf for 1 min minimum. Take out the solution with the help of a pipette.
- To dry the seeds, keep the Eppendorf open for at least 1 h under the laminar flow hood.
- Before storing seeds, check their dryness after sterilization. Seeds can be kept at room temperature, but ideally at 4 °C with low hygrometry.

Sterile seeds can be stored for up to 3 months. When kept for a longer time, the germination rate decreases with time, and seeds do not germinate synchronously.

5.4.4 Sowing Seeds

Seeds are sown onto the agar medium within petri dishes and grown under artificial light in a Peltier-cooled incubator.

5.4.4.1 Agar Preparation

The composition of the agar medium components is shown in Table 5.4.2.

Table 5.4.2: Agar medium.

Components	Amount per liter
Agar Sigma	8 g
Sucrose	5 g
Solution 1	20 mL

- The autoclaved solution 1 is added under sterile conditions.
- Complete to 1 L with deionized water. If a conventional microwave oven is planned for melting the agar, split the preparation into 500 mL bottles.
- Seal the bottles, add a piece of an autoclave tape, and put it in an autoclave for sterilization (120 °C, 30 min).

5.4.4.2 Preparing Agar Plates

Before melting a solidified growth medium in a microwave oven, do not forget to slightly unscrew the lid.[2] Heat the growth medium for around 3 min intermittently or observe until it becomes completely melted.[3] ! Do not forget to wear thermoprotective gloves to handle the bottle. Wait until the agar medium is completely dissolved and melted.[4]

- Take a 50-mL Falcon™ tube and add 15 mL of agar solution.
- Mix in the tube with 300 µL MES buffer (50×) followed by 4 µL $FeCl_2$ (15 mM); 4 µL is the required volume for the desired concentration of iron at 4 µM.
- After a gentle mixing, the preparation is poured in the Petri dish.
- Label the Petri dish with "Fe_0" for zero iron stress, "Fe_4" in case of 4 µM stress, and so on.
- Wait for the agar medium to cool down until the gel is solidified.
- Sow a few seeds carefully into the medium and align them with a sterile wooden stick (see Figure 5.4.1).
- Seal the plate with microtape to prevent the evaporation of the water, yet allowing airflow.

---

2 Without unscrewing the lid, there is a significant risk of explosion during the heating.
3 Do not heat the medium excessively as water will evaporate and thus change the concentration of the nutrients.
4 If needed, the bottle can be stored for a while in an incubator at 60 °C. At this temperature the growth medium will not boil nor evaporate, but still remains in a melting state.

5.4.4.3 Seedling Growth

The agar plate containing the seeds is kept under an artificial LED light box (Indoor Led) in a Peltier-cooled incubator (Memmert IPP + 110) with a 16 h photoperiod at 24 °C/21 °C day/night, respectively. To allow the root to grow on the surface of the medium, incline slightly the plate as shown in Figure 5.4.2. Figure 5.4.1 shows seedlings 4 days after sowing.

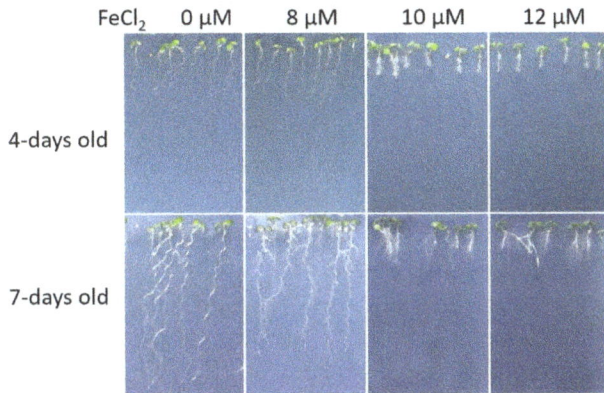

Figure 5.4.1: Four- and Seven-day-old seedlings grown on the solidified agar medium containing 8, 10, or 12 µM $FeCl_2$. After 4 days under 10 and 12 µM $FeCl_2$, a clear root growth arrest is observed; while at day 7, a reduced root growth is also observed under 8 µM $FeCl_2$.

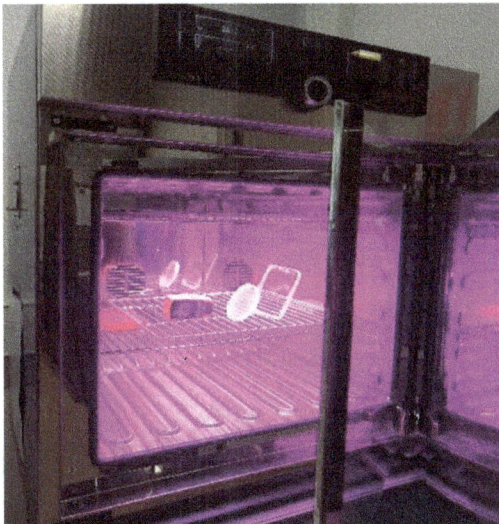

Figure 5.4.2: Slightly inclined squared agar plates in the Peltier-cooled incubator with an artificial LED light box.

5.4.4.4 Seedling Stress

We perform experiments after a 2 h stress by transferring seedling from Fe_0 agar plate to an agar plate with a given stress condition (Fe_0, Fe_5, Fe_{10}, etc.). After 2 h, plants are mounted on a glass slide.

! Because it takes about half an hour to do the experiment on each plant, careful planning of the daily experiment is required.

5.4.5 Mounting Root Samples on Glass Slide

Mounting the root samples on glass slides is the most delicate step in this experiment. The goal is to maintain plant roots immobile on a glass slide but keeping a physiological liquid buffer environment.
– Take two standard glass slides plus two rectangular cover slips.
– Add a superglue drop in the middle of both glass slides.
– Position the cover slips on the glass slides as shown in Figure 5.4.3; make sure that both the coverslips are at the same height in the middle of the slide and that two-thirds of width is touching the slide and one-third of it is outside the slide.

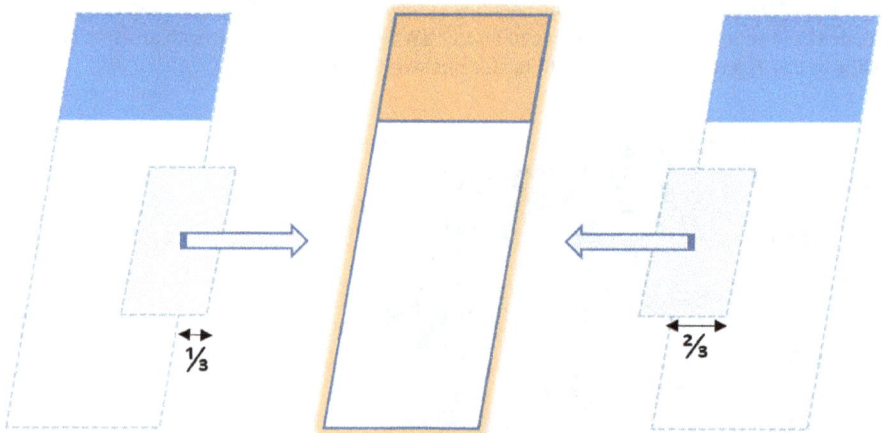

Figure 5.4.3: Principle of assembly of glass slides to mount plant seedlings for further mechanical analysis with AFM.

– Take a third glass slide and place it in the middle of the other two glass slides such that a little rectangular space is left in between (Figure 5.4.3).
– On the middle glass slide, deposit a uniform layer of silicone adhesive (NuSil MED1-1356) approx. 90 µL is spread carefully with the help of a coverslip (square #1, Figure 5.4.4).

- !! The silicon adhesive tends to solidify with time, so it should not be stored in big bottles. Aliquot NuSil in small glass tubes (preferably 2 mL) so that a tube will serve for a single day of work.
- Wait for 20–25 s, so that the adhesive starts drying, yet it is still wet for the sealing purpose. The goal is to place the root on the adhesive when it is semisolid so that it stays on the surface and does not dip into the adhesive.
- Take one seedling from the Petri dish and, using a dedicated tweezer,[5] gently lift it from the shoot area preferably under the cotyledon without pinching the seedling.
- With the help of a binocular, place the plant on the slide such that the root tip is placed first, followed by the shoot part.

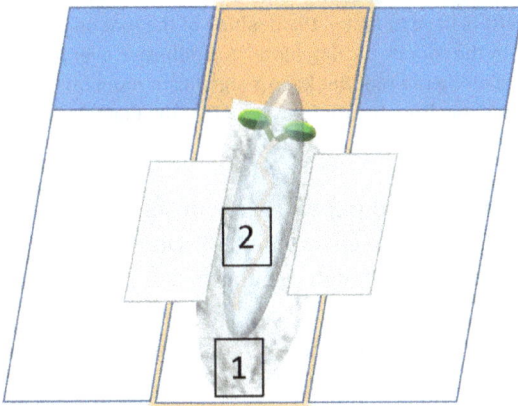

Figure 5.4.4: Principle of mounting plant seedling on glass slides. The adhesive layer in shown in light gray and labeled with the number 1. A long primary root in orange and two green cotyledons schematize the seedling. The seedling is finally covered with a growth medium shown in light transparent blue color and labeled with the number 2.

This is the challenging technical step:
- After placing the root on the glass slide, it is necessary to seal it with the semi-solid adhesive, starting from the top, then the middle and the very end (root tip) to prevent any movement of the root (Figure 5.4.5). !! Sealing by adhesive is done such that the height shall remain constant throughout the slide. With an excess of adhesive, the tip may touch the adhesive during the positioning under the AFM scanner or the edge of cantilever support may also touch the adhesive that will bias nanomechanical measurements.

---

5 NuSil is a glue and it is difficult to fully remove from the tweezer. Thus, the tweezer would not be able to be used for picking cantilevers for instance.

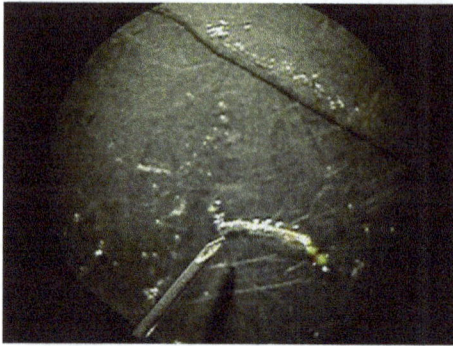

Figure 5.4.5: Optical magnification of deposited seedling roots on adhesive over a glass slide. A thin needle (bright color on the left) is used to fasten the position of the root on the adhesive located all over underneath the root by picking hardening adhesive over the root. Several fastening strips are required along the root but keep a significant room for the investigation zone, which is about 500 μm from the root tip. Note that the diameter of the seedling root is about 120 μm.

– Cover entirely the plant root (at least 200 μL) with growth solution (square #2, Figure 5.4.4) to prevent drying during measurement. !! Do not use water to cover the root as it will generate a stress of high cellular turgor pressure.

The mounted seedling is then positioned under the AFM so that the plant is still alive, and experiment on living root tissues can be performed.

5.4.6 AFM Protocol

Due to the nature of the sample, it is necessary to use an AFM scanner over the top of the sample. In our case, we use a Dimension 3100 AFM (Bruker, Santa Barbara, USA) with a Nanoscope V controller and a Nanoscope 7.3 software. The advantage of this system is the large motorized sample stage, which is perfect for plant root analysis. Before starting measurements, there are several parameters and calibration steps that need to be performed each time.

5.4.6.1 Set Up

When experimenting with AFM cantilever in fluid, the probe holder DTFML-DD (Bruker, Santa Barbara) prevents the scanner tube from coming into contact with the fluid environment. Triangular pyrex nitride cantilevers with pyramidal tips of 10 nm nominal radius and a half-opening angle of 35° were used. The reasons for using this tip are its symmetry, its low sharpness, and its proper spring constant for plant tissues.

5.4.6.2 Laser Alignment

Aligning the laser onto the cantilever is one of the most critical parts of the AFM setup. A good laser alignment helps control the signal-to-noise ratio, the sensitivity of the cantilever deflection or amplitude, and ultimately the quality of any data acquired using the AFM. A poor alignment may cause excessive applied forces or the presence of optical interferences (fringes) from the liquid surface of the sample. Check the instructions of your system for optimal laser alignment. In our case, a SUM value of 3.5–4 V is usually achieved with PNP-TR cantilevers. At the end, the laser signal is centered on the photodetector using proper adjustment knobs.

5.4.6.3 Calibration and Deflection Sensitivity

The deflection sensitivity S is used to convert photodiode measurements (in V) into distances (nm), which in turn can be converted to forces (N) upon further multiplication by the spring constant k (eq. (5.4.1)). The calibration of photodetector sensitivity is first done in air by acquiring a deflection versus distance curve against a hard homogeneous surface (like glass slide), whose effective stiffness is orders of magnitude greater than the cantilever stiffness. Refer to the instructions of your system to perform this step properly. Detailed information can also be found in Chapter 3.1.3 on calibraton issues:

$$F = k^* S^* \Delta x \qquad (5.4.1)$$

where F is the force (N), k is the cantilever-specific spring constant (N/m), S is the sensitivity of the position-sensitive photodetector (m/V), and Δx is deflection (V).

A typical force versus distance curve from which the deflection sensitivity can be obtained is shown in Figure 5.4.6. The relationship between parameters is shown in eq. (5.4.1), an adaptation of Hooke's law in which the deflection is comprised of two terms: the deflection sensitivity as measured in meter per volt and the deflection as measured in volt (Schillers et al., 2017).

The cantilever is ramped over a very stiff surface. Accordingly, all downward piezoelectric motion will be equal to cantilever bending (right part of the graph). The inverse of the measured slope (red line) corresponds to the sought deflection sensitivity (in nm/V) (Gavara, 2017).

Measuring the stiffness of the cantilever is necessary to determine the applied force during the experiment; and this is done using the thermal tuning methods where spring constant is calculated. Finally, the deflection sensitivity is determined

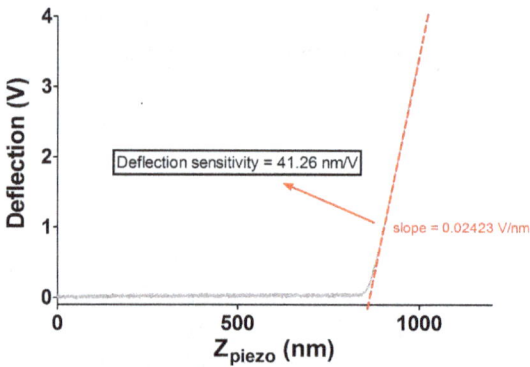

Figure 5.4.6: Calibration of the deflection sensitivity.

again in liquid medium because indentations are performed on samples while they are immersed in the liquid medium and its value is often lower than that in the air. For the case of PNP cantilevers and Nanoscope 7.3 software, it is not possible to determine the spring constant of the cantilevers in liquid medium (low resonance frequency of PNP cantilevers). The value determined in the air is used instead. See Chapter 3.1.3 on calibraton issues for complete description of calibration procedures.

5.4.6.4 Recording the Force–Distance Curves

Before positioning the plant root under the AFM cantilever, make sure that the cantilever is retracted enough to allow a safe distance between the cantilever and the plant root. Lift the scanner head carefully and insert the glass slide with the plant root. Buffer is already on the top of the root, and so carefully lower the scanner head so that the cantilever will not bend upon contact with water (adding a drop of buffer on the cantilever is usually necessary to have a smooth insertion in liquid). Orient the glass slide so that the root orientation is close to vertical, and its tip is on the top of the camera (Figure 5.4.7). The target working area (elongation zone of the root) is located approximatively 500 µm away from the root tip. This distance can be estimated knowing the length of PNP cantilever (about 200 µm). Use the stage positioning controls to move the cantilever on the right zone. Because the root is cylindrical and the tip height rather small (3.5 µm), it is not advised to probe the root beyond the longitudinal median line of the root. Note that there is small setback of the tip on PNP cantilever (4 µm).

During the experiments, the deflection setpoint is kept at 0 V while the initial vertical deflection on the photodiode is set to −2.5 V. For all the analysis, the ramp size for the force–distance curves is kept at 3 µm (maximum z-piezo range of 5 µm). Trigger is set off and no trigger values are used. The later point implies that it is often necessary to adjust the *z*-start value for the ramp each time a new engagement has been done.

Because the AFM tip has a setback of 4 μm, it is not possible to ascertain the exact target area for indentation (edge or middle of cells). A trick is to perform force–distance measurements in 16 different nodes on the elongation zone (Figure 5.4.7). Each time, curves are recorded with the auto ramp feature in 2 × 2 matrix, having four force–distance curves at each node separated by 50 nm in distance between each point. Between each of these 16 nodes, the piezo move in *X*- and *Y*-directions using the offset facility of Nanoscope 7.3 according to the pattern is shown in Figure 5.4.8. Because the plant root is living, even when fixated by nusil on the glass slide, the root continue to grow. Thus, it is important to avoid going up and down along the root during the acquisition. We choose to go down in a single direction as shown in Figure 5.4.8. On our system, force–distance curves are recorded after withdrawing and re-engaging at each matrix node. In rare cases, it is necessary to withdraw and move the cantilever away (5 μm) to probe another area. In practice, it usually takes 25 min to record a full set of force–distance curve for a single plant root. It is important to keep the time for acquisition as short as possible to avoid additional stress effect.

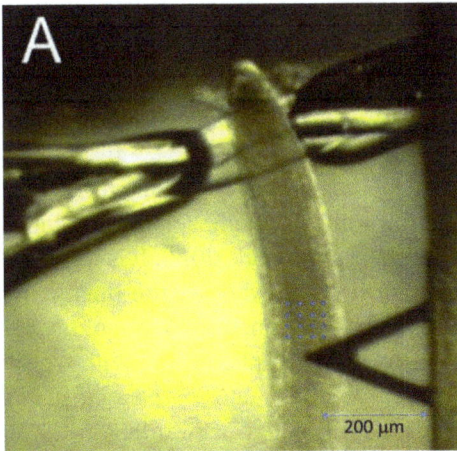

Figure 5.4.7: Camera-taken snapshot of a plant root under an AFM cantilever (triangle on the right). The fastening of the root tip can be seen on the top of the image. The elongation zone where indentation must be performed is located approximately 500 μm from the end of the root tip. This distance can be estimated using the length of the cantilever (200 μm here for PNP-TR). The blue dots represent the redundant measurement performed on the plant root (data points not to scale).

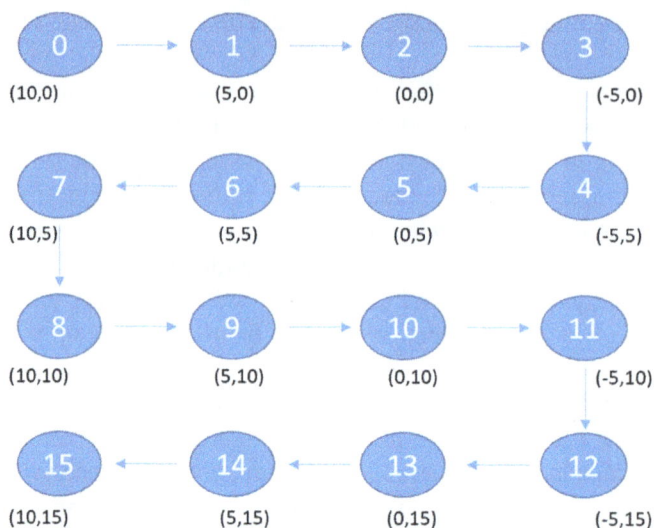

Figure 5.4.8: Schematic diagram used to perform the redundant indentation measurements. An experiment starts at position 0 and goes through the last point 15. The *X,Y* offset (in μm) needed to move from one point to the next one is indicated in parenthesis. Coordinate system is according to the piezo of Dimension 3100.

5.4.7 Plant Phenotype

Detailed results from experiments on plant roots will be presented in Chapter 6.11. In brief, it has been found that the presence of iron (under low phosphate condition) leads to a phenotype of plant root growth arrest (Balzergue et al., 2017). It is therefore an important step in this experimental protocol to check whether the phenotype "root growth arrest" is observed. Experiments were performed on 4-day-old seedlings, and it is necessary to check next day for the expected phenotype. With a pen, label the position of the root tip of the remaining seedlings on the +Fe agar plates and check the next day for the presence or absence of root growth. In Figure 5.4.9, roots transferred from Fe_0 to Fe_0 (left) show a normal growth, that is, the end of the root is beyond the maker line, whereas when roots were transferred from Fe_0 to Fe_{15} (right), no root growth is observed after 1-day post-AFM experiments.

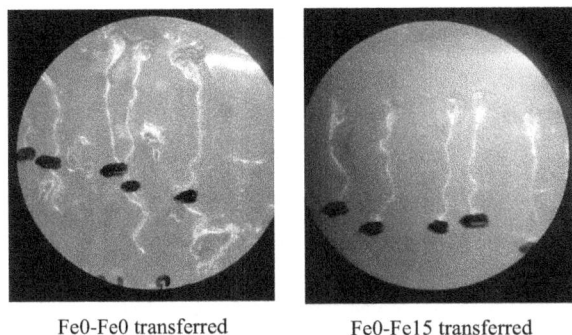

Fe0-Fe0 transferred   Fe0-Fe15 transferred

Figure 5.4.9: Agar plates after mechanical measurements. Several seedlings remained on the plate to confirm the presence of the expected phenotype (here, the root growth arrest under high concentration of Fe and low phosphate). To check the phenotype, a mark is made with a pen on the day of the mechanical experiment, and the growth of plants is checked 1 or 2 days later. Picture taken 24 h after transfer.

References

Arnould, O., D. Siniscalco, A. Bourmaud, A. Le Duigou and C. Baley (2017). "Better insight into the nano-mechanical properties of flax fibre cell walls." Industrial Crops and Products **97**: 224–228.

Balzergue, C., T. Dartevelle, C. Godon, E. Laugier, C. Meisrimler, J.-M. Teulon, A. Creff, M. Bissler, C. Brouchoud, A. Hagège, J. Müller, S. Chiarenza, H. Javot, N. Becuwe-Linka, P. David, B. Péret, E. Delannoy, M.-C. Thibaud, J. Armengaud, S. Abel, J.-L. Pellequer, L. Nussaume and T. Desnos (2017). "Low phosphate activates STOP1-ALMT1 to rapidly inhibit root cell elongation." Nature Communications **8**: 15300.

Gavara, N. (2017). "A beginner's guide to atomic force microscopy probing for cell mechanics." Microscopy Research and Technique **80**: 75–84.

Kozlova, L., A. Petrova, B. Ananchenko and T. Gorshkova (2019). "Assessment of primary cell wall nanomechanical properties in internal cells of non-fixed maize roots." Plants **8**: 172.

Milani, P., M. Gholamirad, J. Traas, A. Arneodo, A. Boudaoud, F. Argoul and O. Hamant (2011). "In vivo analysis of local wall stiffness at the shoot apical meristem in Arabidopsis using atomic force microscopy." The Plant Journal **67**: 1116–1123.

Milani, P., V. Mirabet, C. Cellier, F. Rozier, O. Hamant, P. Das and A. Boudaoud (2014). "Matching patterns of gene expression to mechanical stiffness at cell resolution through quantitative tandem epifluorescence and nanoindentation." Plant Physiology **165**: 1399–1408.

Peaucelle, A., S. Braybrook and H. Hofte (2012). "Cell wall mechanics and growth control in plants: The role of pectins revisited." Frontiers in Plant Science **3**: 121.

Peaucelle, A., S. A. Braybrook, L. Le Guillou, E. Bron, C. Kuhlemeier and H. Hofte (2011). "Pectin-induced changes in cell wall mechanics underlie organ initiation in Arabidopsis." Current Biology **21**: 1720–1726.

Schillers, H., C. Rianna, J. Schäpe, T. Luque, H. Doschke, M. Wälte, J. J. Uriarte, N. Campillo, G. P. Michanetzis, J. Bobrowska, et al. (2017). "Standardized Nanomechanical Atomic force microscopy Procedure (SNAP) for measuring soft and biological samples." Scientific Reports **7**: 5117.

Torode, T. A., R. O'Neill, S. E. Marcus, V. Cornuault, S. Pose, R. P. Lauder, S. K. Kracun, M. G. Rydahl, M. C. F. Andersen, W. G. T. Willats, S. A. Braybrook, B. J. Townsend, M. H. Clausen and J. P. Knox (2018). "Branched pectic galactan in phloem-sieve-element cell walls: Implications for cell mechanics." Plant Physiology **176**: 1547–1558.

Yakubov, G. E., M. R. Bonilla, H. Chen, M. S. Doblin, A. Bacic, M. J. Gidley and J. R. Stokes (2016). "Mapping nano-scale mechanical heterogeneity of primary plant cell walls." Journal of Experimental Botany **67**: 2799–2816.

Zdunek, A. and A. Kurenda (2013). "Determination of the elastic properties of tomato fruit cells with an atomic force microscope." Sensors **13**: 12175–12191.

Zhao, L., D. Schaefer, H. Xu, S. J. Modi, W. R. LaCourse and M. R. Marten (2005). "Elastic properties of the cell wall of Aspergillus nidulans studied with atomic force microscopy." Biotechnology Progress **21**: 292–299.

Mechanics in Diseased Cells and Tissues

Martin Pesl, Jan Pribyl, Petr Dvorak, Zdenek Starek, Petr Skladal,
Kristian Brat and Vladimir Rotrekl
6.1 Cardiovascular Diseases

6.1.1 Introduction

Cardiovascular diseases (CVDs) are the principal cause of death globally contributing
to more than half of the mortality in Europe (42% on males and 51% on females) (Gillespie et al., 2013, Mozaffarian et al., 2015). There is growing interest in studying etiology, hallmarks, progress, and improved therapies for CVDs. Genetic background and
cellular phenotypes are frequent morbidity and mortality causes (Aistrup et al., 2009,
Stienen, 2015, van der Velden and Stienen, 2019), but those are not readily accessible
for diagnostics. Stem-cell-derived cardiomyocytes (CMs) thus stands for readily available and ethically uncompromised model for basic research.

CMs represent the contractile active unit of the heart. Nevertheless, other cardiac
cell populations are attracting research interest, for instance endothelial cells (ECs),
lining of the heart and vessels, cardiac fibroblasts, accounting mainly for extracellular matrix homeostasis, pericardial, adventitial, and smooth muscle cells. The extracellular matrix (ECM) plays an important role in the cardiovascular system, as
mechanical cues are detected and interpreted in a constant cell-matrix interplay
(Nardone, Oliver-De La Cruz et al. 2017), and ECM is discussed in Chapter 4.5.

Phenotype analyses of CMs function allow understanding of complex single-cell
and syncytial properties of excitation–contraction coupling. They represent essential
counterparts of morphologic and genotypic assays. Conventional family of assays,
such as cellular electrophysiology, calcium imaging or mechanical testing, have been
improved and integrated. Quantitative in vitro drug testing has also been expanded
and most importantly standardized (Takeda et al., 2017, Klimovic et al., 2022).

These advancements are in great part due to the introduction of primary animal
and human sources of CMs, but also of human pluripotent stem-cell-derived CMs
(PSC-CMs). Human CMs are not readily available, as myocardial biopsy is an invasive procedure with significant risk (Cooper et al., 2007, Holzmann et al., 2008) and

Martin Pesl, Department of Biology, Faculty of Medicine, Masaryk University, Brno, Czech
Republic; ICRC, St. Anne's University Hospital, Brno, Czech Republic; First Department of Internal
Medicine, Cardio-Angiology, Faculty of Medicine, Masaryk University, Brno, Czech Republic
Jan Pribyl, CEITEC, Masaryk University, Brno, Czech Republic
Petr Dvorak, Vladimir Rotrekl, Department of Biology, Faculty of Medicine, Masaryk University,
Brno, Czech Republic; ICRC, St. Anne's University Hospital, Brno, Czech Republic
Zdenek Starek, ICRC, St. Anne's University Hospital, Brno, Czech Republic
Petr Skladal, Department of Biochemistry, Faculty of Science, Masaryk University, Brno, Czech
Republic
Kristian Brat, Department of Respiratory Diseases, University Hospital, Brno, Czech Republic

https://doi.org/10.1515/9783110989380-011

transplants are rare; thus PSC 2D and 3D models comprised of PSC-CMs have revolutionized the basic CVD research by providing a stable on demand source of CMs which recapitulates patient-specific or imposed genetic background (Acimovic et al., 2014, Pesl et al., 2016b, Borin et al., 2018).

PSC-CMs represent a readily available and renewable cell source for diagnostic, pharma-screening, but also provide a way for basic CMs behavior mechanisms understanding (Liu, Liu et al., 2012, Liang et al., 2013, Caluori et al., 2019a). Through mesodermal differentiation, they express an immature fetal-like phenotype, with protocol-dependent mixed composition of nodal-/atrial-/ventricular-like properties (Hescheler et al., 1997), recapitulated in PSCs (Schweizer et al., 2017, Vestergaard et al., 2017). During in-vivo terminal CMs maturation, parallel arrays of myofibrils are organized into mature sarcomeres, having clearly defined A and I bands, Z disks, but also gap junctions and desmosomes (Westfall et al., 1997). Structural maturity still differs in PSC derived and primary CMs (Koivumäki et al., 2018).

In primary CMs and to certain degree also in PSC-derived CMs, the contractile apparatus unit is a sarcomere; it consists of myosin-containing thick filaments surrounded by a hexagonal array of thin actin filaments with troponin/α-tropomyosin (Tn/Tm) regulatory units (Parmacek and Solaro, 2004). These are responsible for shortening of sarcomere, generating contractile force and slightly differ from other muscle cell types (Léger et al., 1975). However, human CMs are available only after heart transplantation and a few cardiac surgery procedures (Pesl et al., 2020), otherwise animal cells are used with obvious limitations. At the same time, the differentiation of PSC derived CMs is still heavily dependent on surface and ECM elasticity (Jacot et al., 2010, Chung et al., 2013).

Contractility can be defined as the ability of heart muscle to shorten sarcomeres and thus generate force, and it represents the main mechanical function of the heart. Nevertheless, a number of integrins, cadherins, and other membrane proteins mediate ECM cues, which together with stretch activated specific Ca^{2+} channels, connect electrical and mechanical signals (Sachs, 2010, Reed et al., 2014). This complex interaction is resulting in impulse generation and/or contraction (Bers, 2014), or stretch stimulating natriuretic peptide secretion in case of volume overload of the heart (Thibault et al., 1999).

6.1.2 Cardiomyocyte phenotype analysis methods and properties

Patch clamp is almost exclusively based on single cell electrical activity and its ion channels (Sakmann and Neher, 1976). Despite being a gold standard for cellular electrophysiology, it poses significant drawbacks in its invasiveness (opening the cell membrane) and need for transient giga seal formation (Bébarová, 2012) due to which relatively short-term measurement is inherently fatal for the probed CM or neuron.

Table 6.1.1: Methods for analyzing electrophysiology and beating behavior (adapted according to Laurila et al., (2015) related to target substrate and measured parameter).

Method	Biological objects	Parameter
Patch clamp	Single cells	Action potential
Microelectrode arrays	Monolayer/syncytia	Field potential
Fluorescent imaging	Single cell and syncytia	Fluorescent dye intensity (e.g., voltage sensitive dye, displacement)
Video microscopy	Single cells, clusters, monolayers	Displacement – border/texture recognition/optical flow recognition
Traction force microscopy	Single cells and small syncytia	Beads, microposts displacement
Atomic force microscopy	Single cells, clusters, monolayers	Displacement of cell surface, indentation, retraction curve analysis
Impedance assays	Syncytia, monolayers	Electric impedance
Tweezing	Single cell, syncytia	Optical, magnetic acoustic stretching

It requires an experienced operator; the method is low throughput and it provides little or no information on cellular mechanics, as cell contractions are disturbing proper contact (gigaseal). Only highly sophisticated automated patch clamp (Mann et al., 2019) can partially overcome these drawbacks, or similarly scanning patch clamp/or scanning ion conductance microscopy (SICM) (Actis et al., 2014) involving piezo-controlled movements can provide topographical, electrical, and mechanical data together.

Microelectrode arrays (MEAs) are sets of planar electrodes regularly placed on an insulating substrate. Ionic displacement is detected in the cell-electrode cleft during transmembrane voltage changes, described as field potential (Spira and Hai, 2013). Roughly describing action potential (AP) duration and onset, MEAs provide noninvasive and possibly long-term information on arrhythmogenicity or cardiotoxicity of drugs (Braam et al., 2010). They also provide information on electrical propagation with potential to visualize arrhythmic circuits (Natarajan et al., 2011, Gilchrist et al., 2015). The middle- or high-throughput character of the method also prompted stretchable versions of MEA, allowing mechanical analysis either in combination with video (Hayakawa et al., 2014) or impedance-based systems (Hansen et al., 2017).

Fluorescent imaging using calcium or voltage-sensitive dyes allows optical imaging of ion fluxes as well as resulting voltage and contraction changes (Laughner et al., 2012). Most of available dyes are influencing cell metabolism and survival (Herron et al., 2012). In order to allow for contraction quantification, the system must be used on known flexible substrates and further AP quantification is dependent on combined system similarly to MEA (Ahola et al., 2014).

Light microscopic methods involving sequence of still images or video recordings allow analyzing of the mechanical changes of a beating cell or area, which is then analyzed by different computational methods (Kamgoué et al., 2009, Ahola et al., 2014, Chen et al., 2014, Hayakawa et al., 2014, Czirok et al., 2017).

Impedance assays are based on cellular impedance changes, thus similarly to MEA are noninvasive and label-free methods. Cells are plated on an adherent dish, having electrodes on the bottom. Cells' movements and contractions altering the cell membrane properties are monitored by changes in impedance upon application of low-frequency electric current. The system allows for long-term and short-term cell monitoring (Ciambrone et al., 2004, Abassi et al., 2012, Pointon et al., 2015).

Traction force microscopy is an extension of light microscopy. It may be one of the particle tracking techniques enabled by fluorescent beads embedded in cultivation substrate with known mechanical properties (Wang and Lin, 2007). Displacement of the beads related to cell contraction can be tracked with optical microscope and quantified allowing for indirect quantification of the cells' movement. Similar are works based on known mechanical properties of magnetic beads, but also microposts, micropillars, or microthreads (Lin et al., 1995, Kajzar et al., 2008, Kim et al., 2011, Rodriguez et al., 2014).

Atomic force microscopy (AFM), as nonoptical method, relies on a cantilever being in direct contact with measured cardiac cells. During the cell or syncytia contraction cycle, either cantilever follows vertical movement of the cell or remains stable for deflection recoding (locked). Quickly after initial AFM cell-oriented trials (Domke et al., 1999), it was used for topography and submembrane structure imaging studies (Davis et al., 2001). AFM presents wide area of otherwise hard-to-quantify parameters as cellular viscosity and elasticity, contraction, tension, adhesion, friction, and even energy dissipation, moreover contraction and beating profile description is of uttermost importance (Pesl et al., 2016a, Caluori et al., 2018) (see Figure 6.1.1.).

Tweezing force generating and measuring strategies

Tweezers represent a wide variety of physical strategies either with static configurations (i.e., single-bead, two-bead, three-bead) and dynamic configurations (i.e., force clamp, position clamp, dynamic force spectroscopy). Optical, magnetic, as well as acoustic modalities were reviewed by Basoli et al. (2018).

Optical tweezers consist of highly focused laser beam that provides an attractive or repulsive force (Henon et al., 1999, Tan et al., 2012). The laser beam is focused through a microscope objective. The narrowest point of the focused beam (waist), contains a very strong electric field gradient exerting force of 0.1–100 piconewtons (Ashkin et al., 1987) and allows to physically hold and move micrometer-scale objects, similar to tweezers (Mohammed et al., 2019). This allows measurement of the force generated in their native environment on down to single molecules of actin filament (Suzuki et al., 2003) and titin (Leake et al., 2004). Also rheological properties of endothelium (Ayala et al., 2016) and utilization in whole CM activation was proposed

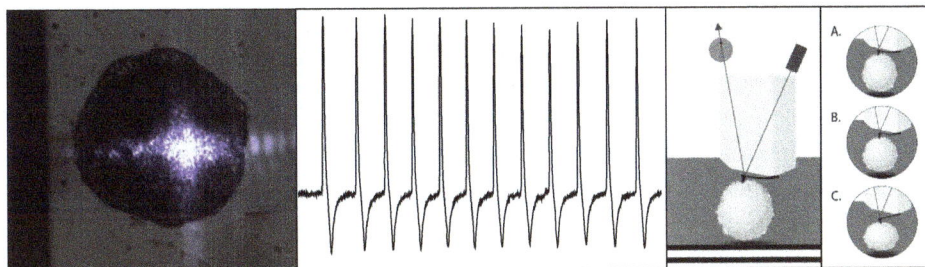

Figure 6.1.1: Microscope top view of the cantilever placed above the EB cluster (10x), recorded mechanocardiogram, the schematic of setup: EB cell cluster, AFM cantilever fixed to glass block, laser detection system, and schematic of contraction cycle A) cluster in "extension" uplifting the cantilever, B) isoline position C) contraction position, cantilever pulled towards the cluster, without loosing contact.

(Gentemann et al., 2017). CMs and similar biological samples are almost transparent to the near-infrared wavelengths normally employed to trap particles by laser traps, thus photodamage could be avoided (Neuman et al., 1999).

Magnetic tweezers (MTs) allow for the simultaneous manipulation and recording of forces in real time using tethered magnetic beads, under a generated magnetic field gradient, having low interference with the specimen (De Vlaminck and Dekker, 2012). Permanent magnets as well as electromagnets with alternating size and shape, magnetization orientation, and varying distance between the magnets and the magnetic beads are used. Thus, allowing the application of stretching/pulling forces perpendicular, but also parallel to the biological sample substrate (Kilinc et al., 2012) and also "magnetic torque tweezers" (Lipfert et al., 2010). Saphirstein et al. (2013) used MTs to investigate the mechanobiology of aortic tissue, describing focal adhesions of the vascular smooth muscle cells (VSMC) as a regulator of aortic stiffness.

Acoustic tweezers (ATs) can stably trap a particle or cluster of cells in a potential well generated by two collimated focused ultrasonic beams propagating along opposite directions in water (Wu, 1991). A stable potential well can be created by radiation pressure at the physical focal point of a focused ultrasonic beam. There is no need for the manipulated cell to undergo surface modifications or labeling so that cells maintain their shape, size, refractive index, charge, and other native properties, resulting in no need of optically purified sample, and a possibility to manipulate large particles or cells and lower damage to biological samples was demonstrated (Hwang et al., 2014). ATs cytometry produces a rapid acoustic radiation force on the microbubbles that provoke the contractility of the intracellular cytoskeleton (Heureaux et al., 2014).

Combination of methods

Each of the abovementioned methods is missing some part of available phenotypic information. For example, 2D optical microscopy cannot grasp three-dimensional contraction displacement or electrophysiological properties of the cells recorded by voltage sensitive dye or MEA application. Yet these complex parameters are essential especially for drug testing studies, but also for physiology and disease description forcing researchers into combining the techniques to sophisticated methods, such as AFM combined with both the MEA and fluorescent microscopy allowing for complex description of the mechanoelectrical transduction and contraction properties (Caluori et al., 2019a, 2019b, Kabanov et al., 2022).

6.1.2.1 Elastic Properties

Viscoelastic properties of ECs, cardiac muscle, and skeletal muscle were quantified early after first topography studies (Mathur et al., 2001), showing that cardiac cells were the stiffest exhibiting elastic modulus reaching up to 100 kPa, the skeletal muscle cells were intermediate (25 kPa) and ECs were the softest with a range of elastic moduli from 1.5 to 6.8 kPa. AFM acting as indenter performs force measurements on cell surface, detecting ample deformation in response to applied force (Burnham and Colton, 1989). These mechanical readouts are particularly useful when the sample experiences changes in mechanical properties with time as during mitosis (Stewart et al., 2011) or CM contraction relaxation cycle (Azeloglu and Costa, 2010).

The myocardium undergoes oscillatory (pulsatile) shear stress, and reflects consequences of turbulent blood flow (Guazzi and Arena, 2009, Ayala et al., 2016). CMs therefore experience stresses and strains, related to own contraction, contraction of surrounding tissue, and the pressure of pumped blood (Jacot et al., 2010). The long-term response of CMs can be proliferation, and physiological hypertrophy, and pathological alteration of the heart geometry as trabeculation and dilatation (Lee et al., 2018). The elastic modulus of ECM surrounding CMs has several ways of signaling. Integrin receptors general function in any cell type is to bind to ECM, focal adhesion is then increased, thus affecting complex cell behavior (Santoro et al., 2019). Integrin-mediated mechano-transduction is thus allowed for signaling mediated by contractile strain. Increasing matrix stiffness therefore causes increasing in CM force. However, substrate stiffness higher than 25 kPa was observed to lower contractile force of neonatal CMs (Jacot et al., 2010).

At long term, the heart tissue properties, such as elasticity, change dramatically during the aging (Hersch et al., 2013) or as an effect of diseases. Stiffness gradually increased during aging of mice (Nance et al., 2015) and rat CMs (30-month-old male animals displayed 42.5 kPa, higher than 4-month-old rats about 35 kPa) (Lieber et al., 2004). These changes are even more pronounced in studied cardiomyopathies (6.1.3) There is an apparent paradox in aging, wherein reparative fibrosis is impaired but interstitial, adverse fibrosis is augmented (Trial and Cieslik, 2018). Stiffness of live CMs isolated from control and diabetic mice also significantly differed, possibly contributing to progressive diastolic left ventricular stiffness (Benech et al., 2014, 2015) corresponding to echocardiographic a cardiac magnetic resonance findings in DCM, muscular dystrophy and diabetic cardiomyopathy (Meluzin et al., 2011, 2017, Panovský et al., 2019, Masarova et al., 2020).

The Hertz model is often applied for quantification of Young's modulus despite nonlinear elastic behavior of CMs (Mathur et al., 2001, Lieber et al., 2004). Depending on the application, the Sneddon model may be more suitable (Sneddon, 1965). Other alternatives are combined models as Johnson–Kendall–Roberts (Johnson et al., 1971) or specific approaches for biological specimens as hyperelastic models (Soufivand et al., 2014). However, despite some drawbacks, most of current studies stick to original Hertzian models.

6.1.2.2 Contraction

Contractility can be defined as the ability of heart muscle to shorten and generate force. It is initiated by electrical stimulus and mediated by calcium levels and flows. Initially, optical analysis studies were based on edge detection and displacement measurement (Claes and Brutsaert, 1971, Steadman et al., 1988, Kawana et al., 1993). Similarly, laser beam dispersion (Krueger et al., 1980), flexible sheets (Balaban et al., 2001) or magnetic beads with optical reading systems (Yin et al., 2005) were used. Nevertheless, these methods inherently suffer from absence of 3D spatial information and are applicable exclusively to flat- and square-shaped adult cells due to reproducibility issues caused by rotation of cells and edge detection errors (Delbridge and Roos, 1997). These methods are thus usually not applicable to PSC derived CMs featuring irregular shape.

An important milestone was application of conditions allowing for measurement of force without actual shortening of cells and fibers, so-called isometric contraction (Sugiura et al., 2003). It was enabled by use of carbon nanofibers of known compliance, but also different elastic structures such as micropillars (Lin et al., 1995) avoiding flat-surface-related formation of stress fibers and focal adhesions (Balaban et al., 2001, Cesa et al., 2007). It was published that CMs show a large scatter of intrinsic amplitudes, APs, and contraction frequencies, most likely because they originate from different areas of the heart (Kajzar et al., 2008, Kim et al., 2011).

Spontaneous contractions were assessed by AFM already in 1999, when beat period and pulse amplitude were quantified. Originally, embryonic chicken CMs were used, where a high degree of irregularity was reported (Domke et al., 1999). Being the most representative of available CMs, PSC-CMs were studied by AFM in single cells (Liu et al., 2012, Sun et al., 2012) as well as in syncytia (Pesl et al., 2014).

6.1.2.3 Adhesivity

Adhesion between individual CMs is mediated by intercalated discs, while their interaction with surrounding environment (ECM) is mediated by integrin receptors. Integrin–ligand interaction is a multistep process resulting in formation of adhesion complexes (Comisar et al., 2011). Integrin receptors are organized in clusters (3–4 integrins/cluster) together forming "nascent adhesions" (Wiseman et al., 2004). Cytoskeleton proteins (e.g., actin and vinculin) and signaling molecules (e.g., focal adhesion kinase and paxillin) are involved in the maturation process of adhesion complexes (Bachir et al., 2017). This is important during development for CMs' differentiation and maturation. Later on, adhesion complexes provide the signals that sustain tissue formation and turnover of ECM proteins (Peters et al., 1994, Humphries et al., 2004). Mimicking those process seems important for myocardial regeneration, which otherwise seems to be very limited in the cardiac tissue, as reviewed in (Leong et al., 2017). At the same time, the relative changes in adhesion, altering connection of CMs to myofibroblasts or ECM adhesion sites during scar formation increases cardiac stiffness, thus constituting the basis for the development of heart failure (Opie et al., 2006).

6.1.3 Cell and Tissue-Based Disease Phenotyping

Acute and chronic changes of heart and vessels have enormous impact on morbidity and mortality all over the world (Abubakar et al., 2015). Globally, CVDs and related mechanical remarks are described in following subchapters.

6.1.3.1 Rheumatic Heart Disease and Valve Involvement

Heart valves are main barriers in the blood flow, facing mechanical stress during heartbeat cycles. Autoimmune inflammatory reactions leading to rheumatic fever (RF) and rheumatic heart disease (RHD) result from untreated infections, and causes increased interstitial cellularity, with consequent increased expression of vimentin and vitronectin, but also changes in valve collagen composition, causing structural and functional valve changes (Tandon et al., 2013, de Oliveira Martins et al., 2017). Similarly, wall shear stress exerted on the valvular ECs influences the inflammatory processes and the signaling between the ECs and the valvular interstitial cells. Equally, these processes are leading to a calcific valve phenotype (Fisher et al., 2013), reported as area of increase strain(Halevi et al., 2015). The consequence is often a reduced valve leaflet movement and valve stenosis, having profound impact on pressure overload of related heart chamber. Even without specific trigger, the valves are stiffening during aging (Sewell-Loftin et al., 2012), which can be described by AFM (Sewell-Loftin et al., 2012).

6.1.3.2 Ischemic Heart Disease

Coronary arteries obstruction results in ischemia of the myocardium. Myocardial infarction (MI) leads to myocardial loss due to necrosis as well as subsequent remodeling with progressive fibrosis (Sutton et al., 1997, Gheorghiade and Bonow, 1998). While elasticities of healthy myocardial tissues range from 10 to 30 kPa, stiffnesses in MI-affected areas increase to values of up to 150 kPa (Berry et al., 2006, Jacot et al., 2010, Van Den Borne et al., 2010). This is by large attributed to loss of CMs and contracting myofibroblasts tissue replacement and interstitial fibrosis following cardiac injury. This uneven process is related to area of ischemia, threatening patient not only by loss of contractile function, but also altered AP propagation resulting in life threatening arrythmias (Liang et al., 2019). Individual CMs in mouse model, early after MI, decreased their Young's modulus, whereas at later stages, cells became stiffer than controls (Dague et al., 2014). This may be attributed to impaired Ca^{2+} management (Kronenbitter et al., 2018) resulting in sarcomere dysfunction, namely changes in ECC (Holt et al., 1998) but also in diastolic relaxation and Ca^{2+} reuptake (Zalvidea et al., 2012). Maintaining the ECM after MI in the scar prevents rupture of the infarct area in the short term, but is leading to increased cardiac stiffness and constitutes the basis of the development of heart failure, possibly related to changes in structural proteins of titin and collagen (Van Den Borne et al., 2010, Zile et al., 2015).

6.1.3.3 Vascular Disease Aortic Aneurysm Cerebral and Atherosclerotic Vessel Disease

Large arteries are composed from collagens I and III, accounting for 60% of the artery wall, and elastin 30% (Rizzo et al., 1989). Due to low elastin turnover during the lifespan, collagen concentration increases with age (Zieman et al., 2005) and as collagen fibers are 100–1,000 times stiffer than elastin, a sharp increase in the incremental elastic modulus at higher levels of circumferential stretch is produced. The changes in the vessel wall ECM composition are leading to gradual arterial stiffening and subsequently to aneurysm formation, expansion and rupture due to repetitive pressure and shear stress (Lasheras, 2007). Similar procesess lead to obstruction or rupture of smaller cerebral vessels (CV) and subsequent brain ischemia. Those states were closely linked to hypertension (Nation et al., 2012), related to arteriolar wall thickening, but also fibrin deposition and functional disruption of vessel wall and even blood brain barrier (Tagami et al., 1987), resulting mainly in atherosclerotic changes, preventable by blood pressure lowering strategies (Hajdu et al., 1991).

Atherosclerotic plaques formation in the large and medium sized arteries is classically driven by systemic factors, such as elevated cholesterol and blood pressure. Starting as vessel wall thickening, plaques form preferentially at vessel branch points, curvatures, and bifurcations exposed to disturbed flow patterns characterized by flow separation, changes in flow direction, and recirculation eddies as reviewed in (Hahn and Schwartz, 2009). At low arterial pressure, elastin fibers absorb the majority of circumferential stretch forces, rigid collagen fibers resist mechanical deformation to high circumferential stretch resulting in elevated strain applied to the smooth muscle layer and promoting proliferation (Yurdagul Jr, Finney et al., 2016). Further cholesterol deposition in the arterial wall promotes a chronic inflammatory response resulting in leukocyte recruitment into the vessel wall (Ramji and Davies, 2015). Definitely altering its mechanical properties and ending in plaque rupture, it becomes the primary cause of thrombotic complications and atherosclerosis-related mortality (Virmani et al., 2006). Large and small vessel changes are closely related to systolic blood pressure.

6.1.3.4 Hypertensive Heart Disease

Change in brachial systolic pressure with age shows a steep rise from age 10 to a plateau when full body height is reached at age 18, then a subsequent rise after age 45 years, as reviewed in (O'Rourke et al., 2002). Arterial stiffening precedes systemic hypertension in condition such as obesity (Weisbrod et al., 2013) or chronic kidney disease (Garnier and Briet, 2016). Age-related stiffening is often linked to the above-described vascular wall ECM composition changes and endothelial dysfunction. Nevertheless, a major role is played by VSMC stiffening and changes of adhesive properties (Sehgel et al., 2013). It is proposed that Ca^{2+} regulates VSMC elasticity and adhesion to the ECM through integrin–actin cytoskeleton axis (Zhu et al., 2018). Rho kinase is the expected regulator (Rho-associated coiled-coil-containing protein kinase, ROCK), through actin/SRF/myocardin pathway effectively lowering alpha smooth muscle actinin expression, reversing aortic stiffening (Zhou et al., 2017).

6.1.3.5 Cardiomyopathy and Myocarditis

Generally, increase in myocardial wall stress stimulates compensatory cardiac myocyte elongation or hypertrophy maintains cardiac output in response to increased mechanical load (Frank–Starling mechanism). Although the heart may functionally tolerate a variety of pathological insults, adaptive responses that aim to maintain function eventually fail. In number of scenarios such as aging or inherited conditions, this response closes and feeds a vicious cycle of CMs exhaustion and apoptosis. Apoptosis may also be triggered by external factors as alcohol or cytostatic therapy (Ikeda et al., 2019). Common for cardiomyopathy subgroups is absence of ischemic or arrhythmic cause.

Cardiac dysfunction in hypertrophic cardiomyopathy (HCM) is attributable to initial increases in heart mass and asymmetric interventricular septal thickening (Maron et al., 1995, Veselka et al., 2017). Most of the cases having genetic background, larger CMs and myofibrillar disarray, due to sarcomeric protein mutations, progressing into perivascular and interstitial fibrosis. Cellular CMs hypertrophy is beyond physiological hypertrophy in response to exercise (McMullen et al., 2003). Upon injury, myofibroblasts appear in the myocardium and are generally believed to arise from resident interstitial and/or adventitial fibroblasts (Powell et al., 1999). At the same time type I collagen increases, together stiffening the ventricles and impeding contraction, torsion, and relaxation (Manabe et al., 2002, Meluzin et al., 2009). A number of genes trigger eventually similar changes, although in some cases the same mechanism is leading to chamber dilation and wall thinning, further described as dilated cardiomyopathy (DCM). Prolonged cardiac insults and a number of cytoskeletal mutations underlie DCM. Generally, disruption of mechano-transduction components prevents the heart from productively responding to both extrinsic and intrinsic stressors. Glaubitz et al. proposed that decrease in single-cell stiffness of left ventricular fibroblasts could trigger left ventricular dilation (Glaubitz et al., 2014). Nevertheless, CMs depletion is also one of the proposed mechanisms involved in muscular dystrophy patients (Michels et al., 1992, Pesl et al., 2020).

Several studies indicate that HCM mutations are associated with increased actin-myosin sliding velocity (Kawana et al., 2017) and DCM with decreased sliding velocity (Robinson et al., 2002), inconsistent reports on altered maximal force development are reviewed in (Eschenhagen and Carrier, 2019), where it was also reported on both HCM and DCM signs of decreased energetic efficiency and altered myofilament calcium sensitivity. In mechanical studies, this could be echocardiographically distinguished as different tension-time integral of the contraction peak, that is, the area under the curve of an averaged contraction peak (Davis et al., 2016).

Restrictive cardiomyopathy presents alterations in tissue flexibility in absence of myofibrillar rearrangement and gross abnormalities, and ventricles filling pressure alterations are responsible for impaired relaxation and commonly atrial enlargement (Huang and Du, 2004, Parvatiyar et al., 2010). Most common cause is extracellular deposition of a misfolded protein – infiltration of amyloid causing CM separation, cellular toxicity, apoptosis, and tissue stiffness (Muchtar et al., 2017). The same may be result of storage disease or endomyocardial fibrosis prevalent in developing world. Nevertheless, both sarcomere as well as cytoskeletal genes could be involved (Tarnovskaya et al., 2017) proposing myofibril multidomain architecture destabilization and increase of myocardial stiffness.

6.1.3.6 Atrial Fibrillation and Flutter

There is increasing recognition that one of the contributing abnormalities to the development of atrial fibrillation (AF) is atrial fibrosis (Boldt et al., 2004), differing in collagen III expression. Progressive atrial myopathy causes disorganized electrical activity, resulting in irregular heartbeat, but also in endothelial dysfunction and blood stasis leading to thrombogenesis and eventual stroke (Melenovsky et al., 2015). A strong role is played by AP alternans referring to a phenomenon whereby a single cell or region of tissue generates APs in a repeated long-short-long-short pattern when stimulated (Weiss et al., 2011); those are heavily dependent on preceding alteration in atrial pressure and its dynamics during exercise (Meluzin et al., 2017). One of the AF consequences is progressive ventricular remodeling, probably due to the adverse hemodynamic effect of loss of coordinated atrial contraction and/or tachycardia-mediated cardiomyopathy from a persistently elevated ventricular rate (Hunter et al., 2014), resulting in microvascular coronary dysfunction and impaired myocardial perfusion, leading to progressive fibrosis similar to the above-described one (ischemic disease, Section 6.1.3.2), but also systemic inflammation and impaired endothelial function (Freestone et al., 2008). Impaired myofibrillar energetics directly impacts myocyte mechanics (Mihm et al., 2001), loss of myofibrils, accumulation of glycogen, changes in mitochondrial shape and size, fragmentation of the sarcoplasmic reticulum, dispersion of nuclear chromatin, and increase in myocyte size (Ausma et al., 1997).

6.1.4 Conclusions

Cardiovascular system mechanics in healthy and disease states are vastly complex areas, studied already for decades. Nevertheless, only recently novel methods as atomic force microscopy or tweezing technologies and their combinations allowed for inclusion of CM mechanics describing methods to the same level as classic and established cellular electrophysiology and optical imaging. Nanotechnology trends are moving ahead not only basic research, but also clinical diagnostic and therapeutic possibilities. A number of areas in cardiovascular medicine remain to be probed, therefore this field is promising not only for expansion of scientific methods, but mainly with high clinical importance, elucidating mechanisms, and timely diagnostics, allowing proper treatments.

Funding information: The work was supported by project Ministry of Health of the Czech Republic, grant NU20-06-00156, project National Institute for Research of Metabolic and Cardiovascular Diseases (Programme EXCELES, No. LX22NPO5104) - Funded by the European Union – Next Generation EU, Project ENOCH (no. CZ.02.1.01/0.0/0.0/16_019/0000868)by the European Regional Development Fund Martin Pešl was supported by the Faculty of Medicine MU to the junior researchers (ROZV/23/LF9/2019 and MUNI/A/1462/2021). We acknowledge CF Nanobiotechnology of CIISB, Instruct-CZ Centre, supported by MEYS CR (LM2018127).

References

Abassi, Y. A., B. Xi, N. Li, W. Ouyang, A. Seiler, M. Watzele, R. Kettenhofen, H. Bohlen, A. Ehlich and E. Kolossov (2012). "Dynamic monitoring of beating periodicity of stem cell-derived cardiomyocytes as a predictive tool for preclinical safety assessment." British Journal of Pharmacology **165**(5): 1424–1441.

Abubakar, I., T. Tillmann and A. Banerjee (2015). "Global, regional, and national age-sex specific all-cause and cause-specific mortality for 240 causes of death, 1990–2013: A systematic analysis for the Global Burden of Disease Study 2013". Lancet **385**(9963): 117–171.

Acimovic I, Vilotic A, Pesl M, Lacampagne A, Dvorak P, Rotrekl V, and Meli AC. Human pluripotent stem cell-derived cardiomyocytes as research and therapeutic tools. Biomed Res Int. 2014;2014:512831. doi: 10.1155/2014/512831. Epub 2014 Apr 2. PMID: 24800237; PMCID: PMC3996932.

Actis, P., S. Tokar, J. Clausmeyer, B. Babakinejad, S. Mikhaleva, R. Cornut, Y. Takahashi, A. López Córdoba, P. Novak and A. I. Shevchuck (2014). "Electrochemical nanoprobes for single-cell analysis". ACS nano **8**(1): 875–884.

Ahola, A., A. L. Kiviaho, K. Larsson, M. Honkanen, K. Aalto-Setälä and J. Hyttinen (2014). "Video image-based analysis of single human induced pluripotent stem cell derived cardiomyocyte beating dynamics using digital image correlation". Biomedical Engineering Online **13**(1): 39.

Aistrup, G. L., Y. Shiferaw, S. Kapur, A. H. Kadish and J. A. Wasserstrom (2009). "Mechanisms underlying the formation and dynamics of subcellular calcium alternans in the intact rat heart". Circulation Research **104**(5): 639–649.

Ashkin, A., J. M. Dziedzic and T. Yamane (1987). "Optical trapping and manipulation of single cells using infrared laser beams". Nature **330**(6150): 769–771.

Ausma, J., M. Wijffels, F. Thoné, L. Wouters, M. Allessie and M. Borgers (1997). "Structural changes of atrial myocardium due to sustained atrial fibrillation in the goat". Circulation **96**(9): 3157–3163.

Ayala, Y. A., B. Pontes, D. S. Ether, L. B. Pires, G. R. Araujo, S. Frases, L. F. Romão, M. Farina, V. Moura-Neto and N. B. Viana (2016). "Rheological properties of cells measured by optical tweezers". BMC Biophysics **9**(1): 5.

Azeloglu, E. U. and K. D. Costa (2010). "Cross-bridge cycling gives rise to spatiotemporal heterogeneity of dynamic subcellular mechanics in cardiac myocytes probed with atomic force microscopy". American Journal of Physiology-Heart and Circulatory Physiology **298**(3): H853–H860.

Bachir, A. I., A. R. Horwitz, W. J. Nelson and J. M. Bianchini (2017). "Actin-based adhesion modules mediate cell interactions with the extracellular matrix and neighboring cells". Cold Spring Harbor Perspectives in Biology **9**(7): a023234.

Balaban, N. Q., U. S. Schwarz, D. Riveline, P. Goichberg, G. Tzur, I. Sabanay, D. Mahalu, S. Safran, A. Bershadsky and L. Addadi (2001). "Force and focal adhesion assembly: A close relationship studied using elastic micropatterned substrates". Nature Cell Biology **3**(5): 466–472.

Basoli, F., S. M. Giannitelli, M. Gori, P. Mozetic, A. Bonfanti, M. Trombetta and A. Rainer. (2018). "Biomechanical characterization at the cell scale: Present and prospects". Frontiers in Physiology **9**: 1449.

Bébarová, M. (2012). "Advances in patch clamp technique: Towards higher quality and quantity". General Physiology and Biophysics **31**(2): 131–140.

Benech, J. C., N. Benech, A. I. Zambrana, I. Rauschert, V. Bervejillo, N. Oddone, A. Alberro and J. P. Damián. (2015). "Intrinsic nanomechanical changes in live diabetic cardiomyocytes". Cardiovascular Regenerative Medicine **2**: e893.

Benech, J. C., N. Benech, A. I. Zambrana, I. Rauschert, V. Bervejillo, N. Oddone and J. P. Damián (2014). "Diabetes increases stiffness of live cardiomyocytes measured by atomic force microscopy nanoindentation". American Journal of Physiology. Cell Physiology **307**(10): C910–919.

Berry, M. F., A. J. Engler, Y. J. Woo, T. J. Pirolli, L. T. Bish, V. Jayasankar, K. J. Morine, T. J. Gardner, D. E. Discher and H. L. Sweeney (2006). "Mesenchymal stem cell injection after myocardial infarction improves myocardial compliance". American Journal of Physiology-Heart and Circulatory Physiology **290**(6): H2196–H2203.

Bers, D. M. (2014). Cardiac electrophysiology: From cell to bedside. Sixth Edition edited by. Zipes, D. P. and Jalife, J., Philadelphia, W.B. Saunders, 161–169.

Boldt, A., U. Wetzel, J. Lauschke, J. Weigl, J. Gummert, G. Hindricks, H. Kottkamp and S. Dhein (2004). "Fibrosis in left atrial tissue of patients with atrial fibrillation with and without underlying mitral valve disease". Heart **90**(4): 400–405.

Borin, D., I. Pecorari, B. Pena and O. Sbaizero (2018). Novel insights into cardiomyocytes provided by atomic force microscopy. Seminars in cell & developmental biology, Elsevier.

Braam, S. R., L. Tertoolen, A. van de Stolpe, T. Meyer, R. Passier and C. L. Mummery (2010). "Prediction of drug-induced cardiotoxicity using human embryonic stem cell-derived cardiomyocytes". Stem Cell Research **4**(2): 107–116.

Burnham, N. A. and R. J. Colton (1989). "Measuring the nanomechanical properties and surface forces of materials using an atomic force microscope". Journal of Vacuum Science & Technology A: Vacuum, Surfaces, and Films **7**(4): 2906–2913.

Caluori, G., J. Pribyl, V. Cmiel, M. Pesl, T. Potocnak, I. Provaznik, P. Skladal and V. Rotrekl (2019a). "Simultaneous study of mechanobiology and calcium dynamics on hESC-derived cardiomyocytes clusters". Journal of Molecular Recognition **32**(2): e2760.

Caluori, G., J. Pribyl, M. Pesl, S. Jelinkova, V. Rotrekl, P. Skladal and R. Raiteri. (2019b). "Non-invasive electromechanical cell-based biosensors for improved investigation of 3D cardiac models". Biosensors & Bioelectronics **124**: 129–135.

Caluori, G., J. Pribyl, M. Pesl, G. Nardone, P. Skladal and G. Forte. (2018). "Advanced and rationalized atomic force microscopy analysis unveils specific properties of controlled cell mechanics". Frontiers in Physiology **9**: 1121.

Cesa, C. M., N. Kirchgeßner, D. Mayer, U. S. Schwarz, B. Hoffmann and R. Merkel (2007). "Micropatterned silicone elastomer substrates for high resolution analysis of cellular force patterns". Review of Scientific Instruments **78**(3): 034301.

Chen, A., E. Lee, R. Tu, K. Santiago, A. Grosberg, C. Fowlkes and M. Khine (2014). "Integrated platform for functional monitoring of biomimetic heart sheets derived from human pluripotent stem cells". Biomaterials 35(2): 675–683.

Chung, C., B. L. Pruitt and S. C. Heilshorn (2013). "Spontaneous cardiomyocyte differentiation of mouse embryoid bodies regulated by hydrogel crosslink density". Biomaterials Science 1(10): 1082–1090.

Ciambrone, G. J., V. F. Liu, D. C. Lin, R. P. McGuinness, G. K. Leung and S. Pitchford (2004). "Cellular dielectric spectroscopy: A powerful new approach to label-free cellular analysis". Journal of Biomolecular Screening 9(6): 467–480.

Claes, V. and D. Brutsaert (1971). "Infrared-emitting diode and optic fibers for underwater force measurement in heart muscle". Journal of Applied Physiology 31(3): 497–498.

Comisar, W., D. Mooney and J. Linderman (2011). "Integrin organization: Linking adhesion ligand nanopatterns with altered cell responses". Journal of Theoretical Biology 274(1): 120–130.

Cooper, L. T., K. L. Baughman, A. M. Feldman, A. Frustaci, M. Jessup, U. Kuhl, G. N. Levine, J. Narula, R. C. Starling and J. Towbin (2007). "The role of endomyocardial biopsy in the management of cardiovascular disease: A scientific statement from the American Heart Association, the American College of Cardiology, and the European Society of Cardiology Endorsed by the Heart Failure Society of America and the Heart Failure Association of the European Society of Cardiology". Journal of the American College of Cardiology 50(19): 1914–1931.

Czirok, A., D. G. Isai, E. Kosa, S. Rajasingh, W. Kinsey, Z. Neufeld and J. Rajasingh (2017). "Optical-flow based non-invasive analysis of cardiomyocyte contractility". Scientific Reports 7(1): 1–11.

Dague, E., G. Genet, V. Lachaize, C. Guilbeau-Frugier, J. Fauconnier, C. Mias, B. Payré, L. Chopinet, D. Alsteens and S. Kasas. (2014). "Atomic force and electron microscopic-based study of sarcolemmal surface of living cardiomyocytes unveils unexpected mitochondrial shift in heart failure". Journal of Molecular and Cellular Cardiology 74: 162–172.

Davis, J., L. C. Davis, R. N. Correll, C. A. Makarewich, J. A. Schwanekamp, F. Moussavi-Harami, D. Wang, A. J. York, H. Wu and S. R. Houser (2016). "A tension-based model distinguishes hypertrophic versus dilated cardiomyopathy". Cell 165(5): 1147–1159.

Davis, J. J., H. A. O. Hill and T. Powell (2001). "High resolution scanning force microscopy of cardiac myocytes". Cell Biology International 25(12): 1271–1277.

de Oliveira Martins, C., L. Demarchi, F. M. Ferreira, P. M. A. Pomerantzeff, C. Brandao, R. O. Sampaio, G. S. Spina, J. Kalil, E. Cunha-Neto and L. Guilherme (2017). "Rheumatic heart disease and myxomatous degeneration: Differences and similarities of valve damage resulting from autoimmune reactions and matrix disorganization". PloS one 12(1).

De Vlaminck, I. and C. Dekker. (2012). "Recent advances in magnetic tweezers". Annual Review of Biophysics 41: 453–472.

Delbridge, L. M. and K. P. Roos (1997). "Optical methods to evaluate the contractile function of unloaded isolated cardiac myocytes". Journal of Molecular and Cellular Cardiology 29(1): 11–25.

Domke, J., W. J. Parak, M. George, H. E. Gaub and M. Radmacher (1999). "Mapping the mechanical pulse of single cardiomyocytes with the atomic force microscope". European Biophysics Journal 28(3): 179–186.

Eeva, L., A. Antti, H. Jari and A.-S. Katriina (2015). "Biochim". Biophysica Acta 7.

Eschenhagen, T. and L. Carrier (2019). "Cardiomyopathy phenotypes in human-induced pluripotent stem cell-derived cardiomyocytes – a systematic review". Pflügers Archiv-European Journal of Physiology 471(5): 755–768.

Fisher, C. I., J. Chen and W. D. Merryman (2013). "Calcific nodule morphogenesis by heart valve interstitial cells is strain dependent". Biomechanics and Modeling in Mechanobiology **12**(1): 5–17.

Freestone, B., A. Y. Chong, S. Nuttall and G. Y. Lip (2008). "Impaired flow mediated dilatation as evidence of endothelial dysfunction in chronic atrial fibrillation:: Relationship to plasma von Willebrand factor and soluble E-selectin levels". Thrombosis Research **122**(1): 85–90.

Garnier, A. and M. Briet (2016). "Arterial Stiffness and Chronic Kidney Disease". Pulse Basel, Switzerland **3**(3–4): 229–241.

Gentemann, L., S. Kalies, M. Coffee, H. Meyer, T. Ripken, A. Heisterkamp, R. Zweigerdt and D. Heinemann (2017). "Modulation of cardiomyocyte activity using pulsed laser irradiated gold nanoparticles". Biomedical Optics Express **8**(1): 177–192.

Gheorghiade, M. and R. O. Bonow (1998). "Chronic heart failure in the United States: A manifestation of coronary artery disease". Circulation **97**(3): 282–289.

Gilchrist, K. H., G. F. Lewis, E. A. Gay, K. L. Sellgren and S. Grego (2015). "High-throughput cardiac safety evaluation and multi-parameter arrhythmia profiling of cardiomyocytes using microelectrode arrays". Toxicology and Applied Pharmacology **288**(2): 249–257.

Gillespie, C. D., C. Wigington and Y. Hong (2013). "Coronary heart disease and stroke deaths-United States, 2009". MMWR Supply **62**(3): 157–160.

Glaubitz, M., S. Block, J. Witte, K. Empen, S. Gross, R. Schlicht, K. Weitmann, K. Klingel, R. Kandolf and W. Hoffmann (2014). "Stiffness of left ventricular cardiac fibroblasts is associated with ventricular dilation in patients with recent-onset nonischemic and nonvalvular cardiomyopathy". Circulation Journal **78**(7): 1693–1700.

Guazzi, M. and R. Arena (2009). "Endothelial dysfunction and pathophysiological correlates in atrial fibrillation". Heart **95**(2): 102–106.

Hahn, C. and M. A. Schwartz (2009). "Mechanotransduction in vascular physiology and atherogenesis". Nature Reviews. Molecular Cell Biology **10**(1): 53–62.

Hajdu, M. A., D. D. Heistad and G. L. Baumbach (1991). "Effects of antihypertensive therapy on mechanics of cerebral arterioles in rats". Hypertension **17**(3): 308–316.

Halevi, R., A. Hamdan, G. Marom, M. Mega, E. Raanani and R. Haj-Ali (2015). "Progressive aortic valve calcification: Three-dimensional visualization and biomechanical analysis". Journal of Biomechanics **48**(3): 489–497.

Hansen, K. J., J. T. Favreau, J. R. Gershlak, M. A. Laflamme, D. R. Albrecht and G. R. Gaudette (2017). "Optical method to quantify mechanical contraction and calcium transients of human pluripotent stem cell-derived cardiomyocytes". Tissue Engineering. Part C, Methods **23**(8): 445–454.

Hayakawa, T., T. Kunihiro, T. Ando, S. Kobayashi, E. Matsui, H. Yada, Y. Kanda, J. Kurokawa and T. Furukawa. (2014). "Image-based evaluation of contraction–relaxation kinetics of human-induced pluripotent stem cell-derived cardiomyocytes: Correlation and complementarity with extracellular electrophysiology". Journal of Molecular and Cellular Cardiology **77**: 178–191.

Henon, S., G. Lenormand, A. Richert and F. Gallet (1999). "A new determination of the shear modulus of the human erythrocyte membrane using optical tweezers". Biophysical Journal **76**(2): 1145–1151.

Herron, T. J., P. Lee and J. Jalife (2012). "Optical imaging of voltage and calcium in cardiac cells & tissues". Circulation Research **110**(4): 609–623.

Hescheler, J., B. Fleischmann, S. Lentini, V. Maltsev, J. Rohwedel, A. Wobus and K. Addicks (1997). "Embryonic stem cells: A model to study structural and functional properties in cardiomyogenesis". Cardiovascular Research **36**(2): 149–162.

Heureaux, J., D. Chen, V. L. Murray, C. X. Deng and A. P. Liu (2014). "Activation of a bacterial mechanosensitive channel in mammalian cells by cytoskeletal stress". Cellular and Molecular Bioengineering **7**(3): 307–319.

Holt, E., T. Tønnessen, P. K. Lunde, S. O. Semb, J. A. Wasserstrom, O. M. Sejersted and G. Christensen (1998). "Mechanisms of cardiomyocyte dysfunction in heart failure following myocardial infarction in rats". Journal of Molecular and Cellular Cardiology **30**(8): 1581–1593.

Holzmann, M., A. Nicko, U. Kühl, M. Noutsias, W. Poller, W. Hoffmann, A. Morguet, B. Witzenbichler, C. Tschöpe and H. Schultheiss (2008). "Complication rate of right ventricular endomyocardial biopsy via the femoral approach: A retrospective and prospective study analyzing 3048 diagnostic procedures over an 11-year period". Circulation **118**(17): 1722.

Huang, X.-P. and J.-F. Du. (2004). "Troponin I, cardiac diastolic dysfunction and restrictive cardiomyopathy". Acta Pharmacologica Sinica **25**: 1569–1575.

Humphries MJ, Travis MA, Clark K, Mould AP. Mechanisms of integration of cells and extracellular matrices by integrins. Biochem Soc Trans. 2004 Nov;**32**(Pt 5): 822–825. doi: 10.1042/BST0320822. PMID: 15494024.

Hunter, R. J., T. J. Berriman, I. Diab, R. Kamdar, L. Richmond, V. Baker, F. Goromonzi, V. Sawhney, E. Duncan and S. P. Page (2014). "A randomized controlled trial of catheter ablation versus medical treatment of atrial fibrillation in heart failure (the CAMTAF trial)". Circulation. Arrhythmia and Electrophysiology **7**(1): 31–38.

Hwang, J. Y., H. G. Lim, C. W. Yoon, K. H. Lam, S. Yoon, C. Lee, C. T. Chiu, B. J. Kang, H. H. Kim and K. K. Shung (2014). "Non-contact high-frequency ultrasound microbeam stimulation for studying mechanotransduction in human umbilical vein endothelial cells". Ultrasound in Medicine & Biology **40**(9): 2172–2182.

Ikeda, S., S. Matsushima, K. Okabe, M. Ikeda, A. Ishikita, T. Tadokoro, N. Enzan, T. Yamamoto, M. Sada and H. Deguchi (2019). "Blockade of L-type Ca 2+ channel attenuates doxorubicin-induced cardiomyopathy via suppression of CaMKII-NF-κB pathway". Scientific Reports **9**(1): 1–14.

Jacot, J. G., J. C. Martin and D. L. Hunt (2010). "Mechanobiology of cardiomyocyte development". Journal of Biomechanics **43**(1): 93–98.

Johnson, K. L., K. Kendall and A. Roberts (1971). "Surface energy and the contact of elastic solids." Proceedings of the royal society of London. A. mathematical and physical sciences **324**(1558): 301–313.

Kabanov D, Klimovic S, Rotrekl V, Pesl M, Pribyl J. Atomic force spectroscopy is a promising tool to study contractile properties of cardiac cells. Micron. 2022 Apr;155:103199. doi: 10.1016/j.micron.2021.103199. Epub 2021 Dec 18. PMID: 35140035.

Kajzar, A., C. Cesa, N. Kirchgessner, B. Hoffmann and R. Merkel (2008). "Toward physiological conditions for cell analyses: Forces of heart muscle cells suspended between elastic micropillars". Biophysical Journal **94**(5): 1854–1866.

Kamgoué, A., J. Ohayon, Y. Usson, L. Riou and P. Tracqui (2009). "Quantification of cardiomyocyte contraction based on image correlation analysis". Cytometry Part A: The Journal of the International Society for Advancement of Cytometry **75**(4): 298–308.

Kawana, M., S. S. Sarkar, S. Sutton, K. M. Ruppel and J. A. Spudich (2017). "Biophysical properties of human β-cardiac myosin with converter mutations that cause hypertrophic cardiomyopathy". Science Advances **3**(2): e1601959.

Kawana, S., H. Kimura, A. Miyamoto, H. Ohshika and A. Namiki (1993). "Application of a Fotonic Sensor for measurement of chronotropy and contractility in cultured rat cardiac myocytes". Nihon yakurigaku zasshi. Folia pharmacologica Japonica **102**(4): 279–286.

Klimovic S, Scurek M, Pesl M, Beckerova D, Jelinkova S, Urban T, Kabanov D, Starek Z, Bebarova M, Pribyl J, Rotrekl V, Brat K. Aminophylline Induces Two Types of Arrhythmic Events in Human

Pluripotent Stem Cell-Derived Cardiomyocytes. Front Pharmacol. 2022 Jan 17;**12**:789730. doi: 10.3389/fphar.2021.789730. PMID: 35111056; PMCID: PMC8802108.

Kilinc, D., A. Blasiak, J. J. O'Mahony, D. M. Suter and G. U. Lee (2012). "Magnetic tweezers-based force clamp reveals mechanically distinct apCAM domain interactions". Biophysical Journal **103**(6): 1120–1129.

Kim, K., R. Taylor, J. Sim, S.-J. Park, J. Norman, G. Fajardo, D. Bernstein and B. Pruitt (2011). "Calibrated micropost arrays for biomechanical characterisation of cardiomyocytes". Micro & Nano Letters **6**(5): 317–322.

Koivumäki, J. T., N. Naumenko, T. Tuomainen, J. Takalo, M. Oksanen, K. A. Puttonen, Š. Lehtonen, J. Kuusisto, M. Laakso and J. Koistinaho. (2018). "Structural immaturity of human iPSC-derived cardiomyocytes: In silico investigation of effects on function and disease modeling". Frontiers in Physiology **9**: 80.

Kronenbitter, A., F. Funk, K. Hackert, S. Gorreßen, D. Glaser, P. Boknik, G. Poschmann, K. Stühler, M. Isić and M. Krüger. (2018). "Impaired Ca2+ cycling of nonischemic myocytes contributes to sarcomere dysfunction early after myocardial infarction". Journal of Molecular and Cellular Cardiology **119**: 28–39.

Krueger, J. W., D. Forletti and B. A. Wittenberg (1980). "Uniform sarcomere shortening behavior in isolated cardiac muscle cells". The Journal of General Physiology **76**(5): 587–607.

Laurila E, Ahola A, Hyttinen J, Aalto-Setälä K. Methods for in vitro functional analysis of iPSC derived cardiomyocytes - Special focus on analyzing the mechanical beating behavior. Biochim Biophys Acta. 2016 Jul;1863(7Pt B):1864-72. doi: 10.1016/j.bbamcr.2015.12.013. Epub 2015

Lasheras, J. C. (2007). "The biomechanics of arterial aneurysms". Annual Review of Fluid Mechanics **39**: 293–319.

Laughner, J. I., F. S. Ng, M. S. Sulkin, R. M. Arthur and I. R. Efimov (2012). "Processing and analysis of cardiac optical mapping data obtained with potentiometric dyes". American Journal of Physiology-Heart and Circulatory Physiology **303**(7): H753–H765.

Leake, M. C., D. Wilson, M. Gautel and R. M. Simmons (2004). "The elasticity of single titin molecules using a two-bead optical tweezers assay". Biophysical Journal **87**(2): 1112–1135.

Lee, J., V. Vedula, K. I. Baek, J. Chen, J. J. Hsu, Y. Ding, -C.-C. Chang, H. Kang, A. Small and P. Fei (2018). "Spatial and temporal variations in hemodynamic forces initiate cardiac trabeculation". JCI Insight **3**(13): e96672. doi: 10.1172/jci.insight.96672. PMID: 29997298; PMCID: PMC6124527.

Léger, J., G. Berson, C. Delcayre, C. Klotz, K. Schwartz, J. Léger, M. Stephens and B. Swynghedauw (1975). "Heart contractile proteins". Biochimie **57**(11–12): 1249.

Leong, Y. Y., W. H. Ng, G. M. Ellison-Hughes and J. J. Tan. (2017). "Cardiac stem cells for myocardial regeneration: They are not alone". Frontiers in Cardiovascular Medicine **4**: 47.

Liang, C., K. Wang, Q. Li, J. Bai and H. Zhang (2019). "Influence of the distribution of fibrosis within an area of myocardial infarction on wave propagation in ventricular tissue". Scientific Reports **9**(1): 1–14.

Liang, P., F. Lan, A. S. Lee, T. Gong, V. Sanchez-Freire, Y. Wang, S. Diecke, K. Sallam, J. W. Knowles and P. J. Wang (2013). "Drug screening using a library of human induced pluripotent stem cell–derived cardiomyocytes reveals disease-specific patterns of cardiotoxicity". Circulation **127**(16): 1677–1691.

Lieber, S. C., N. Aubry, J. Pain, G. Diaz, S.-J. Kim and S. F. Vatner (2004). "Aging increases stiffness of cardiac myocytes measured by atomic force microscopy nanoindentation". American Journal of Physiology-Heart and Circulatory Physiology **287**(2): H645–H651.

Lin, G., K. Pister and K. Roos (1995). "Novel microelectromechanical system force transducer to quantify contractile characteristics from isolated cardiac muscle cells". Journal of the Electrochemical Society **142**(3): L31.

Lipfert, J., S. Klijnhout and N. H. Dekker (2010). "Torsional sensing of small-molecule binding using magnetic tweezers". Nucleic Acids Research **38**(20): 7122–7132.

Liu, J., N. Sun, M. A. Bruce, J. C. Wu and M. J. Butte (2012). "Atomic force mechanobiology of pluripotent stem cell-derived cardiomyocytes". PloS one **7**(5): e37559. doi: 10.1371/journal. pone.0037559. Epub 2012 May 18. PMID: 22624048; PMCID: PMC3356329.

Manabe, I., T. Shindo and R. Nagai (2002). "Gene expression in fibroblasts and fibrosis: Involvement in cardiac hypertrophy". Circulation Research **91**(12): 1103–1113.

Mann, S. A., J. Heide, T. Knott, R. Airini, F. B. Epureanu, A.-F. Deftu, A.-T. Deftu, B. M. Radu and B. Amuzescu (2019). Journal of Pharmacological and Toxicological Methods **100**: 106599. doi: 10.1016/j.vascn.2019.106599. Epub 2019 Jun 20. PMID: 31228558.

Maron, B. J., J. M. Gardin, J. M. Flack, S. S. Gidding, T. T. Kurosaki and D. E. Bild (1995). "Prevalence of hypertrophic cardiomyopathy in a general population of young adults: Echocardiographic analysis of 4111 subjects in the CARDIA study". Circulation **92**(4): 785–789.

Martins CO, Demarchi L, Ferreira FM, Pomerantzeff PM, Brandao C, Sampaio RO, Spina GS, Kalil J, Cunha-Neto E, Guilherme L. Rheumatic Heart Disease and Myxomatous Degeneration: Differences and Similarities of Valve Damage Resulting from Autoimmune Reactions and Matrix Disorganization. PLoS One. 2017 Jan 25;**12**(1):e0170191. doi: 10.1371/journal. pone.0170191. PMID: 28121998; PMCID: PMC5266332.

Masarova, L., M. Mojica-Pisciott, R. Panovsky, V. Kincl, M. Pesl, L. Opatril, J. Machal, J. Novak, T. Holecek, L. Jurikova and V. Feitova (2020). "JACC". Cardiovascular Imaging doi.org/10.1016/ j.jcmg.2020.09.016.

Masárová L, Mojica-Pisciotti ML, Panovský R, Kincl V, Pešl M, Opatřil L, Máchal J, Novák J, Holeček T, Juříková L, Feitová V. Decreased Global Strains of LV in Asymptomatic Female Duchenne Muscular Dystrophy Gene Carriers Using CMR-FT. JACC Cardiovasc Imaging. 2021 May;**14** (5):1070–1072. doi: 10.1016/j.jcmg.2020.09.016. Epub 2020 Nov 18. PMID: 33221218.

Mathur, A. B., A. M. Collinsworth, W. M. Reichert, W. E. Kraus and G. A. Truskey (2001). "Endothelial, cardiac muscle and skeletal muscle exhibit different viscous and elastic properties as determined by atomic force microscopy". Journal of Biomechanics **34**(12): 1545–1553.

McMullen, J. R., T. Shioi, L. Zhang, O. Tarnavski, M. C. Sherwood, P. M. Kang and S. Izumo (2003). "Phosphoinositide 3-kinase (p110α) plays a critical role for the induction of physiological, but not pathological, cardiac hypertrophy." Proceedings of the National Academy of Sciences **100** (21): 12355–12360.

Melenovsky, V., S.-J. Hwang, M. M. Redfield, R. Zakeri, G. Lin and B. A. Borlaug (2015). "Left atrial remodeling and function in advanced heart failure with preserved or reduced ejection fraction". Circulation. Heart Failure **8**(2): 295–303.

Meluzin, J., L. Spinarova, P. Hude, J. Krejci, H. Poloczkova, H. Podrouzkova, M. Pesl, M. Orban, L. Dusek and J. Korinek (2009). "Left ventricular mechanics in idiopathic dilated cardiomyopathy: Systolic-diastolic coupling and torsion". Journal of the American Society of Echocardiography **22**(5): 486–493.

Meluzin J, Spinarova L, Hude P, Krejci J, Podrouzkova H, Pesl M, Orban M, Dusek L, Jarkovsky J, Korinek J. Estimation of left ventricular filling pressures by speckle tracking echocardiography in patients with idiopathic dilated cardiomyopathy. Eur J Echocardiogr. 2011 Jan;**12**(1):11–18. doi: 10.1093/ejechocard/jeq088. Epub 2010 Aug 4. PMID: 20688766.

Meluzin, J., L. Spinarova, P. Hude, J. Krejci, H. Podrouzkova, M. Pesl, M. Orban, L. Dusek, J. Jarkovsky and J. Korinek (2011). European Journal of Echocardiography **12**: 11–18.

Meluzin, J., Z. Starek, T. Kulik, J. Jez, F. Lehar, J. Tomandl, L. Dusek, J. Wolf, P. Leinveber and M. Novak (2017). "Improvement in the prediction of exercise-induced elevation of left ventricular filling pressure in patients with normal left ventricular ejection fraction". Echocardiography 34(1): 78–86.

Michels, V. V., P. P. Moll, F. A. Miller, A. J. Tajik, J. S. Chu, D. J. Driscoll, J. C. Burnett, R. J. Rodeheffer, J. H. Chesebro and H. D. Tazelaar (1992). "The frequency of familial dilated cardiomyopathy in a series of patients with idiopathic dilated cardiomyopathy". New England Journal of Medicine 326(2): 77–82.

Mihm, M. J., F. Yu, C. A. Carnes, P. J. Reiser, P. M. McCarthy, D. R. Van Wagoner and J. A. Bauer (2001). "Impaired myofibrillar energetics and oxidative injury during human atrial fibrillation". Circulation 104(2): 174–180.

Mohammed, D., M. Versaevel, C. Bruyère, L. Alaimo, M. Luciano, E. Vercruysse, A. Procès and S. Gabriele (2019). "Innovative tools for mechanobiology: Unraveling outside-in and inside-out mechanotransduction". Frontiers in Bioengineering and Biotechnology 7:162. doi: 10.3389/fbioe.2019.00162. PMID: 31380357; PMCID: PMC6646473.

Mozaffarian, D., E. J. Benjamin, A. S. Go, D. K. Arnett, M. J. Blaha, M. Cushman, S. De Ferranti, J.-P. Després, H. J. Fullerton and V. J. Howard (2015). "Executive summary: Heart disease and stroke statistics – 2015 update: A report from the American Heart Association". circulation 131(4): 434–441.

Muchtar, E., L. A. Blauwet and M. A. Gertz (2017). "Restrictive cardiomyopathy: Genetics, pathogenesis, clinical manifestations, diagnosis, and therapy". Circulation Research 121(7): 819–837.

Nance, M. E., J. T. Whitfield, Y. Zhu, A. K. Gibson, L. M. Hanft, K. S. Campbell, G. A. Meininger, K. S. McDonald, S. S. Segal and T. L. Domeier (2015). "Attenuated sarcomere lengthening of the aged murine left ventricle observed using two-photon fluorescence microscopy". American Journal of Physiology-Heart and Circulatory Physiology 309(5): H918–H925.

Nardone, G., J. Oliver-De, L. Cruz, J. Vrbsky, C. Martini, J. Pribyl, P. Skládal, M. Pešl, G. Caluori, S. Pagliari and F. Martino (2017). "YAP regulates cell mechanics by controlling focal adhesion assembly". Nature Communications 8(1): 1–13.

Natarajan, A., M. Stancescu, V. Dhir, C. Armstrong, F. Sommerhage, J. J. Hickman and P. Molnar (2011). "Patterned cardiomyocytes on microelectrode arrays as a functional, high information content drug screening platform". Biomaterials 32(18): 4267–4274.

Nation, D. A., L. Delano-Wood, K. J. Bangen, C. E. Wierenga, A. J. Jak, L. A. Hansen, D. R. Galasko, D. P. Salmon and M. W. Bondi (2012). "Antemortem pulse pressure elevation predicts cerebrovascular disease in autopsy-confirmed Alzheimer's disease". Journal of Alzheimer's Disease 30(3): 595–603.

Neuman, K. C., E. H. Chadd, G. F. Liou, K. Bergman and S. M. Block (1999). "Characterization of photodamage to Escherichia coli in optical traps". Biophysical Journal 77(5): 2856–2863.

O'Rourke, M. F., J. A. Staessen, C. Vlachopoulos and D. Duprez (2002). "Clinical applications of arterial stiffness; definitions and reference values". American Journal of Hypertension 15(5): 426–444.

Opie, L. H., P. J. Commerford, B. J. Gersh and M. A. Pfeffer (2006). "Controversies in ventricular remodelling". The Lancet 367(9507): 356–367.

Panovský, R., M. Pešl, T. Holeček, J. Máchal, V. Feitová, L. Mrázová, J. Haberlová, A. Slabá, P. Vít, V. Stará and V. Kincl (2019). Orphanet Journal of Rare Diseases 14(1): 10. doi: 10.1186/s13023-018-0986-0. PMID: 30626423; PMCID: PMC6327529.

Parmacek, M. S. and R. J. Solaro (2004). "Biology of the troponin complex in cardiac myocytes". Progress in Cardiovascular Diseases 47(3): 159–176.

Parvatiyar, M. S., J. R. Pinto, D. Dweck and J. D. Potter (2010). "Cardiac troponin mutations and restrictive cardiomyopathy". BioMed Research International 2010.

Pesl, M., I. Acimovic, J. Pribyl, R. Hezova, A. Vilotic, J. Fauconnier, J. Vrbsky, P. Kruzliak, P. Skladal and T. Kara (2014). "Forced aggregation and defined factors allow highly uniform-sized embryoid bodies and functional cardiomyocytes from human embryonic and induced pluripotent stem cells". Heart and Vessels **29**(6): 834–846.

Pesl, M., J. Pribyl, I. Acimovic, A. Vilotic, S. Jelinkova, A. Salykin, A. Lacampagne, P. Dvorak, A. C. Meli and P. Skladal. (2016a). "Atomic force microscopy combined with human pluripotent stem cell derived cardiomyocytes for biomechanical sensing". Biosensors & Bioelectronics **85**: 751–757.

Pesl, M., J. Pribyl, G. Caluori, V. Cmiel, I. Acimovic, S. Jelinkova, P. Dvorak, Z. Starek, P. Skladal and V. Rotrekl (2016b). "Phenotypic assays for analyses of pluripotent stem cell–derived cardiomyocytes". Journal of Molecular Recognition: JMR e2602.

Pesl M, Pribyl J, Caluori G, Cmiel V, Acimovic I, Jelinkova S, Dvorak P, Starek Z, Skladal P, Rotrekl V. Phenotypic assays for analyses of pluripotent stem cell-derived cardiomyocytes. J Mol Recognit. 2017 Jun;30(6). doi: 10.1002/jmr.2602. Epub 2016 Dec 20. PMID: 27995655.

Pesl, M., S. Jelinkova, G. Caluori, M. Holicka, J. Krejci, P. Nemec, A. Kohutova, V. Zampachova, P. Dvorak and V. Rotrekl (2020). "Cardiovascular progenitor cells and tissue plasticity are reduced in a myocardium affected by Becker muscular dystrophy". Orphanet Journal of Rare Diseases **15**(1): 1–8.

Peters, N. S., N. J. Severs, S. M. Rothery, C. Lincoln, M. H. Yacoub and C. R. Green (1994). "Spatiotemporal relation between gap junctions and fascia adherens junctions during postnatal development of human ventricular myocardium". Circulation **90**(2): 713–725.

Pointon, A., A. R. Harmer, I. L. Dale, N. Abi-Gerges, J. Bowes, C. Pollard and H. Garside (2015). "Assessment of cardiomyocyte contraction in human-induced pluripotent stem cell-derived cardiomyocytes". Toxicological Sciences **144**(2): 227–237.

Powell, D., R. Mifflin, J. Valentich, S. Crowe, J. Saada and A. West (1999). "Myofibroblasts. I. Paracrine cells important in health and disease". American Journal of Physiology-Cell Physiology **277**(1): C1–C19.

Ramji, D. P. and T. S. Davies (2015). "Cytokines in atherosclerosis: Key players in all stages of disease and promising therapeutic targets". Cytokine & Growth Factor Reviews **26**(6): 673–685.

Reed, A., P. Kohl and R. Peyronnet (2014). "Molecular candidates for cardiac stretch-activated ion channels". Global Cardiology Science and Practice **2014**(2): 19.

Rizzo, R. J., W. J. McCarthy, S. N. Dixit, M. P. Lilly, V. P. Shively, W. R. Flinn and J. S. Yao (1989). "Collagen types and matrix protein content in human abdominal aortic aneurysms". Journal of Vascular Surgery **10**(4): 365–373.

Robinson, P., M. Mirza, A. Knott, H. Abdulrazzak, R. Willott, S. Marston, H. Watkins and C. Redwood (2002). "Alterations in thin filament regulation induced by a human cardiac troponin T mutant that causes dilated cardiomyopathy are distinct from those induced by troponin T mutants that cause hypertrophic cardiomyopathy". Journal of Biological Chemistry **277**(43): 40710–40716.

Rodriguez, M. L., B. T. Graham, L. M. Pabon, S. J. Han, C. E. Murry and N. J. Sniadecki (2014). "Measuring the contractile forces of human induced pluripotent stem cell-derived cardiomyocytes with arrays of microposts". Journal of Biomechanical Engineering **136**(5): 051005. doi: 10.1115/1.4027145. PMID: 24615475; PMCID: PMC4158804.

Sachs, F. (2010). "Stretch-activated ion channels: What are they?". Physiology **25**(1): 50–56.

Sakmann, B. and E. Neher (1976). "Single channel currents recorded from membrane of denervated frog muscle fibers". Nature **260**(799–802): 7.

Santoro, R., G. L. Perrucci, A. Gowran and G. Pompilio (2019). "Unchain My Heart: Integrins at the Basis of iPSC Cardiomyocyte Differentiation". Stem Cells International 2019: 8203950. doi: 10.1155/2019/8203950. PMID: 30906328; PMCID: PMC6393933.

Saphirstein, R. J., Y. Z. Gao, M. H. Jensen, C. M. Gallant, S. Vetterkind, J. R. Moore and K. G. Morgan (2013). "The focal adhesion: A regulated component of aortic stiffness". PloS one **8**(4): e62461. doi: 10.1371/journal.pone.0062461. PMID: 23626821; PMCID: PMC3633884.

Schweizer, P. A., F. F. Darche, N. D. Ullrich, P. Geschwill, B. Greber, R. Rivinius, C. Seyler, K. Müller-Decker, A. Draguhn and J. Utikal (2017). "Subtype-specific differentiation of cardiac pacemaker cell clusters from human induced pluripotent stem cells". Stem Cell Research & Therapy **8**(1): 229.

Sehgel, N. L., Y. Zhu, Z. Sun, J. P. Trzeciakowski, Z. Hong, W. C. Hunter, D. E. Vatner, G. A. Meininger and S. F. Vatner (2013). "Increased vascular smooth muscle cell stiffness: A novel mechanism for aortic stiffness in hypertension". American Journal of Physiology-Heart and Circulatory Physiology **305**(9): H1281–H1287.

Sewell-Loftin, M.-K., C. B. Brown, H. S. Baldwin and W. D. Merryman (2012). "Novel technique for quantifying mouse heart valve leaflet stiffness with atomic force microscopy". The Journal of Heart Valve Disease **21**(4): 513.

Sneddon, I. N. (1965). "The relation between load and penetration in the axisymmetric Boussinesq problem for a punch of arbitrary profile". International Journal of Engineering Science **3**(1): 47–57.

Soufivand, A., M. Navidbakhsh and M. Soleimani (2014). "Is it appropriate to apply hertz model to describe cardiac myocytes' mechanical properties by atomic force microscopy nanoindentation?". Micro & Nano Letters **9**(3): 153–156.

Spira, M. E. and A. Hai (2013). "Multi-electrode array technologies for neuroscience and cardiology". Nature Nanotechnology **8**(2): 83.

Steadman, B., K. Moore, K. Spitzer and J. Bridge (1988). "A video system for measuring motion in contracting heart cells". IEEE Transactions on Biomedical Engineering **35**(4): 264–272.

Stewart, M. P., J. Helenius, Y. Toyoda, S. P. Ramanathan, D. J. Muller and A. A. Hyman (2011). "Hydrostatic pressure and the actomyosin cortex drive mitotic cell rounding". Nature **469** (7329): 226–230.

Stienen, G. (2015). "Pathomechanisms in heart failure: The contractile connection". Journal of Muscle Research and Cell Motility **36**(1): 47–60.

Sugiura, S., S.-I. Yasuda, H. Yamashita, K. Kato, Y. Saeki, H. Kaneko, Y. Suda, R. Nagai and H. Sugi (2003). Measurement of force developed by a single cardiac myocyte using novel carbon fibers. Molecular and cellular aspects of muscle contraction. Springer, 381–387. doi: 10.1007/ 978-1-4419-9029-7_35. PMID: 15098684.

Sun, N., M. Yazawa, J. Liu, L. Han, V. Sanchez-Freire, O. J. Abilez, E. G. Navarrete, S. Hu, L. Wang and A. Lee (2012). "Patient-specific induced pluripotent stem cells as a model for familial dilated cardiomyopathy". Science Translational Medicine **4**(130): 130ra147–130ra147.

Sutton, M. S. J., M. A. Pfeffer, L. Moye, T. Plappert, J. L. Rouleau, G. Lamas, J. Rouleau, J. O. Parker, M. O. Arnold and B. Sussex (1997). "Cardiovascular death and left ventricular remodeling two years after myocardial infarction: Baseline predictors and impact of long-term use of captopril: Information from the Survival and Ventricular Enlargement (SAVE) trial". Circulation **96**(10): 3294–3299.

Suzuki, M., H. Fujita and S. I. Ishiwata (2003). Bio-nanomuscle project: Contractile properties of single actin filaments in an a-band motility assay system. molecular and cellular aspects of muscle contraction. Springer, **538**:103–110.

Tagami, M., Y. Nara, A. Kubota, T. Sunaga, H. Maezawa, H. Fujino and Y. Yamori (1987). "Ultrastructural characteristics of occluded perforating arteries in stroke-prone spontaneously hypertensive rats". Stroke **18**(4): 733–740.

Takeda, M., S. Miyagawa, S. Fukushima, A. Saito, E. Ito, A. Harada, R. Matsuura, H. Iseoka, N. Sougawa and N. Mochizuki-Oda (2017). "Development of In Vitro Drug-Induced Cardiotoxicity Assay by Using Three-Dimensional Cardiac Tissues Derived from Human Induced Pluripotent Stem Cells". Tissue Engineering. Part C, Methods **24**(1): 56–67.

Tan, Y., C.-W. Kong, S. Chen, S. H. Cheng, R. A. Li and D. Sun (2012). "Probing the mechanobiological properties of human embryonic stem cells in cardiac differentiation by optical tweezers". Journal of Biomechanics **45**(1): 123–128.

Tandon, R., M. Sharma, Y. Chandrasekhar, M. Kotb, M. H. Yacoub and J. Narula (2013). "Revisiting the pathogenesis of rheumatic fever and carditis". Nature Reviews Cardiology **10**(3): 171–177.

Tarnovskaya, S., A. Kiselev, A. Kostareva and D. Frishman (2017). "Structural consequences of mutations associated with idiopathic restrictive cardiomyopathy". Amino Acids **49**(11): 1815–1829.

Thibault, G., F. Amiri and R. Garcia (1999). "Regulation of natriuretic peptide secretion by the heart". Annual Review of Physiology **61**(1): 193–217.

Trial, J. and K. A. Cieslik. (2018). "Am". Journal of Physiology-Heart and Circulatory Physiology **315**: H745–H755.

Van Den Borne, S. W., J. Diez, W. M. Blankesteijn, J. Verjans, L. Hofstra and J. Narula (2010). "Myocardial remodeling after infarction: The role of myofibroblasts". Nature Reviews Cardiology **7**(1): 30.

van der Velden, J. and G. J. Stienen (2019). "Cardiac disorders and pathophysiology of sarcomeric proteins". Physiological Reviews **99**(1): 381–426.

Veselka, J., N. S. Anavekar and P. Charron (2017). "Hypertrophic obstructive cardiomyopathy". The Lancet **389**(10075): 1253–1267.

Vestergaard, M. L., S. Grubb, K. Koefoed, Z. Anderson-Jenkins, K. Grunnet-Lauridsen, K. Calloe, C. Clausen, S. T. Christensen, K. Møllgård and C. Y. Andersen (2017). "Human embryonic stem cell-derived cardiomyocytes self-arrange with areas of different subtypes during differentiation". Stem Cells and Development **26**(21): 1566–1577.

Virmani, R., A. P. Burke, A. Farb and F. D. Kolodgie (2006). "Pathology of the vulnerable plaque". Journal of the American College of Cardiology **47** 8 Supplement: C13–C18.

Wang, J. H. and J.-S. Lin (2007). "Cell traction force and measurement methods". Biomechanics and Modeling in Mechanobiology **6**(6): 361.

Weisbrod, R. M., T. Shiang, L. Al Sayah, J. L. Fry, S. Bajpai, C. A. Reinhart-King, H. E. Lob, L. Santhanam, G. Mitchell and R. A. Cohen (2013). "Arterial stiffening precedes systolic hypertension in diet-induced obesity". Hypertension **62**(6): 1105–1110.

Weiss, J. N., M. Nivala, A. Garfinkel and Z. Qu (2011). "Alternans and arrhythmias: From cell to heart". Circulation Research **108**(1): 98–112.

Westfall, M. V., K. A. Pasyk, D. I. Yule, L. C. Samuelson and J. M. Metzger (1997). "Ultrastructure and cell-cell coupling of cardiac myocytes differentiating in embryonic stem cell cultures". Cell Motility and the Cytoskeleton **36**(1): 43–54.

Wiseman, P. W., C. M. Brown, D. J. Webb, B. Hebert, N. L. Johnson, J. A. Squier, M. H. Ellisman and A. Horwitz (2004). "Spatial mapping of integrin interactions and dynamics during cell migration by image correlation microscopy". Journal of Cell Science **117**(23): 5521–5534.

Wu, J. (1991). "Acoustical tweezers". The Journal of the Acoustical Society of America **89**(5): 2140–2143.

Yin, S., X. Zhang, C. Zhan, J. Wu, J. Xu and J. Cheung (2005). "Measuring single cardiac myocyte contractile force via moving a magnetic bead". Biophysical Journal **88**(2): 1489–1495.

Yurdagul, A. Jr, A. C. Finney, M. D. Woolard and A. W. Orr (2016). "The arterial microenvironment: The where and why of atherosclerosis". Biochemical Journal **473**(10): 1281–1295.

Zalvidea, S., L. Andre, X. Loyer, C. Cassan, Y. Sainte-Marie, J. Thireau, I. Sjaastad, C. Heymes, J.-L. Pasquie and O. Cazorla (2012). "ACE inhibition prevents diastolic Ca2+ overload and loss of myofilament Ca2+ sensitivity after myocardial infarction". Current Molecular Medicine **12**(2): 206–217.

Zhou, N., -J.-J. Lee, S. Stoll, B. Ma, K. D. Costa and H. Qiu (2017). "Rho kinase regulates aortic vascular smooth muscle cell stiffness via actin/SRF/myocardin in hypertension". Cellular Physiology and Biochemistry **44**(2): 701–715.

Zhu, Y., L. He, J. Qu and Y. Zhou (2018). "Regulation of vascular smooth muscle cell stiffness and adhesion by [Ca 2+] i: An atomic force microscopy-based study". Microscopy and Microanalysis **24**(6): 708–712.

Zieman, S. J., V. Melenovsky and D. A. Kass (2005). "Mechanisms, pathophysiology, and therapy of arterial stiffness". Arteriosclerosis, Thrombosis, and Vascular Biology **25**(5): 932–943.

Zile, M. R., C. F. Baicu, J. S. Ikonomidis, R. E. Stroud, P. J. Nietert, A. D. Bradshaw, R. Slater, B. M. Palmer, P. Van Buren and M. Meyer (2015). "Myocardial stiffness in patients with heart failure and a preserved ejection fraction: Contributions of collagen and titin". Circulation **131**(14): 1247–1259.

Ramon Farré, Daniel Navajas

6.2 Respiratory Diseases

6.2.1 Outline

This chapter is focused on the mechanical properties of cells and tissues in the lungs, with different respiratory diseases investigated in vitro in animal models and in patients with respiratory pathologies. The chapter is aimed at providing a general perspective on the current knowledge available with the application of micro/nanomechanical techniques for a better understanding of lung pathophysiology. First, respiratory mechanics is briefly described and the mechanical role played by lung cells and tissues is presented. Subsequently, the main micro-scale techniques employed to measure the lung cell and tissue mechanics are discussed. Finally, the current knowledge available on cell and tissue mechanics in most relevant respiratory diseases is summarized. Potential developments of advanced micromechanical tools for diagnosis, for patient's follow-up, and future high-throughput system for drug testing are also discussed.

6.2.2 Respiratory System Mechanics

The main physiological function of the respiratory system is to allow gas exchange in the blood. Indeed, the lung is the organ where blood receives the O_2 required for supplying the whole-body metabolism, where the CO_2 produced in all organs and tissues is eliminated. This process of gas exchange is carried out by diffusion through the very thin alveolar-capillary membrane that separates the air in the lung alveoli from the blood in the pulmonary capillaries. This passive process of O_2 and CO_2 diffusion requires both a continuous flow of blood through the lung capillaries and renewal of the air in the alveoli. While blood circulates trough a two-port vascular system, with an inlet (the pulmonary artery) and outlet (the pulmonary vein), air circulates to-from

Acknowledgments: This book chapter was funded in part by the Spanish Ministry of Sciences, Innovation and Universities (PID2020-113910RB-I00-AEI/10.13039/501100011033), and by the Marie Sklodowska-Curie Action, Innovative Training Networks 2018, EU grant agreement no. 812772.

Ramon Farré, Unitat de Biofísica i Bioenginyeria, Facultat de Medicina i Ciències de la Salut, Universitat de Barcelona, Barcelona, Spain; CIBER de Enfermedades Respiratorias, Madrid, Spain; Institut d'Investigacions Biomediques August Pi Sunyer, Barcelona, Spain
Daniel Navajas, Unitat de Biofísica i Bioenginyeria, Facultat de Medicina i Ciències de la Salut, Universitat de Barcelona, Institute for Bioengineering of Catalonia, Barcelona, Spain, dnavajas@ub.edu

https://doi.org/10.1515/9783110989380-012

the airway tree through a one-port circuit having a unique conduit that is open to the atmosphere (the trachea). Such a mechanical feature of air circulation in the lung during breathing requires that this organ is compliant. Indeed, it cyclically increases its volume during inspiration to inhale fresh air and subsequently recovers the original volume to exhale the previously inhaled air after gas exchange has increased/decreased the concentrations of CO_2/O_2 in the exhaled air. The mechanical properties of the lung are determined by the contribution of its anatomical components and by their mutual interaction. On the one hand, the airways compartment is mainly resistive, with a resistance that depends on the lung volume (because inspiration enlarges the airway diameter and thus reduces resistance) and on flow (since there are nonlinearities induced by turbulent air circulation), on the other hand, lung tissues present viscoelastic properties that also depend on volume, typically exhibiting strain-hardening behavior. Remarkably, lung mechanical properties may depend on history because some breathing maneuvers modify the degree of cell contraction. For instance, lung resistance at the spontaneous end-expiration volume (achieved when breathing muscles are relaxed) is considerably modified after performing a deep inspiration, and the baseline value is not recovered after several cycles of normal breathing.

The classical methodology to study lung mechanics was developed many decades ago, at a time when there were no suitable tools to measure the mechanical properties at the microscopic scale. Hence, the respiratory system had to be analyzed by measuring the only macroscopic variables available, which described the mechanics of the whole organ (pressure, volume, and flow). Even when advanced pathophysiological research was carried out by using invasive methods (using oesophagal and gastric balloons to assess pleural and trans-diaphragmatic pressures), the mechanical variables measured were still macroscopic. This classical approach has been and is still of great importance to understand the pathophysiology of respiratory diseases. Remarkably, this approach is currently a fundamental clinical tool for the diagnosis and follow-up of patients at lung function labs and to monitor patients with lung diseases at intensive care units. However, recent developments in techniques to measure the mechanical properties of samples at the microscopic level have allowed us to investigate lung mechanics through a more basic window. Indeed, the mechanical properties of the lung can currently be probed at the cell scale, which has enormous interest since the mechanics of cells, of their surrounding extracellular matrix (ECM), and the cell-ECM mechanical cross talk are the fundamental determinants of the mechanical properties of the whole organ in both health and disease.

6.2.3 Mechanical Role of Lung Cells and Tissues

6.2.3.1 Cells

The lung has a high variety of cell types (up to 60) (Franks et al., 2008) and although it is still partially unknown, most of them experience functional alterations during respiratory diseases. Specifically, it is currently clear that mechanical changes experienced by some lung cell types are relevant in prevalent and severe respiratory diseases, potentially compromising the normal function of the organ. The cells that present different mechanical behavior in healthy and diseased lungs are widely distributed at the different structural parts of the organ: the conducting airways and vessels, the alveolar-capillary membrane, the lung parenchymal tissue, and the blood circulating through the lung capillaries.

The cells in the airways that play a more relevant mechanical role in respiratory diseases are epithelial and smooth muscle cells. Bronchial epithelial cells form a monolayer, covering the luminal surface of the airway, and are crucial for maintaining a functional and safe barrier, separating the internal part of the airway from the circulating air. Smooth muscle cells are contractile components within the airway walls, which, depending on their degree of contraction, regulate the section of the conduit lumen and hence its air flow resistance. The alveolar-capillary membrane has two main cell types playing a relevant mechanical role. Alveolar epithelial cells form a monolayer at the side of the membrane, which is in contact with the alveolar gas. On the other side of the membrane, there is a monolayer of endothelial cells in contact with the blood circulating through the capillaries. Both monolayers are important since their mechanical integrity determines the homeostatic permeability of the membrane for water, different types of molecules, and cells. As such, this barrier function is crucial to protect the lung from toxic pollutants and microbes transported by the inhaled air. It is interesting to note that the integrity of the alveolar-capillary membrane is mechanically challenged continuously, since it is subjected to the cyclic stretch associated with inspiration and expiration. The type of cells that play a more relevant mechanical role in lung tissue are fibroblasts (and differentiated myofibroblasts), since, when pathologically activated, these cells may substantially remodel and stiffen the lung tissue by secreting ECM components, such as collagen. Circulating cells in the blood, namely erythrocytes and leukocytes, have a size that is slightly higher than the diameter of lung capillaries. Therefore, they should present a physiological stiffness, allowing them to deform for circulating through the lung vascular circuit. Accordingly, stiffening of these cells in lung diseases may compromise blood circulation and thus the gas exchange.

As discussed in more detail later in this chapter, mechanical alterations in the previously mentioned cells are key in the most prevalent and severe respiratory diseases: acute lung injury (ALI), asthma, lung fibrosis, chronic obstructive pulmonary disease (COPD), and pulmonary arterial hypertension (PAH). It should be noted that in addition to the structural cells that normally participate in the pulmonary homeostatic balance, neoplastic cells should also be considered in lung cancer, which is among the most devastating malignancies in terms of prevalence and mortality. As in other types of cancer, the mechanical properties of lung cells are crucial since they may regulate the migration and metastatic potential of malignant cells.

6.2.3.2 Tissues

Tissues in each of the lung anatomical components – airways, alveoli, and vessels – have a well-defined architecture built with cells and ECM. Whereas in the past, the ECM was viewed as just a passive component aimed at providing structural support to cells, it is now clear that the ECM is an active biomechanical component. Indeed, a constant bidirectional cross talk between the ECM and the cells contributes to regulate lung homeostasis (Zhou et al., 2018). From a biomechanical viewpoint, it is interesting to note that the response of lung tissues to external challenges – either mechanical, such as cyclic stretch, or biochemical, such as histamine provocation – should be considered at different time scales. Whereas the short-time mechanical response is mainly due to ECM properties, with small cell contribution, the longer time response is principally determined by cells since they respond to the external stimuli by modifying the ECM composition and thus its mechanical properties (Suki and Bates, 2008). ECM biomechanics is mainly determined by three types of molecular components: collagen, elastin, and proteoglycans. Collagen is the most abundant of them and is of major relevance for the biomechanical homeostasis of the lung (Cavalcante et al., 2005). The lung ECM is continuously being remodeled by lung cells in a process that involves a physiological balance between synthesis, deposition, degradation, and clearance of ECM components (Haak et al., 2018). To highlight how dynamic is the lung ECM, it is worth noting that 10% of lung collagen is newly synthetized every day and 40% of it is immediately degraded. Given the anatomical and functional variety of the structural components across the lung, the composition and mechanical properties of the ECM are not uniform. Indeed, the ECM is adapted to provide the specific biomechanical properties required by each anatomical compartment, for example, the relatively stiff airway walls to maintain their dimensions under different stresses as compared to the compliant alveolar walls to allow breathing (Haak et al., 2018).

Interestingly, alteration of the composition and biomechanics of lung ECM also occurs as a physiological process during ageing. Indeed, a recent study using human lung samples from healthy subjects covering a wide range of ages (11–60 years old) reported that. In contrast, cell mechanics did not change with age; ageing showed a trend towards increasing ECM deposition and traction forces, suggesting that tissue mechanics may contribute to the well-known age-related alterations of lung function (Sicard et al., 2018). Despite ageing, alteration of the normal ECM remodeling, and thus changes in its mechanical properties is one of the main characteristics in most relevant respiratory diseases. For instance, increase in parenchymal collagen content in lung fibrosis (Bidan et al., 2015), abnormal deposition and cross-link of collagen and the related breakdown of elastic laminae in vessel walls in PAH (Thenappan et al., 2018), alveolar wall disruption in COPD emphysema (Bidan et al., 2015), or fibrotic thickening of the airway wall in asthma (Seow, 2013). Given that the different lung diseases present specific alterations, it is important to investigate the mechanics of lung tissues from a dual perspective (Tschumperlin et al., 2010) – on the one hand, to use a microscopic approach to understand the mechanical microenvironment sensed by each individual cell at focal adhesion level, because this micron-size scale is the one cell determining mechanosensing and mechanoresponse, and on the other hand, to use a macroscopic perspective for featuring how changes in ECM and tissue mechanics impact whole lung mechanics and thus in ventilation and effectiveness of gas exchange. As will be discussed later in this chapter, techniques for such a multiscale assessment of lung tissue and ECM mechanics are currently available.

6.2.4 Techniques to Measure the Mechanics of Lung Cells and Tissues

6.2.4.1 Cells

The very different nature of cells modulating lung mechanics and the complexity of lung tissues where they are residing in vivo require the use of a variety of techniques to explore lung cell mechanical properties in healthy conditions and in different respiratory diseases. Among the available mechanical techniques for adherent cells, atomic force microscopy (AFM) and magnetic twisting cytometry (MTC) allow us to measure cell stiffness/viscoelasticy, whereas traction force microscopy (TFM) provides an assessment of cell contraction. Typical techniques for measuring stiffness/deformability in nonadherent cells are micropipette aspiration and microchannel circulation. However, it should be mentioned that each of these techniques, which are widely used in mechanobiology, presents specific advantages and limitations when applied to measure lung cell mechanics.

MTC was an early technique used to investigate the viscoelasticity of lung cells (Berrios et al., 2001, Fabry et al., 2001, Puig-de-Morales et al., 2004). The method is based on adhering magnetic microbeads, coated with ligands (such as RGD), to the cytoskeleton through focal adhesions.

Once attached to the cell surface, beads are magnetized into a horizontal direction and subsequently subjected to a perpendicular low-amplitude oscillatory magnetic field that induces an oscillatory twisting torque to the beads. As each bead is attached to the cell at one area of its surface, beads cannot freely rotate but only oscillate around the attachment zone with an amplitude that is determined by the stiffness of the cell.

Since the amplitude of bead oscillations can be optically measured and the magnetic torque applied is known, cell stiffness can be derived. Since each MTC measurement is carried out on several cells simultaneously in a same culture plate, and beads are randomly attached to different points of the cell surface, an advantage of the technique is that the biological variability of the measurements is reduced as compared with techniques, such as AFM, in which a given measurement explores just one specific site in a single cell. Nevertheless, a remarkable potential limitation of MTC is that attaching beads to the cell remodels the cytoskeleton (Deng et al., 2004), hence potentially modifying the baseline mechanical status of cells and their response to pathophysiological challenges. Moreover, as the contact area of the bead-cell attachment is unknown, the actual magnitude of Young's modulus of the cell cannot be obtained. However, regardless of their limitations, MTC has provided important information on the mechanical properties of lung smooth muscle cells (Deng et al., 2004, Puig-de-Morales et al., 2004, Stamenovic et al., 2004), lung endothelial cells and neutrophils (Wang et al., 2001, Suresh et al., 2019), alveolar epithelial cells (Berrios et al., 2001, Trepat et al., 2004–2006, Puig et al., 2009, 2013, Lan et al., 2018), and lung cancer cells (Coughlin and Fredberg, 2013). It should be mentioned that MTC shares limitations with AFM (which was mentioned in Chapter 3.1) regarding the stiffness of the substrate and the 2D nature of measurements.

The theoretical principles and general methodological issues of AFM have been described in detail in previous chapters. Since this technique is based on contacting and indenting the cell surface, it is particularly well suited for measuring the mechanics of cells that in vivo form interface monolayers, such as lung epithelial and endothelial cells. Moreover, as the contact area between the tip and the cell surface can be accurately estimated, AFM provides a measurement of the Young's modulus of the cell. Accordingly, application of AFM to conventional in vitro 2D cultures provides a model that mimics the native cell environment reasonably well when the aim is to study the alveolar-capillary barrier, the bronchial epithelium, and the vessel endothelium in respiratory diseases. There is, however, a limitation in the technique since AFM requires that the substrate to which the cell is adhered is much stiffer than cell stiffness. This is not a problem in conventional AFM measurements carried out in plastic or glass culture plates, even when coated with ECM, since in that case, the cell substrate is virtually rigid. Nevertheless, such a substrate has a stiffness that is in orders of magnitude higher than the stiffness of the natural lung ECM substrate to which cells are placed in vivo (Melo et al., 2014a). Given that lung alveolar and endothelial cells, as virtually all cells, are mechanosensitive, the mechanical properties of cells obtained in conventional AFM measurements (and also in MTC) may differ from actual in vivo cell stiffness. It should also be considered that AFM (as MTC) has a limitation to assess the mechanics of lung cells that in vivo are in a 3D microenvironment (e.g., smooth muscle cells, fibroblasts, and cancer cells). Indeed, as cells sense whether their microenvironment is 2D or 3D and behave accordingly, results from measuring cell mechanics in 2D culture plates may differ from the actual in vivo mechanical properties, and their response to experimental interventions may also be different. The limitations of substrate stiffness and 2D culture are usually addressed by arguing that their effects are minor in experiments investigating the consequences of a given biological challenge (e.g., inflammatory cytokines) or drug treatment on cell mechanics, since both pre- and post-intervention data are obtained under the same measuring conditions. Nevertheless, this reasoning holds true as far as the mechanical response of the cell to the specific intervention under test is not inhibitory or synergistically mediated by substrate stiffness or 2D/3D microenvironment. Regardless of their limitations, AFM is an extremely useful technique to provide insight into the mechanical properties of the relevant types of lung cells: epithelial cells (Alcaraz et al., 2003, Azeloglu et al., 2008, Waters et al., 2012, Wilhelm et al., 2014, Oliveira et al., 2019a, 2019b), endothelial cells (Whitlock et al., 2008, Birukova et al., 2009, Wiesinger et al., 2013, Job et al., 2016, Viswanathan et al., 2016, Wang et al., 2017, Merna et al., 2018, Schimmel et al., 2018), smooth muscle cells (Smith et al., 2005), fibroblasts (Gabasa et al., 2017, Jaffar et al., 2018), and cancer cells (Rotsch et al., 2001, Bulk et al., 2017, Iida et al., 2017, Lartey et al., 2017, Sobiepanek et al., 2017, Prina-Mello et al., 2018, Zhang et al., 2018, Bobrowska et al., 2019). It is noteworthy that a recent technical development (Jorba et al., 2019) allows AFM to measure the mechanics of cells subjected to stretch, which is not possible with conventional settings.

Such measurements under stretching conditions are fundamental when investigating lung cells since in vivo they are continuously subjected to cyclic deformations, owing to breathing, and it has been shown that stretch is a key mediator of lung cell mechanics (Trepat et al., 2006, Gavara et al., 2008, Lan et al., 2018).

While MTC and AFM measure cell stiffness, TFM is aimed at assessing the force that cells exert on their substrate. This is obviously an important variable for lung cells with specific contractile function such as smooth muscle cells, but it is also important for other lung cells (epithelial, endothelial, fibroblasts, and cancer cells), since all of them have actomyosin machinery. Indeed, traction forces in such lung cells modulate important physiological functions, such as mechanosensing, keeping thigh cell-cell and cell-substrate adhesions in the epithelial and endothelial monolayers, and in migration and regulating the ECM composition of lung parenchyma. Although it has been reported that contractile forces are correlated with cell stiffness (Schierbaum et al., 2019), TFM is of much interest to specifically measure the mechanical effects of the cell contractile machinery. The technique is based on culturing cells on a soft substrate, with embedded fluorescent microbeads. The position of each microbead is determined with a fluorescence microscope at different times: at baseline, after challenging cells (e.g., cytokine or drug), and finally after removing the cells from the substrate (e.g., by trypsinization). Such sequential bead imaging allows us to observe how cell tractions on the substrate have locally deformed it at each measuring time, thereby obtaining deformation maps at microscopic scale. As the stiffness of the substrate (ideally linear, isotropic, and homogeneous) is well known (usually assessed by AFM), application of a suitable algorithm allows us to transform the bead displacement maps into quantitative traction force maps.

Although early TFM implementation required that measured cells are isolated (no cell-cell interaction), current TFM developments are suited for confluent cells also (Serra-Picamal et al., 2015). A remarkable interest of TFM is that cells are cultured, not in plates with nonphysiological stiffness but in soft substrates with a stiffness that can theoretically mimic the actual stiffness of the cell niche in vivo. To this end, the only technical limitation is that the substrate stiffness should be matched with the magnitude of cell forces so that substrate deformation remains in the linear range (i.e., sufficiently small deformations) and can be measured with an acceptable resolution (i.e., sufficiently high deformations). Interestingly, given that conventional TFM (such as MTC and AFM) is designed for 2D cultured cells, the technique is particularly adequate for investigating traction forces in lung epithelial and endothelial cells (Gavara et al., 2006, 2008, Puig et al., 2009, Oliveira et al., 2019b). Moreover, TFM can be used to investigate the contraction of endothelial cells when subjected to shear stress, mimicking physiological conditions (Perrault et al., 2015). However, when TFM is applied in cells whose physiological environment is not 2D – such as smooth muscle cells (Stamenovic et al., 2004, Lan et al., 2018, Lin et al., 2018) or lung fibroblasts (Caporarello et al., 2019, Reed et al., 2019), or lung cancer cells (Mierke et al., 2011, Kraning-Rush et al., 2012, Vizoso et al., 2015) – the results obtained must be interpreted carefully, taking into account that a change in the dimension of the microenvironment could modify cell forces. Remarkably, recent efforts are focused on developing a 3D version of TFM, which will allow us to measure the forces exerted by cells embedded in realistic ECM matrices (Koch et al., 2012, Colin-York et al., 2019, Holenstein et al., 2019).

The stiffness/deformability of cells circulating in lung vessels (e.g., leukocytes and erythrocytes) could be measured by MTC (Wang et al., 2001) and AFM (Roca-Cusachs et al., 2006, Berthold et al., 2015) using procedures to slightly adhere the cells on the substrate. However, to better mimic the nonadherent in vivo nature of these cells, their mechanics may be measured by techniques requiring no cell adhesion (and hence avoiding a certain degree of cytoskeleton remodeling). Ektacytometry uses laser diffraction viscometry to measure erythrocyte deformability under increasing shear stress or under osmotic gradient at a given shear stress (Parrow et al., 2018). Micropipette aspiration was an early technique to directly assess circulating cells deformability (Skoutelis et al., 2000). This technique consists in approaching the cell and contacting it with a pipette, with a diameter smaller than cell dimensions. When the pipette is in contact with the cell surface, application of a negative pressure induces a curvature of the cell surface that is aspired. Local deformability can be quantified from the relationship between the deformation radius measured optically and the negative pressure applied. The technique also allows us to characterize whole cell deformability by measuring the negative pressure required to achieve full aspiration of the cell into the micropipette. While this technique is physiologically realistic since it subjects the cell to a challenge similar to the one experienced when it should circulate through a narrow lung capillary, it is time consuming and requires precise single cell manipulation. A variant of this early technique has been developed more recently thanks to the advancement in fabricating microchannels (Tanaka et al., 2001, Inoue et al., 2006, Morikawa et al., 2014, Preira et al., 2016). With this approach, a number of circulating cells are directed to microchannels with different diameters and lengths, and the time required to cross the microchannel – which mimic the dimension of lung capillaries – provides an index of cell deformability. This setting can be implemented in microchips and data observed by computer-controlled microscopy, potentially providing automatic high-throughput settings for different clinical applications. For instance, the deformability of blood circulating cells, freshly isolated from a patient, could be tested either in a conventional medium, in healthy plasma, or in the plasma of the patient. Also, the capability of the plasma from a given patient to modify the deformability of healthy circulating cells can be determined (Preira et al., 2016).

6.2.4.2 Tissues

Native lung tissues and ECM (decellularized lung scaffold) can be explored from both microscopic and macroscopic perspectives (Polio et al., 2018). Macroscopic viscoelasticity is typically measured by applying tensile stretch to strips or rings of tissue (e.g., lung parenchyma and pulmonary vessels) (Jorba et al., 2019). Microscopic assessment of lung tissue samples is usually carried out by AFM (Liu and Tschumperlin, 2011). Since the main features of this technique are discussed in other chapters, here, we only mention specific issues regarding application of AFM to lung tissues.

The lung is extremely nonhomogeneous, with components such as cells, airspaces, and ECM that present very different mechanical properties and that are placed at close distance from each other. Accordingly, the dimensions of the cantilever tip indenting the sample are particularly relevant. Specifically, pyramidal tips provide more precise local resolution whereas spherical tips allow for sensing a wider, nonhomogeneous area of the sample at the cost of lower resolution. The pros and cons of the different AFM tip shapes in lung measurements has been discussed (Sicard et al., 2017, Jorba et al., 2019) Another issue that particularly affects the measured stiffness in lung samples, also related to their inhomogeneity, concerns the comparison and interpretation of the stiffness values obtained from macroscopic and microscopic measurements. Indeed, while a technique such as AFM measures the local stiffness of the sample, techniques such as tensile stretch applied to macroscopic strips provide the bulk stiffness of the whole network, which depends both on the 3D arrangement of fibers and on the local stiffness of the material that constitutes the fibers. An example of the disparity between the stiffness data obtained in lung ECM is depicted in Figure 6.2.4 where it can be observed that the elastic modulus obtained by tensile stretch is one magnitude lower than the values obtained by AFM (regardless of the tip geometry) (Jorba et al., 2019). This observation applies also when stiffness data obtained from AFM are compared with the elastic properties of the whole lung in vivo, as measured clinically from the pressure-volume relationship (lung elastance). (Uriarte et al., 2016).

A very relevant issue to optimally characterize lung tissue mechanics by AFM is pre-stress. Indeed, in vivo whole lung tissues (and thus the ECM) are pre-stressed since they are subjected to a baseline tension as a result of the equilibrium between the chest wall and the lung elastic forces at physiological lung volumes. By contrast, lung samples probed with conventional AFM settings are explored in the absence of pre-stress, which may considerably affect the measured results because of the strain-hardening behavior of the lung tissues. In particular, it is important to note that the Young's modulus of the lung tissue and the ECM is commonly measured at levels of stretch different from those experienced during normal breathing. To address this technical issue, a setting has been recently designed to allow AFM measurements at different levels of tissue stretch, showing that both macroscopic and microscopic stiffness increase dramatically with tissue stretch (Jorba et al., 2019).

6.2.5 Cell and Tissue Mechanics in Respiratory Diseases

6.2.5.1 Acute Lung Injury

ALI and the associated acute respiratory distress syndrome (ARDS) are very severe conditions, deeply compromising patient survival in intensive care units. Accordingly, much effort is devoted to better understand the mechanisms involved and to find potential therapies. Among the multidisciplinary research carried out, cell mechanics is relevant since the disease is characterized by a loss of mechanical integrity in the alveolar-capillary membrane, resulting in abnormal increased permeability for water (lung edema), proteins, and inflammatory cells. Loss of membrane integrity is potentiated by systemic inflammatory mediators, bacterial LPS from the pneumonia/ sepsis (usually associated with ALI/ARDS), and also by the considerable stretch caused by the high level of mechanical ventilation required for achieving enough gas exchange in the lungs of these patients. The mechanical effects of stretch on alveolar cell viscoelasticity was investigated using MTC and TFM (Trepat et al., 2004, 2006, Gavara et al., 2008). These techniques have also been used to understand how inflammatory mediators, which play a relevant role in ALI/ARDS, such as thrombin and histamine, modify alveolar epithelial cell mechanics (Trepat et al., 2005, Gavara et al., 2006, Suresh et al., 2019).

AFM has been employed to assess the effect of hyperoxia – a potential consequence of O_2-enriched mechanical ventilation in patients with ALI/ARDS – on alveolar epithelial cells (Wilhelm et al., 2014). Moreover, recent AFM data show that bacterial LPS considerably modulates the stiffness in these lung cells (Oliveira et al., 2019a). Interestingly, all these four challenges present in ALI/ARDS (stretch, inflammatory mediators, hyperoxia, and LPS) result in stiffening/contraction of the alveolar epithelial cells, potentially contributing to imbalance the cell-cell and cell-matrix force equilibrium that maintains cell monolayer integrity along the breathing cycle.

The potential effect on cell mechanics of drugs to alleviate ALI/ARDS has also been investigated. For instance, dexamethasone, a widely used anti-inflammatory drug, reduced the stiffening/contraction induced by thrombin in alveolar epithelial cells (Puig et al., 2009), and a similar effect was observed when, before thrombin challenge, cells were pretreated with activated protein C (Puig et al., 2013). Moreover, data from AFM in the pulmonary endothelium report significant mechanical effects of the barrier-disrupting thrombin and barrier-enhancing sphingosine 1-phosphate (Whitlock et al., 2008) and of sepsis-associated mediators (thrombin, LPS, TNF-α) (Wiesinger et al., 2013). The effects of treatment drugs on the mechanics of the lung epithelium have also been further investigated by AFM. On the one hand, Abl kinase inhibition by imatinib diminished the elastic modulus at both the cytoplasm and cell periphery (Wang et al., 2017). On the other hand, how lung endothelial glycocalyx stiffness is modulated by different concentrations of resuscitation colloids (albumin and hydroxyethyl starch) has been investigated (Job et al., 2016).

Measuring the mechanics of circulating (i.e., nonadherent) cells in ALI/ARDS is of interest since stiffening of these cells may compromise correct lung perfusion, and hence gas exchange, in patients. The early observation that neutrophils in patients with ARDS were stiffer than in healthy controls (Skoutelis et al., 2000) was followed by more detailed studies, indicating that in ALI patients a drug-inhibiting neutrophil elastase (sivelestat) reduced leukocyte stiffness and increased lung oxygenation (Inoue et al., 2006). More recently, it has been reported that serum from patients with ARDS was able to rapidly stiffen the control of neutrophils and monocytes as a result of the cytokines in the patient serum (Preira et al., 2016). Indeed, using antibodies, blocking IL-1β, IL-8, or TNF-α significantly reduced the stiffening effect of the patient's serum, which points to potential therapeutic applications (Preira et al., 2016). In a similar line, it has been reported that in patients with sepsis (which is common in ALI/ARDS), the deformability of leukocytes correlates with disease severity and, therefore, this cell mechanics index could be a biomarker for patient follow-up (Morikawa et al., 2014).

Given the important role that cell mechanics plays in ALI/ARDS, most studies on micromechanics have been focused on cells and little is known on changes in tissue mechanics in this lung disease. However, a relatively recent paper has used

AFM to investigate how the ECM is remodeled, and thus how tissue stiffness is changed, in living lung slices, from an animal model of ALI, induced by LPS (Meng et al., 2015). This bacterial agent increased perivascular stiffness and boosted expression of main ECM proteins (e.g., fibronectin and collagen) and cross-linkers (e.g., lysyl oxidase). The authors also explored the effects of an analogue of lipoxin in reducing the LPS-induced hardening and found that this drug had softening local effects and, importantly, restored lung compliance in vivo (Meng et al., 2015).

6.2.5.2 Asthma

Asthma is a very prevalent disease associated with an increase in airway obstruction due to abnormal reduction in bronchial lumen section. Regardless of the cause inducing bronchial hyperreactivity and thus triggering asthma attacks (e.g., exercise, allergy, drugs, and pollutants) the main mechanical process that increases airway resistance is the abnormal contraction of the smooth muscle cells in the airway wall. Accordingly, measuring the mechanics of these cells provides insight into the mechanisms of asthma and on possible treatments. How the rheological properties of airway smooth muscle cells are associated with cytoskeletal contractile stress was investigated by combining the stiffness and contraction forces data obtained by MTC and TFM, respectively (Puig-de-Morales et al., 2004, Stamenovic et al., 2004). In addition, the complex elastic modulus of airway smooth muscle cells was measured with AFM, providing information on the fast cell stiffening induced by the contractile agonist 5-hydroxytryptamine (Smith et al., 2005). This study showed that dynamic actin polymerization plays a key role in cell stiffness and contraction, both in resting baseline and during activation in these cells (Smith et al., 2005). Very recently, TFM data have shown that after subjecting airway smooth muscle cells to a transient stretch simulating a deep inspiration, the cytoskeleton experienced fluidification, and that this phenomenon was strongly modified by cofilin – an actin-severing protein (Lan et al., 2018). Accordingly, cofilin plays an important role in stretch-induced cytoskeletal fluidization and, therefore, may partially account for bronchodilation after deep inhalation (Lan et al., 2018).

Also, using TFM, it has been recently reported that smooth muscle cell mechanics may be substantially altered by a pollutant agent, such as ZnO nanoparticles (Lin et al., 2018). Specifically, cell contraction was reduced as the concentration of ZnO increased. both in baseline and after stimulation with KCl (Lin et al., 2018). Interestingly, these results are consistent with the reduction of stiffness (AFM) and contraction (TFM) that was recently reported when alveolar epithelial cells were challenged with nanoparticles of Fe_2O_3 and of TiO_2 (Oliveira et al., 2019b). These data indicate that nanoparticle exposure may induce alterations in the lung through changes in cell mechanics associated with cytoskeleton remodeling. It is worth noting that asthma does not only alter the mechanics of airway-resident cells but also of the

circulating neutrophils. In particular, assessing neutrophil chemotaxis (which involves activation of the contractile machinery for migration) with a microfluidic-
based handheld device allowed discriminating asthma from allergic rhinitis, within
minutes from a simple drop of a patient's whole blood (Sackmann et al., 2014).

Tissue mechanics in asthma has also been investigated by means of different
techniques. An early study employed tensile stretch tests to analyze the resistance,
elastance, and hysteresivity of guinea pig subpleural lung strips, subjected to two
well-known challenges in asthma: step stretch and histamine (Romero et al., 2001).
Interestingly, the mechanical changes induced by both challenges were towards tissue
stiffening, but with some differences among them. The frequency dependence analysis
of the data with a constant-phase model allowed the authors to conclude that pneumoconstriction significantly alters the structure of the connective matrix (Romero et al.,
2001). Much more recently, AFM has been used to explore the micromechanics of bronchial tissue obtained from biopsy in asthma patients (Zemła et al., 2018). The main
finding of this methodological study was that freezing–thawing these human samples
preserved tissue mechanical properties, which may facilitate the clinical application of
AFM for routine testing of bronchial biopsies (Zemła et al., 2018).

6.2.5.3 Lung Fibrosis

Clinically interesting data on the micromechanics of cells from patients with lung fibrosis have been reported very recently. In an AFM study on lung fibroblasts, from patients with pulmonary fibrosis and normal controls, which were cultured on profibrotic
conditions (stiff substrate and TGF-β), the authors found that myofibroblasts from lung
fibrosis patients were stiffer and expressed more fibrillar collagen than control fibroblasts (Gabasa et al., 2017). As these cells also exhibited enhanced focal adhesion kinase (FAK) activity, the authors inhibited FAK and found decreased fibroblast stiffness
and collagen expression, suggesting that FAK could be a target to recover the physiological mechanobiology in human lung fibrosis (Gabasa et al., 2017). In another AFM
study, the authors also found that fibroblasts from patients with pulmonary fibrosis
were stiffer, when compared with healthy controls (Jaffar et al., 2018). The fibroblasts
from fibrotic patients were not mechanoresponsive when cultured on soft and hard surfaces, which was in contrast to healthy fibroblasts (Jaffar et al., 2018). However, as fibrotic fibroblasts augmented their cytoskeleton responses to TGF-β1, the authors
suggested that defective cytoskeletal machinery was not a cause for the lack of mechanical response in the fibrotic patient cells (Jaffar et al., 2018).

The mechanical properties of tissue and ECM in lung fibrosis were investigated.
Alterations in the ECM and its mechanics were first described in rodent models of
bleomycin-induced pulmonary fibrosis. Using tensile-stretch testing of fibrotic lung
strips with different oscillation frequencies, the increase in both tissue resistance and
elastance was observed, as compared to healthy controls (Ebihara et al., 2000). This

study also showed a significant correlation between bi-glycan and all mechanical parameters, suggesting that changes in proteoglycans are important in modifying the lung tissue mechanics in fibrosis (Ebihara et al., 2000). More recently, a study using AFM in lung tissue slices of rodent models of fibrosis has reported a 6-fold stiffening as compared to healthy controls (Liu et al., 2010). Interestingly, culturing fibroblast on the substrates of controlled stiffness, which were within the physiological range, considerably activated cells from a quiescent state to increased matrix synthesis and also reduced the expression of matrix proteolytic genes (Liu et al., 2010). Inhomogeneous and considerable stiffening has also been observed when focusing on the ECM of fibrotic lungs with AFM. Indeed, decellularized lungs from mice, with bleomycin-induced fibrosis, showed increased ECM stiffness, which was dependent on fibrosis severity, as classified by histologic criteria, in the different sites of the lung scaffold (alveolar septa, visceral pleura, and vessels tunicae adventitia and intima) (Melo et al., 2014b). Consistently, AFM data from decellularized human lungs were stiffer and less homogeneous in patients with lung fibrosis than in healthy subject lungs (Booth et al., 2012).

6.2.5.4 COPD

Data currently available on lung cell and tissue micromechanics in COPD are scarce. A recent report on cell mechanics has clinical interest since it described that erythrocyte deformability, measured by ektacytometry, in COPD patients was lower than in healthy controls, and red blood cells deformability was improved after patients were treated for COPD exacerbations (Ugurlu et al., 2017). Regarding lung tissue mechanics, an early study applied tensile stretch to lung strips of rats with elastase-induced emphysema (a phenotype in COPD), while imaging with immunofluorescent-labeled elastin-collagen deformations (Kononov et al., 2001). As compared to healthy lungs, strips challenged with elastase showed thickened elastin and collagen fibers. The new fibers experienced larger deformations than in controls, with reduced threshold for collagen failure, thereby reducing the mechanical stability of the lung. The authors concluded that tissue forces during breathing may induce failure of the remodeled ECM, thus contributing to emphysema progression (Kononov et al., 2001). Very recently, it has been proposed to test lung tissue mechanics with an approach based on precision-cut slices from patients with mild to moderate COPD and the controls undergoing surgery for lung cancer (using tumor-free tissue far from the tumor site) (Hiorns et al., 2016). Following this approach, lung tissue samples were challenged with methacholine (a drug inducing airway hyper-responsiveness) and with LPS (Maarsingh et al., 2019). A mathematical model used to interpret tissue images from lung slices indicated that COPD weakens matrix mechanics, enhancing stiffness differences between airways and lung parenchyma. The authors concluded that in COPD, there is a relationship between the small airway hyperresponsiveness and reduction in retraction forces in the parenchyma and biomechanical changes in the airway wall (Maarsingh et al., 2019).

6.2.5.5 Pulmonary Arterial Hypertension

Application of micromechanical techniques in PAH has started very recently. AFM was applied to investigate the stiffness of the vascular tissue in lung vessel samples from both a PAH rat model (monocrotaline) and from patients (Bertero et al., 2016, Liu et al., 2016), which showed that pulmonary vascular tissue stiffness was increased as compared to their corresponding controls in both rats and humans. Moreover, distal vascular matrix stiffening was an early mechanobiological regulator of the disease progression, and it was found that cyclooxygenase-2 (COX-2) is a determinant in the mechanisms regulating stiffness-dependent vascular cell activation (Liu et al., 2016). In another study, it was also found that pulmonary arteriolar tissue was increased in the PAH rat model as compared with controls (Bertero et al., 2016). Moreover, the authors reported that ECM stiffening resulted in mechanoactivation of the transcriptional coactivators, YAP and TAZ (Bertero et al., 2016). A more recent study has also employed AFM to investigate the role of Deleted in liver cancer 1 (DLC-1) – a regulator for cell proliferation, adhesion, and migration involved in PAH – on the stiffness of endothelial cells, showing that silencing of endothelial DLC-1 significantly reduced cellular stiffness (Schimmel et al., 2018). Interestingly, a study has focused on the mechanics of circulating cells, specifically erythrocyte deformability, in PAH patients and in healthy controls. Patient erythrocytes were significantly stiffer than in controls, and blood rheological parameters correlated with conventional biomarkers of PAH severity (Yaylali et al., 2019).

6.2.5.6 Lung Cancer

The mechanical properties of tumor cells, ECM, and the tissues in lung cancer exhibit similar basic features as malignancies in other organs (Chapter 6.3) (Rianna et al., 2018). At a macroscopic scale, the mechanical properties of lung tumors can be measured by elastography, as described for other cancers (Chapter 6.3). Indeed, recent data suggest that elastography can be incorporated into endobronchial ultrasound-guided transbronchial needle aspiration (EBUS-TBNA) – which is a widely used clinical tool for diagnosing and staging mediastinal lymph nodes to assess tumor tissue compressibility, and thus stiffness. Preliminary studies strongly suggest that elastography – combined with EBUS – is feasible to classify mediastinal lymph nodes (Gu et al., 2017) and to help in predicting lymph node malignancy in patients with lung cancer (Verhoeven et al., 2019). Moreover, a recent proof-of-concept study suggests that transthoracic elastography can be a noninvasive tool to differentiate benign and malign subpleural masses (Ozgokce et al., 2018). Regarding other noninvasive techniques for assessing tissue stiffness in lung cancer, it is interesting to note that in rats, dynamic CT imaging of pulmonary nodules correlates with physical measurements of stiffness. Specifically, the volumetric ratios of pulmonary nodules, derived from breathing-gated CT images, exhibited a significant correlation with the Young's modulus of the same nodules, measured by AFM after excision (Lartey et al., 2017). Although still at the stage of preliminary development and testing, it is expected that determining the macroscopic stiffness of lung cancer tissue by minimally-invasive techniques could be clinically useful in the future.

However, most information currently available on the mechanics of cells and ECM in lung tumors has been obtained at the research level using mainly AFM. A first important fact to highlight is the relationship between cell softening and malignant progression of human lung cancer cells. A study focused on six human non-small cell lung cancer (NSCLC) cells – the most common in lung cancers (Chen et al., 2014) – showed that low cell motility was strongly associated with high cell stiffness, and opposite results were found in high motility cancer cells (Iida et al., 2017).

Interestingly, this study reported that activation of AXL receptor tyrosine-kinase induced cell softening and promoted malignant progression in these lung cancer cells, thus playing a key biophysical role (Iida et al., 2017). A recent work has investigated the mechanisms involved in the elevation of rigidity and invasiveness of TGFβ-stimulated NSCLC cells, and its correlation with upregulation of cytoskeletal and motor proteins, suggesting that mediators of elevated cell stiffness and migratory activity could be promising agents for pharmaceutically reducing lung cancer progression (Gladilin et al., 2019). It should be noted, however, that stiffness of lung cancer cell depends on whether its measurement is carried out when cells are cultured on conventional 2D culture plates or in 3D soft environments. This fact was recently reported by measuring the stiffness of lung adenocarcinoma cells (A549) by AFM nanoindentation (Prina-Mello et al., 2018). The authors found that lung cancer cells were stiffer when cultured on 2D glass substrates, compared to 3D soft gels. Moreover, in addition to stiffness, other important cell properties, such as morphology, topography, and biochemical signatures, were strongly influenced by culturing in 2D or 3D microenvironments. These findings stress the importance of improving cell culture conditions to more realistically mimic the in vivo lung cancer cell microenvironment. In this connection, a second important fact in the biomechanics in lung cancer cells is that the stiffness of cell substrate plays an important role in modulating cellular response in lung tumors (Tilghman et al., 2010). Accordingly, human adenocarcinoma cells (A549) cultured on collagen-coated PDMS substrates with increased rigidity exhibited increased cells stiffness and slower migration (Shukla et al., 2016).

Several studies have investigated the mechanisms involved in the effects of substrate stiffness in lung cancer cells. For instance, slower cell migration correlated with decreased levels of phosphorylated focal adhesion kinase (FAK) and paxillin, indicating that substrate stiffness modulates lung cancer cell migration via focal adhesion signaling (Shukla et al., 2016). Moreover, by using polyacrylamide hydrogels with stiffness of 2 and 25 kPa, it was also shown that a stiff substrate enhanced PD-L1 expression in HCC827 lung adenocarcinoma cells and that this occurred via actin-dependent mechanisms (Miyazawa et al., 2018). Furthermore, tumor suppressor RASSF1A has recently been shown to modulate deposition of ECM, tumor stiffness, and metastatic dissemination in vitro and in vivo, and has been identified as a clinical biomarker associated with ECM mechanics that increases cancer stemness and risk of metastasis in lung cancer (Pankova et al., 2019). Interestingly, tumor-associated fibroblasts (TAF) – which are known to play an important role in the progression of NSCLC cells – are also modulated by substrate stiffness. Indeed, when culturing fibroblasts on substrata with normal- or tumor-like stiffness, it was shown that in control fibroblasts from nonmalignant tissue matrix, stiffening alone increased fibroblast accumulation (Figure 6.2.10) in a process driven by β1 integrin, mechanosensing through FAK (Puig et al., 2015). Moreover, matrix stiffening induced a larger TAF accumulation in squamous cells carcinoma–TAFs (>50%) compared with adenocarcinoma–TAFs (10– 20%). This study suggested that treatments aimed at restoring normal lung elasticity may be useful in lung cancer therapy. The important role played by integrins α11β1 in TAFs was also investigated in two patient-derived NSCLC cells in vitro and in xenograft animal models, confirming the importance of α11 signaling pathway in TAFs, promoting tumor growth and metastatic potential of NSCLC cells, and being closely associated with collagen cross-linking and stiffness (Navab et al., 2016).

6.2.5.6.1 Advanced Therapies for Lung Diseases

Research on potential application of cell therapies for chronic and severe respiratory diseases has increased considerably in the last years, with a considerable number of clinical trials already finished or on going. The use of bone marrow-derived mesenchymal (stromal) cells (MSCs) for ARDS (Laffey and Matthay, 2017) is the application having more background from preclinical studies (Chimenti et al., 2012, Nonaka et al., 2020) and from pilot clinical trials (Matthay et al., 2019). Regardless of the application route (tracheal instillation or venous perfusion), MSCs from donors are administered in suspension (i.e., nonadherent) within a liquid vehicle, thereby running the risk of microvascular sequestration in the lung and hence potentially hindering suitable distribution within the target organ sites. Given that circulation of cells through capillaries that are narrower than cell diameter requires cell deformation, characterization of the mechanics of MSCs may provide useful information for optimizing the therapy. Indeed, real-time deformability cytometry has already been used to characterize different populations of stem cells from the bone marrow, with potential application for cell type selection (Xavier et al., 2016). Also, a compressive strain technique has been used for investigating how vimentin intermediate filaments modulate MSCs deformability (Sharma et al., 2017). Similarly, high-throughput real-time deformability cytometry, based on a microfluidic setting, has been used to characterize the mechanical properties of MSCs and their aggregation (Sarem et al., 2019). The phenotypic variability in MSCs deformability has been studied with a micropore/microchannel setting and the results suggest that selecting fractions of MSCs population, according to cell deformability, could be useful to control cell sequestration in the microvasculature, thereby potentially optimizing lung cell therapy (Lipowsky et al., 2018).

Tissue engineering preclinical research is progressing toward designing future advanced therapies for respiratory diseases (Farré et al., 2018), with biomechanics playing a potential role in optimizing them (Nonaka et al., 2016). For instance, new tissue engineering materials, such as hydrogels, can be used as vehicles with optimal viscoelastic properties to administer therapeutic drugs into the airway (Wu et al., 2017, Shamskhou et al., 2019). Interestingly, the first description of a protocol for obtaining hydrogels made from lung ECM (Pouliot et al., 2016) opened the possibility of using this stiffness-tunable, physio-mimicking material as a substrate for culturing lung cells in microenvironments that more realistically approach the in vivo cell niche (Farre et al., 2019). Such a lung ECM hydrogel could be used under cyclic stretch in future high-throughput microfluidic platforms for more realistic drug testing or for physiologically preconditioning MSCs to optimize cell therapy for respiratory diseases (Nonaka et al., 2020).

6.2.6 Conclusion and Future Perspective

The lung is an organ involving continuous cyclic deformations, and air and blood circulation through the airway tree and lung vascular bed, respectively. The basic biological components of the lung – cells and ECM – are therefore subjected to mechanical stimuli in a bidirectional cross talk. Indeed, cells sense and respond to the mechanical cues from the ECM and, in turn, cells remodel the ECM. It is therefore expected that a better knowledge of the mechanisms involved in such micromechanical cross talk will improve our understanding of the most relevant respiratory diseases (ALI/ARDS, asthma, lung fibrosis, COPD, PAH, and lung cancer). Fortunately, development of micro/nano techniques (e.g., MTC, AFM, and TFM) to measure cell and tissue mechanics has made it possible that in the last 20 years, we have gained considerable knowledge on the micromechanics of the lung. Application of these techniques is useful from different perspectives. First, to better understand the pathophysiology of major respiratory diseases. Second, to investigate how potential therapies may counteract the cell and tissue mechanics alterations that are induced by lung diseases. While these two classical research aims are of much interest, another perspective is already possible nowadays, and with great innovation potential, Indeed, it is in principle feasible to integrate the principles of techniques, such as AFM and TFM (now extremely complex, since they are implemented by means of big all-purpose devices), into small chips (Jang et al., 2019) with micro-channeling, to produce straightforward point-of-care devices for diagnosis and treatment follow-up of respiratory disease at the patient bedside. While such a solution seems to be technologically feasible, more innovation will be required to find out what patient samples (cells and tissues) are realistic candidates for such clinical applications. Circulating cells are the most obvious candidates because they are easily accessible, and they carry mechanical information for diagnosis and disease follow-up. Moreover, developing minimally invasive techniques to obtain cells/ECM from mini-bronchoalveolar lavages (Hendrickson et al., 2017) or lung microbiopsies (Herath and Cooper, 2017, Righi et al., 2017) would boost the clinical application of micromechanics in respiratory diseases.

References

Alcaraz, J., L. Buscemi, M. Grabulosa, X. Trepat, B. Fabry, R. Farré and D. Navajas (2003). "Microrheology of human lung epithelial cells measured by atomic force microscopy." Biophysical Journal **84**(3): 2071–2079.

Arce, F. T., J. L. Whitlock, A. A. Birukova, K. G. Birukov, M. F. Arnsdorf, R. Lal, J. G. Garcia and S. M. Dudek (2008). "Regulation of the micromechanical properties of pulmonary endothelium by S1P and thrombin: Role of cortactin." Biophysical Journal **95**(2): 886–894.

Azeloglu, E. U., J. Bhattacharya and K. D. Costa (2008). "Atomic force microscope elastography reveals phenotypic differences in alveolar cell stiffness." Journal of Applied Physiology **105**(2): 652–661.

Berrios, J. C., M. A. Schroeder and R. D. Hubmayr (2001). "Mechanical properties of alveolar epithelial cells in culture." Journal of Applied Physiology **91**(1): 65–73.

Bertero, T., W. M. Oldham, K. A. Cottrill, S. Pisano, R. R. Vanderpool, Q. Yu, J. Zhao, Y. Tai, Y. Tang and -Y.-Y. Zhang (2016). "Vascular stiffness mechanoactivates YAP/TAZ-dependent glutaminolysis to drive pulmonary hypertension." The Journal of Clinical Investigation **126**(9): 3313–3335.

Berthold, T., M. Glaubitz, S. Muschter, S. Groß, R. Palankar, A. Reil, C. A. Helm, T. Bakchoul, H. Schwertz and J. Bux (2015). "Human neutrophil antigen-3a antibodies induce neutrophil stiffening and conformational activation of CD11b without shedding of L-selectin." Transfusion **55**(12): 2939–2948.

Bidan, C. M., A. C. Veldsink, H. Meurs and R. Gosens (2015). "Airway and extracellular matrix mechanics in COPD." Frontiers in Physiology **6**: 346.

Birukova, A. A., F. T. Arce, N. Moldobaeva, S. M. Dudek, J. G. Garcia, R. Lal and K. G. Birukov (2009). "Endothelial permeability is controlled by spatially defined cytoskeletal mechanics: Atomic force microscopy force mapping of pulmonary endothelial monolayer." Nanomedicine: Nanotechnology, Biology and Medicine **5**(1): 30–41.

Bobrowska, J., K. Awsiuk, J. Pabijan, P. Bobrowski, J. Lekki, K. M. Sowa, J. Rysz, A. Budkowski and M. Lekka (2019). "Biophysical and biochemical characteristics as complementary indicators of melanoma progression." Analytical Chemistry **91**(15): 9885–9892.

Booth, A. J., R. Hadley, A. M. Cornett, A. A. Dreffs, S. A. Matthes, J. L. Tsui, K. Weiss, J. C. Horowitz, V. F. Fiore and T. H. Barker (2012). "Acellular normal and fibrotic human lung matrices as a culture system for in vitro investigation." American Journal of Respiratory and Critical Care Medicine **186**(9): 866–876.

Bulk, E., N. Kramko, I. Liashkovich, F. Glaser, H. Schillers, H.-J. Schnittler, H. Oberleithner and A. Schwab (2017). "KCa3. 1 channel inhibition leads to an ICAM-1 dependent increase of cell-cell adhesion between A549 lung cancer and HMEC-1 endothelial cells." Oncotarget **8**(68): 112268.

Caporarello, N., J. A. Meridew, D. L. Jones, Q. Tan, A. J. Haak, K. M. Choi, L. J. Manlove, Y. Prakash, D. J. Tschumperlin and G. Ligresti (2019). "PGC1α repression in IPF fibroblasts drives a pathologic metabolic, secretory and fibrogenic state." Thorax **74**(8): 749–760.

Cavalcante, F. S., S. Ito, K. Brewer, H. Sakai, A. M. Alencar, M. P. Almeida, J. S. Andrade, Jr, A. Majumdar, E. P. Ingenito and B. Suki (2005). "Mechanical interactions between collagen and proteoglycans: Implications for the stability of lung tissue." Journal of Applied Physiology **98**(2): 672–679.

Chen, Z., C. M. Fillmore, P. S. Hammerman, C. F. Kim and -K.-K. Wong (2014). "Non-small-cell lung cancers: A heterogeneous set of diseases." Nature Reviews Cancer **14**(8): 535–546.

Chimenti, L., T. Luque, M. R. Bonsignore, J. Ramírez, D. Navajas and R. Farré (2012). "Pre-treatment with mesenchymal stem cells reduces ventilator-induced lung injury." European Respiratory Journal **40**(4): 939–948.

Colin-York, H., Y. Javanmardi, L. Barbieri, D. Li, K. Korobchevskaya, Y. Guo, C. Hall, A. Taylor, S. Khuon and G. K. Sheridan (2019). "Spatiotemporally super-resolved volumetric traction force microscopy." Nano Letters **19**(7): 4427–4434.

Coughlin, M. F. and J. J. Fredberg (2013). "Changes in cytoskeletal dynamics and nonlinear rheology with metastatic ability in cancer cell lines." Physical Biology **10**(6): 065001.

Deng, L., N. J. Fairbank, B. Fabry, P. G. Smith and G. N. Maksym (2004). "Localized mechanical stress induces time-dependent actin cytoskeletal remodeling and stiffening in cultured airway smooth muscle cells." American Journal of Physiology-Cell Physiology **287**(2): C440–C448.

Ebihara, T., N. Venkatesan, R. Tanaka and M. S. Ludwig (2000). "Changes in extracellular matrix and tissue viscoelasticity in bleomycin–induced lung fibrosis: Temporal aspects." American Journal of Respiratory and Critical Care Medicine **162**(4): 1569–1576.

Fabry, B., G. N. Maksym, J. P. Butler, M. Glogauer, D. Navajas and J. J. Fredberg (2001). "Scaling the microrheology of living cells." Physical Review Letters **87**(14): 148102.

Farré, R., J. Otero, I. Almendros and D. Navajas (2018). "Bioengineered lungs: A challenge and an opportunity." Archivos de Bronconeumologia **54**(1): 31–38.

Farre, R., J. Otero, B. Falcones, E. Marhuenda, I. Almendros and D. Navajas (2019). Bioprinted 3D Model to Study the Cross-talk Between Lung Mesenchymal Stem Cells and Lung Extracellular Matrix. A68. New Techniques, Methodologies, and Mathematical Modelling, American Thoracic Society: A2244–A2244.

Franks, T. J., T. V. Colby, W. D. Travis, R. M. Tuder, H. Y. Reynolds, A. R. Brody, W. V. Cardoso, R. G. Crystal, C. J. Drake and J. Engelhardt (2008). "Resident cellular components of the human lung: Current knowledge and goals for research on cell phenotyping and function." Proceedings of the American Thoracic Society **5**(7): 763–766.

Gabasa, M., P. Duch, I. Jorba, A. Giménez, R. Lugo, I. Pavelescu, F. Rodríguez-Pascual, M. Molina-Molina, A. Xaubet and J. Pereda (2017). "Epithelial contribution to the profibrotic stiff microenvironment and myofibroblast population in lung fibrosis." Molecular Biology of the Cell **28**(26): 3741–3755.

Gavara, N., P. Roca-Cusachs, R. Sunyer, R. Farré and D. Navajas (2008). "Mapping cell-matrix stresses during stretch reveals inelastic reorganization of the cytoskeleton." Biophysical Journal **95**(1): 464–471.

Gavara, N., R. Sunyer, P. Roca-Cusachs, R. Farré, M. Rotger and D. Navajas (2006). "Thrombin-induced contraction in alveolar epithelial cells probed by traction microscopy." Journal of Applied Physiology **101**(2): 512–520.

Gladilin, E., S. Ohse, M. Boerries, H. Busch, C. Xu, M. Schneider, M. Meister and R. Eils (2019). "TGFβ-induced cytoskeletal remodeling mediates elevation of cell stiffness and invasiveness in NSCLC." Scientific Reports **9**(1): 1–12.

Gu, Y., H. Shi, C. Su, X. Chen, S. Zhang, W. Li, F. Wu, G. Gao, H. Wang and H. Chu (2017). "The role of endobronchial ultrasound elastography in the diagnosis of mediastinal and hilar lymph nodes." Oncotarget **8**(51): 89194.

Haak, A. J., Q. Tan and D. J. Tschumperlin (2018). "Matrix biomechanics and dynamics in pulmonary fibrosis." Matrix Biology **73**: 64–76.

Hendrickson, C. M., J. Abbott, H. Zhuo, K. D. Liu, C. S. Calfee, M. A. Matthay and N. A. Network (2017). "Higher mini-BAL total protein concentration in early ARDS predicts faster resolution of lung injury measured by more ventilator-free days." American Journal of Physiology-Lung Cellular and Molecular Physiology **312**(5): L579–L585.

Herath, S. and W. A. Cooper (2017). "The novel 19G endobronchial USS (EBUS) needle samples processed as tissue "core biopsies" facilitate PD-L1 and other biomarker testing in lung cancer specimens: Case report and the view point from the Respiratory Physician and the Pathologist." Respirology Case Reports **5**(6): e00271.

Hiorns, J. E., C. M. Bidan, O. E. Jensen, R. Gosens, L. E. Kistemaker, J. J. Fredberg, J. P. Butler, R. Krishnan and B. S. Brook (2016). "Airway and parenchymal strains during bronchoconstriction in the precision cut lung slice." Frontiers in Physiology **7**: 309.

Holenstein, C., C. Lendi, N. Wili and J. Snedeker (2019). "Simulation and evaluation of 3D traction force microscopy." Computer Methods in Biomechanics and Biomedical Engineering **22**(8): 853–860.

Iida, K., R. Sakai, S. Yokoyama, N. Kobayashi, S. Togo, H. Y. Yoshikawa, A. Rawangkan, K. Namiki and M. Suganuma (2017). "Cell softening in malignant progression of human lung cancer cells by activation of receptor tyrosine kinase AXL." Scientific Reports **7**(1): 1–11.

Inoue, Y., H. Tanaka, H. Ogura, I. Ukai, K. Fujita, H. Hosotsubo, R. Shimazu and H. Sugimoto (2006). "A neutrophil elastase inhibitor, sivelestat, improves leukocyte deformability in patients with acute lung injury." Journal of Trauma and Acute Care Surgery **60**(5): 936–943.

Jaffar, J., S.-H. Yang, S. Y. Kim, H.-W. Kim, A. Faiz, W. Chrzanowski and J. K. Burgess (2018). "Greater cellular stiffness in fibroblasts from patients with idiopathic pulmonary fibrosis." American Journal of Physiology-Lung Cellular and Molecular Physiology **315**(1): L59–L65.

Jang, H., J. Kim, J. H. Shin, J. J. Fredberg, C. Y. Park and Y. Park (2019). "Traction microscopy with integrated microfluidics: Responses of the multi-cellular island to gradients of HGF." Lab on a Chip **19**(9): 1579–1588.

Job, K. M., R. O'Callaghan, V. Hlady, A. Barabanova and R. O. Dull (2016). "Biomechanical effects of resuscitation colloids on the compromised lung endothelial glycocalyx." Anesthesia and Analgesia **123**(2): 382.

Jorba, I., G. Beltrán, B. Falcones, B. Suki, R. Farré, J. M. García-Aznar and D. Navajas (2019). "Nonlinear elasticity of the lung extracellular microenvironment is regulated by macroscale tissue strain." Acta Biomaterialia **92**: 265–276.

Koch, T. M., S. Münster, N. Bonakdar, J. P. Butler and B. Fabry (2012). "3D traction forces in cancer cell invasion." PLoS ONE **7**(3): e33476.

Kononov, S., K. Brewer, H. Sakai, F. S. Cavalcante, C. R. Sabayanagam, E. P. Ingenito and B. Suki (2001). "Roles of mechanical forces and collagen failure in the development of elastase-induced emphysema." American Journal of Respiratory and Critical Care Medicine **164**(10): 1920–1926.

Kraning-Rush, C. M., J. P. Califano and C. A. Reinhart-King (2012). "Cellular traction stresses increase with increasing metastatic potential." PLoS ONE **7**(2): e32572.

Laffey, J. G. and M. A. Matthay (2017). "Fifty years of research in ARDS. Cell-based therapy for acute respiratory distress syndrome. Biology and potential therapeutic value." American Journal of Respiratory and Critical Care Medicine **196**(3): 266–273.

Lan, B., R. Krishnan, C. Y. Park, R. A. Watanabe, R. Panganiban, J. P. Butler, Q. Lu, W. C. Cole and J. J. Fredberg (2018). "Transient stretch induces cytoskeletal fluidization through the severing action of cofilin." American Journal of Physiology-Lung Cellular and Molecular Physiology **314**(5): L799–L807.

Lartey, F. M., M. Rafat, M. Negahdar, A. V. Malkovskiy, X. Dong, X. Sun, M. Li, T. Doyle, J. Rajadas and E. E. Graves (2017). "Dynamic CT imaging of volumetric changes in pulmonary nodules correlates with physical measurements of stiffness." Radiotherapy and Oncology **122**(2): 313–318.

Lin, F., H. Zhang, J. Huang and C. Xiong (2018). "Contractility of airway smooth muscle cell in response to zinc oxide nanoparticles by traction force microscopy." Annals of Biomedical Engineering **46**(12): 2000–2011.

Lipowsky, H. H., D. T. Bowers, B. L. Banik and J. L. Brown (2018). "Mesenchymal stem cell deformability and implications for microvascular sequestration." Annals of Biomedical Engineering **46**(4): 640–654.

Liu, F., C. M. Haeger, P. B. Dieffenbach, D. Sicard, I. Chrobak, A. M. F. Coronata, M. M. S. Velandia, S. Vitali, R. A. Colas and P. C. Norris (2016). "Distal vessel stiffening is an early and pivotal mechanobiological regulator of vascular remodeling and pulmonary hypertension." JCI Insight **1**(8): e86987.

Liu, F., J. D. Mih, B. S. Shea, A. T. Kho, A. S. Sharif, A. M. Tager and D. J. Tschumperlin (2010). "Feedback amplification of fibrosis through matrix stiffening and COX-2 suppression." Journal of Cell Biology **190**(4): 693–706.

Liu, F. and D. J. Tschumperlin (2011). "Micro-mechanical characterization of lung tissue using atomic force microscopy." JoVE (Journal of Visualized Experiments) (54): e2911.

Luque, T., E. Melo, E. Garreta, J. Cortiella, J. Nichols, R. Farré and D. Navajas (2013). "Local micromechanical properties of decellularized lung scaffolds measured with atomic force microscopy." Acta biomaterialia **9**(6): 6852–6859.

Maarsingh, H., C. M. Bidan, B. S. Brook, A. B. Zuidhof, C. R. Elzinga, M. Smit, A. Oldenburger, R. Gosens, W. Timens and H. Meurs (2019). "Small airway hyperresponsiveness in COPD: Relationship between structure and function in lung slices." American Journal of Physiology-Lung Cellular and Molecular Physiology **316**(3): L537–L546.

Matthay, M. A., C. S. Calfee, H. Zhuo, B. T. Thompson, J. G. Wilson, J. E. Levitt, A. J. Rogers, J. E. Gotts, J. P. Wiener-Kronish and E. K. Bajwa (2019). "Treatment with allogeneic mesenchymal stromal cells for moderate to severe acute respiratory distress syndrome (START study): A randomised phase 2a safety trial." The Lancet Respiratory Medicine **7**(2): 154–162.

Melo, E., E. Garreta, T. Luque, J. Cortiella, J. Nichols, D. Navajas and R. Farré (2014a). "Effects of the decellularization method on the local stiffness of acellular lungs." Tissue Engineering Part C Methods **20**: 412–422.

Melo, E., N. Cárdenes, E. Garreta, T. Luque, M. Rojas, D. Navajas and R. Farré (2014b). "Inhomogeneity of local stiffness in the extracellular matrix scaffold of fibrotic mouse lungs." Journal of the Mechanical Behavior of Biomedical Materials **37**: 186–19.

Meng, F., I. Mambetsariev, Y. Tian, Y. Beckham, A. Meliton, A. Leff, M. L. Gardel, M. J. Allen, K. G. Birukov and A. A. Birukova (2015). "Attenuation of lipopolysaccharide-induced lung vascular stiffening by lipoxin reduces lung inflammation." American Journal of Respiratory Cell and Molecular Biology **52**(2): 152–161.

Merna, N., A. K. Wong, V. Barahona, P. Llanos, B. Kunar, B. Palikuqi, M. Ginsberg, S. Rafii and S. Y. Rabbany (2018). "Laminar shear stress modulates endothelial luminal surface stiffness in a tissue-specific manner." Microcirculation **25**(5): e12455.

Mierke, C. T., N. Bretz and P. Altevogt (2011). "Contractile forces contribute to increased glycosylphosphatidylinositol-anchored receptor CD24-facilitated cancer cell invasion." Journal of Biological Chemistry **286**(40): 34858–34871.

Miyazawa, A., S. Ito, S. Asano, I. Tanaka, M. Sato, M. Kondo and Y. Hasegawa (2018). "Regulation of PD-L1 expression by matrix stiffness in lung cancer cells." Biochemical and Biophysical Research Communications **495**(3): 2344–2349.

Morikawa, M., Y. Inoue, Y. Sumi, Y. Kuroda and H. Tanaka (2014). "Leukocyte deformability is a novel biomarker to reflect sepsis-induced disseminated intravascular coagulation." Acute Medicine & Surgery **2**(1): 13.

Navab, R., D. Strumpf, C. To, E. Pasko, K. Kim, C. Park, J. Hai, J. Liu, J. Jonkman and M. Barczyk (2016). "Integrin α11β1 regulates cancer stromal stiffness and promotes tumorigenicity and metastasis in non-small cell lung cancer." Oncogene **35**(15): 1899–1908.

Nonaka, P. N., J. J. Uriarte, N. Campillo, V. R. Oliveira, D. Navajas and R. Farré (2016). "Lung bioengineering: Physical stimuli and stem/progenitor cell biology interplay towards biofabricating a functional organ." Respiratory Research **17**(1): 161.

Nonaka, P. N., B. Falcones, R. Farre, A. Artigas, I. Almendros and D. Navajas (2020). "Biophysically Preconditioning Mesenchymal Stem Cells Improves Treatment of Ventilator-Induced Lung Injury." Archivos de Bronconeumología (Engl Ed) **56**(3): 179–181.

Oliveira, V. R., J. J. Uriarte, B. Falcones, I. Jorba, W. A. Zin, R. Farré, D. Navajas and I. Almendros (2019a). "Biomechanical response of lung epithelial cells to iron oxide and titanium dioxide nanoparticles." Frontiers in Physiology **10**: 1047.

Oliveira, V. R., J. J. Uriarte, B. Falcones, W. A. Zin, D. Navajas, R. Farré and I. Almendros (2019b). "Escherichia coli lipopolysaccharide induces alveolar epithelial cell stiffening." Journal of Biomechanics **83**: 315–318.

Ozgokce, M., A. Yavuz, I. Akbudak, F. Durmaz, I. Uney, Y. Aydin, H. Yildiz, A. Batur, H. Arslan and I. Dundar (2018). "Usability of transthoracic shear wave elastography in differentiation of subpleural solid masses." Ultrasound Quarterly **34**(4): 233–237.

Pankova, D., Y. Jiang, M. Chatzifrangkeskou, I. Vendrell, J. Buzzelli, A. Ryan, C. Brown and E. O'Neill (2019). "RASSF1A controls tissue stiffness and cancer stem-like cells in lung adenocarcinoma." The EMBO Journal **38**(13): e100532.

Parrow, N. L., P.-C. Violet, H. Tu, J. Nichols, C. A. Pittman, C. Fitzhugh, R. E. Fleming, N. Mohandas, J. F. Tisdale and M. Levine (2018). "Measuring deformability and red cell heterogeneity in blood by ektacytometry." JoVE (Journal of Visualized Experiments (131): e56910.

Perrault, C. M., A. Brugues, E. Bazellieres, P. Ricco, D. Lacroix and X. Trepat (2015). "Traction forces of endothelial cells under slow shear flow." Biophysical Journal **109**(8): 1533–1536.

Polio, S. R., A. N. Kundu, C. E. Dougan, N. P. Birch, D. E. Aurian-Blajeni, J. D. Schiffman, A. J. Crosby and S. R. Peyton (2018). "Cross-platform mechanical characterization of lung tissue." PloS one **13**(10): e0204765.

Pouliot, R. A., P. A. Link, N. S. Mikhaiel, M. B. Schneck, M. S. Valentine, F. J. Kamga Gninzeko, J. A. Herbert, M. Sakagami and R. L. Heise (2016). "Development and characterization of a naturally derived lung extracellular matrix hydrogel." Journal of Biomedical Materials Research Part A **104**(8): 1922–1935.

Preira, P., J.-M. Forel, P. Robert, P. Nègre, M. Biarnes-Pelicot, F. Xeridat, P. Bongrand, L. Papazian and O. Theodoly (2016). "The leukocyte-stiffening property of plasma in early acute respiratory distress syndrome (ARDS) revealed by a microfluidic single-cell study: The role of cytokines and protection with antibodies." Critical Care **20**, 8 (2015).

Prina-Mello, A., N. Jain, B. Liu, J. I. Kilpatrick, M. A. Tutty, A. P. Bell, S. P. Jarvis, Y. Volkov and D. Movia (2018). "Culturing substrates influence the morphological, mechanical and biochemical features of lung adenocarcinoma cells cultured in 2D or 3D." Tissue and Cell **50**: 15–30.

Puig, F., G. Fuster, M. Adda, L. Blanch, R. Farre, D. Navajas and A. Artigas (2013). "Barrier-protective effects of activated protein C in human alveolar epithelial cells." PLoS one **8**(2): e56965.

Puig, F., N. Gavara, R. Sunyer, A. Carreras, R. Farré and D. Navajas (2009). "Stiffening and contraction induced by dexamethasone in alveolar epithelial cells." Experimental Mechanics **49**(1): 47–55.

Puig, M., R. Lugo, M. Gabasa, A. Giménez, A. Velásquez, R. Galgoczy, J. Ramírez, A. Gómez-Caro, Ó. Busnadiego and F. Rodríguez-Pascual (2015). "Matrix stiffening and β1 integrin drive subtype-specific fibroblast accumulation in lung cancer." Molecular Cancer Research **13**(1): 161–173.

Puig-de-Morales, M., E. Millet, B. Fabry, D. Navajas, N. Wang, J. P. Butler and J. J. Fredberg (2004). "Cytoskeletal mechanics in adherent human airway smooth muscle cells: Probe specificity and scaling of protein-protein dynamics." American Journal of Physiology-Cell Physiology **287**(3): C643–C654.

Reed, E. B., S. Ard, J. La, C. Y. Park, L. Culligan, J. J. Fredberg, L. V. Smolyaninova, S. N. Orlov, B. Chen and R. Guzy (2019). "Anti-fibrotic effects of tannic acid through regulation of a sustained TGF-beta receptor signaling." Respiratory Research **20**(1): 168.

Rianna, C., P. Kumar and M. Radmacher (2018). "The role of the microenvironment in the biophysics of cancer." Seminars in Cell & Developmental Biology **73**: 107–114.

Righi, L., F. Franzi, F. Montarolo, G. Gatti, M. Bongiovanni, F. Sessa and S. La Rosa (2017). "Endobronchial ultrasound-guided transbronchial needle aspiration (EBUS-TBNA) – from morphology to molecular testing." Journal of Thoracic Disease **9**(Suppl 5): S395.

Roca-Cusachs, P., I. Almendros, R. Sunyer, N. Gavara, R. Farré and D. Navajas (2006). "Rheology of passive and adhesion-activated neutrophils probed by atomic force microscopy." Biophysical Journal **91**(9): 3508–3518.

Romero, P. V., W. A. Zin and J. Lopez-Aguilar (2001). "Frequency characteristics of lung tissue strip during passive stretch and induced pneumoconstriction." Journal of Applied Physiology **91**(2): 882–890.

Rotsch, C., K. Jacobson, J. Condeelis and M. Radmacher (2001). "EGF-stimulated lamellipod extension in adenocarcinoma cells." Ultramicroscopy **86**(1–2): 97–106.

Sackmann, E. K.-H., E. Berthier, E. A. Schwantes, P. S. Fichtinger, M. D. Evans, L. L. Dziadzio, A. Huttenlocher, S. K. Mathur and D. J. Beebe (2014). "Characterizing asthma from a drop of blood using neutrophil chemotaxis." Proceedings of the National Academy of Sciences **111**(16): 5813–5818.

Sarem, M., O. Otto, S. Tanaka and V. P. Shastri (2019). "Cell number in mesenchymal stem cell aggregates dictates cell stiffness and chondrogenesis." Stem Cell Research & Therapy **10**(1): 10.

Schierbaum, N., J. Rheinlaender and T. E. Schäffer (2019). "Combined atomic force microscopy (AFM) and traction force microscopy (TFM) reveals a correlation between viscoelastic material properties and contractile prestress of living cells." Soft Matter **15**(8): 1721–1729.

Schimmel, L., M. van der Stoel, C. Rianna, A.-M. van Stalborch, A. de Ligt, M. Hoogenboezem, S. Tol, J. van Rijssel, R. Szulcek and H. J. Bogaard (2018). "Stiffness-induced endothelial DLC-1 expression forces leukocyte spreading through stabilization of the ICAM-1 adhesome." Cell Reports **24**(12): 3115–3124.

Seow, C. Y. (2013). "Passive stiffness of airway smooth muscle: The next target for improving airway distensibility and treatment for asthma?." Pulmonary Pharmacology & Therapeutics **26**(1): 37–41.

Serra-Picamal, X., V. Conte, R. Sunyer, J. J. Muñoz and X. Trepat (2015). "Mapping forces and kinematics during collective cell migration." Methods in Cell Biology, Elsevier **125**: 309–330.

Shamskhou, E. A., M. J. Kratochvil, M. E. Orcholski, N. Nagy, G. Kaber, E. Steen, S. Balaji, K. Yuan, S. Keswani and B. Danielson (2019). "Hydrogel-based delivery of Il-10 improves treatment of bleomycin-induced lung fibrosis in mice." Biomaterials **203**: 52–62.

Sharma, P., Z. T. Bolten, D. R. Wagner and A. H. Hsieh (2017). "Deformability of human mesenchymal stem cells is dependent on vimentin intermediate filaments." Annals of Biomedical Engineering **45**(5): 1365–1374.

Shukla, V., N. Higuita-Castro, P. Nana-Sinkam and S. Ghadiali (2016). "Substrate stiffness modulates lung cancer cell migration but not epithelial to mesenchymal transition." Journal of Biomedical Materials Research Part A **104**(5): 1182–1193.

Sicard, D., L. E. Fredenburgh and D. J. Tschumperlin (2017). "Measured pulmonary arterial tissue stiffness is highly sensitive to AFM indenter dimensions." Journal of the Mechanical Behavior of Biomedical Materials **74**: 118–127.

Sicard, D., A. J. Haak, K. M. Choi, A. R. Craig, L. E. Fredenburgh and D. J. Tschumperlin (2018). "Aging and anatomical variations in lung tissue stiffness." American Journal of Physiology-Lung Cellular and Molecular Physiology **314**(6): L946–L955.

Skoutelis, A. T., V. Kaleridis, G. M. Athanassiou, K. I. Kokkinis, Y. F. Missirlis and H. P. Bassaris (2000). "Neutrophil deformability in patients with sepsis, septic shock, and adult respiratory distress syndrome." Critical Care Medicine **28**(7): 2355–2359.

Smith, B. A., B. Tolloczko, J. G. Martin and P. Grütter (2005). "Probing the viscoelastic behavior of cultured airway smooth muscle cells with atomic force microscopy: Stiffening induced by contractile agonist." Biophysical Journal **88**(4): 2994–3007.

Sobiepanek, A., M. Milner-Krawczyk, M. Lekka and T. Kobiela (2017). "AFM and QCM-D as tools for the distinction of melanoma cells with a different metastatic potential." Biosensors and Bioelectronics **93**: 274–281.

Stamenovic, D., B. Suki, B. Fabry, N. Wang, J. J. Fredberg and J. E. Buy (2004). "Rheology of airway smooth muscle cells is associated with cytoskeletal contractile stress." Journal of Applied Physiology **96**(5): 1600–1605.

Suki, B. and J. H. Bates (2008). "Extracellular matrix mechanics in lung parenchymal diseases." Respiratory Physiology & Neurobiology **163**(1–3): 33–43.

Suresh, K., K. Carino, L. Johnston, L. Servinsky, C. E. Machamer, T. M. Kolb, H. Lam, S. M. Dudek, S. S. An and M. J. Rane (2019). "A nonapoptotic endothelial barrier-protective role for caspase-3." American Journal of Physiology-Lung Cellular and Molecular Physiology **316**(6): L1118–L1126.

Tanaka, H., M. Nishino, Y. Nakamori, H. Ogura, K. Ishikawa, T. Shimazu and H. Sugimoto (2001). "Granulocyte colony-stimulating factor (G-CSF) stiffens leukocytes but attenuates inflammatory response without lung injury in septic patients." Journal of Trauma and Acute Care Surgery **51**(6): 1110–1116.

Thenappan, T., S. Y. Chan and E. K. Weir (2018). "Role of extracellular matrix in the pathogenesis of pulmonary arterial hypertension." American Journal of Physiology-Heart and Circulatory Physiology **315**(5): H1322–H1331.

Tilghman, R. W., C. R. Cowan, J. D. Mih, Y. Koryakina, D. Gioeli, J. K. Slack-Davis, B. R. Blackman, D. J. Tschumperlin and J. T. Parsons (2010). "Matrix rigidity regulates cancer cell growth and cellular phenotype." PloS one **5**(9): e12905.

Trepat, X., M. Grabulosa, L. Buscemi, F. Rico, R. Farré and D. Navajas (2005). "Thrombin and histamine induce stiffening of alveolar epithelial cells." Journal of Applied Physiology **98**(4): 1567–1574.

Trepat, X., M. Grabulosa, F. Puig, G. N. Maksym, D. Navajas and R. Farré (2004). "Viscoelasticity of human alveolar epithelial cells subjected to stretch." American Journal of Physiology-Lung Cellular and Molecular Physiology **287**(5): L1025–L1034.

Trepat, X., F. Puig, N. Gavara, J. J. Fredberg, R. Farre and D. Navajas (2006). "Effect of stretch on structural integrity and micromechanics of human alveolar epithelial cell monolayers exposed to thrombin." American Journal of Physiology-Lung Cellular and Molecular Physiology **290**(6): L1104–L1110.

Tschumperlin, D. J., F. Boudreault and F. Liu (2010). "Recent advances and new opportunities in lung mechanobiology." Journal of Biomechanics **43**(1): 99–107.

Ugurlu, E., E. Kilic-Toprak, I. Can, O. Kilic-Erkek, G. Altinisik and M. Bor-Kucukatay (2017) "Impaired hemorheology in exacerbations of COPD." Canadian Respiratory Journal **2017**: 1286263.

Uriarte, J. J., T. Meirelles, D. G. Del Blanco, P. N. Nonaka, N. Campillo, E. Sarri, D. Navajas, G. Egea and R. Farre (2016). "Early impairment of lung mechanics in a murine model of Marfan syndrome." PloS one **11**(3): e0152124.

Verhoeven, R. L., C. L. de Korte and E. H. van der Heijden (2019). "Optimal endobronchial ultrasound strain elastography assessment strategy: An explorative study." Respiration **97**(4): 337–347.

Viswanathan, P., Y. Ephstein, J. G. Garcia, M. Cho and S. Dudek (2016). "Differential elastic responses to barrier-altering agonists in two types of human lung endothelium." Biochemical and Biophysical Research Communications **478**(2): 599–605.

Vizoso, M., M. Puig, F. J. Carmona, M. Maqueda, A. Velásquez, A. Gómez, A. Labernadie, R. Lugo, M. Gabasa and L. G. Rigat-Brugarolas (2015). "Aberrant DNA methylation in non-small cell lung cancer-associated fibroblasts." Carcinogenesis **36**(12): 1453–1463.

Wang, Q., E. T. Chiang, M. Lim, J. Lai, R. Rogers, P. A. Janmey, D. Shepro and C. M. Doerschuk (2001). "Changes in the biomechanical properties of neutrophils and endothelial cells during adhesion." Blood, the Journal of the American Society of Hematology **97**(3): 660–668.

Wang, X., R. Bleher, L. Wang, J. G. Garcia, S. Dudek, G. Shekhawat and V. Dravid (2017). "Imatinib alters agonists-mediated cytoskeletal biomechanics in lung endothelium." Scientific Reports **7**(1): 1–14.

Waters, C. M., E. Roan and D. Navajas (2012). "Mechanobiology in lung epithelial cells: Measurements, perturbations, and responses." Comprehensive Physiology **2**(1): 1.

Wiesinger, A., W. Peters, D. Chappell, D. Kentrup, S. Reuter, H. Pavenstädt, H. Oberleithner and P. Kümpers (2013). "Nanomechanics of the endothelial glycocalyx in experimental sepsis." PloS one **8**(11): e80905.

Wilhelm, K. R., E. Roan, M. C. Ghosh, K. Parthasarathi and C. M. Waters (2014). "Hyperoxia increases the elastic modulus of alveolar epithelial cells through Rho kinase." The FEBS Journal **281**(3): 957–969.

Wu, J., P. Ravikumar, K. T. Nguyen, C. C. Hsia and Y. Hong (2017). "Lung protection by inhalation of exogenous solubilized extracellular matrix." PloS one **12**(2): e0171165.

Xavier, M., P. Rosendahl, M. Herbig, M. Kräter, D. Spencer, M. Bornhäuser, R. O. Oreffo, H. Morgan, J. Guck and O. Otto (2016). "Mechanical phenotyping of primary human skeletal stem cells in heterogeneous populations by real-time deformability cytometry." Integrative Biology **8**(5): 616–623.

Yaylali, Y. T., E. Kilic-Toprak, Y. Ozdemir, H. Senol and M. Bor-Kucukatay (2019). "Impaired Blood Rheology in Pulmonary Arterial Hypertension." Heart, Lung and Circulation **28**(7): 1067–1073.

Zemła, J., T. Stachura, I. Gross-Sondej, K. Górka, K. Okoń, G. Pyka-Fościak, J. Soja, K. Sładek and M. Lekka (2018). "AFM-based nanomechanical characterization of bronchoscopic samples in asthma patients." Journal of Molecular Recognition **31**(12): e2752.

Zhang, H., L. Xiao, Q. Li, X. Qi and A. Zhou (2018). "Microfluidic chip for non-invasive analysis of tumor cells interaction with anti-cancer drug doxorubicin by AFM and Raman spectroscopy." Biomicrofluidics **12**(2): 024119.

Zhou, Y., J. C. Horowitz, A. Naba, N. Ambalavanan, K. Atabai, J. Balestrini, P. B. Bitterman, R. A. Corley, B.-S. Ding and A. J. Engler (2018). "Extracellular matrix in lung development, homeostasis and disease." Matrix Biology **73**: 77–104.

Małgorzata Lekka, Manfred Radmacher, Arnaud Millet,
Massimo Alfano

6.3 Cancer

6.3.1 Introduction

Despite considerable progress in cancer research leading to the identification of various key molecules and processes, cancer is still a leading cause of the worldwide deaths. GLOBOCAN database presenting global cancer statistics shows an increasing incidence of cancer from 18.1 million in 2018 to 29.5 million cases in 2040 (http://gco.iarc.fr). Cancer development and metastasis are highly complex processes. Many cancer-related deaths are associated with the metastatic phase; therefore, understanding of mechanisms governing them will help fight the disease. Metastatic cancers develop during years or even decades after primary tumor diagnosis by accumulation of a huge number of mutations leading to various phenotypes of cancer cells. Therefore, metastasis is characterized by a large degree of heterogeneity present in the structure, properties, and functioning of cells and affected organs. A severe consequence is that mechanisms discovered for one type of cancer may not necessarily be valid for others. Over one decade of research performed on various aspects of cancer development has delivered evidence that one of the main features of cancer progression is the alteration of cells, tissues, and tumor microenvironment deformability.

6.3.2 Metastasis – Cancer Cell Characteristics

Cancer can be considered as a collection of multiple changes characterized by the aberrant growth of cells that have collected mutations in genes controlling cell proliferation and survival. Metastasis, being the dissemination of cancer cells to distinct organs, is mostly responsible for most cancer-related deaths. During this process, after acquiring an invasive cell phenotype, individual cancer cells, or a small cluster of cells invade the microenvironment surrounding the primary tumor site (Figure 6.3.1). They change from an anchorage-dependent to an anchorage-independent phenotype, resulting in altered adhesive properties of cells. Penetrating cells enter

Małgorzata Lekka, Institute of Nuclear Physics, Polish Academy of Sciences, Kraków, Poland
Manfred Radmacher, Institute of Biophysics, University of Bremen, Bremen, Germany
Arnaud Millet, Institute for Advanced Biosciences, Grenoble-Alpes University, Inserm U1209 –CNRS
UMR 5309, and Research Department University Hospital of Grenoble Alpes, Grenoble, France
Massimo Alfano, Division of Experimental Oncology/Unit of Urology, URI, IRCCS Ospedale San
Raffaele, Milan, Italy

https://doi.org/10.1515/9783110989380-013

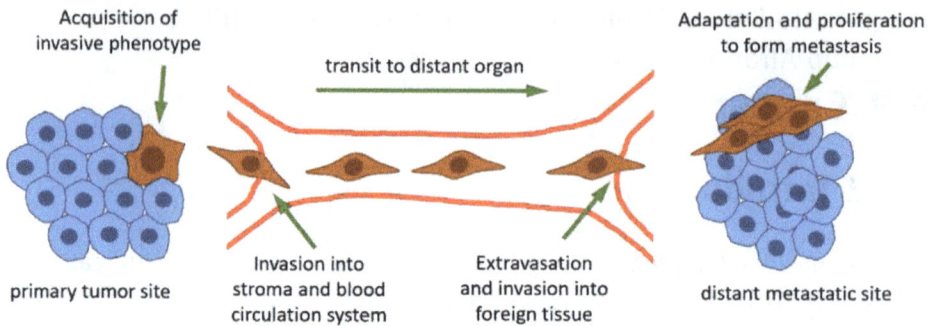

Figure 6.3.1: Dissemination of cancerous cells. Metastasis starts in the primary tumor site. Cells that acquire an invasive phenotype detach from the cells detach from the cells forming a primary tumor site forming a primary tumor site and invade into the surrounding extracellular matrix, reaching blood or lymph circulations systems. Here, cells display properties of anchorage-independent survival that enable them to travel through the circulation system up to certain distant location, where they begin invading the microenvironment of the foreign tissue. To form metastasis, the cancer cells must be able to adapt to this microenvironment and begin proliferating.

the vascular system of the lymph and blood. Inside the microvessels, cancer cells circulate and translocate to distant tissues. Next, cells escape from the microvessels and invade the microenvironment of distant tissues. Finally, they adapt to this microenvironment and start colonizing the tissue by extensive proliferation, leading to the formation of a secondary tumor (Fouad and Aanei, 2017, Hanahan and Weinberg, 2011).

Transformation of cancerous cells to an aggressive and migratory phenotype is a key step leading to the formation of metastasis. Cancer cells differ from normal cells at structural and functional levels, forming rather disorganized structures both in cell culture and at the tissue levels. Structurally, cancer cells possess large and irregularly shaped cell nuclei, occupying a relatively large volume of the cytoplasm. Within cell nuclei, multiple nucleoli and coarse chromatin are frequently observed.

Cancer cells are variable in size and shape, which implicates alterations in the organization of the cytoskeleton. It has already been reported that abnormal behavior of cancer cells, including uncontrolled growth and accumulation of a migratory phenotype, affects the organization of the cell cytoskeleton (Ben-Ze'ev, 1985, Hall, 2009). The cell cytoskeleton is composed of three main elements: microtubules, intermediate filaments, and actin filaments. All these components are highly dynamic and participate in numerous cellular processes. Their dysregulation, observed in various pathological processes, results in severe consequences as they are crucial for controlling cell behavior (Gaspar et al, 2015a, Hanahan and Weinberg, 2011). Microtubules mainly serve as transporting tracks for vesicles and organelles within cells, but they are essential for the formation of mitotic spindles during cell division. The structural instability of microtubules that affects the equilibrium of its assembly and disassembly is regulated by microtubule-associated proteins. Accumulation of multiple mutations in mitotic cells alters normal cell mitosis, leading to uncontrolled proliferation of cancer cells. Intermediate filaments are relatively conserved cytoskeletal structures. They provide mechanical strength and distribute stresses generated inside the cell and in the cell microenvironment. Most cancers arise from the epithelium. In such cells, enhanced keratin and vimentin expression is a potential biomarker of the disease (Satelli and Li, 2011, Trivedi et al., 2017). Microfilaments are the thinnest cytoskeletal structures made from polymerized F-actin. Apart from being responsible for cell shape and mechanical resistance, these filaments participate in various functional processes such as cell proliferation, adhesion, and migration. Disorganization of actin filaments is one of the most visible features of cancer transformation (Li et al., 2008).

6.3.3 Significance of Nanomechanics in Cancer

Dissemination of cancer cells and colonization of distant organs are highly inefficient processes, but the results are a huge number of deaths. The current hypothesis is that dissemination is initiated by circulating tumor cells (CTCs) found in the blood of patients (Osmulski et al., 2014). The number of CTCs is significantly smaller than the total number of cancer cells present within the primary tumor site. Yet, the question of how molecular and physicochemical properties of cells contribute to the colonization of distant sites by a single cancer cell remains obscure.

Research gathered over the last two decades delivers evidence that cell functioning relies not only on molecular bases but also involves nanomechanics. Cells must resist physiologically relevant deformations and stresses present both inside the cell and in the surrounding environment. Tissues are formed by cells and by the noncellular environment, mainly ECM, in which cells are embedded. ECM does not serve as a scaffold for cells. It provides an active platform on which the cells grow, influencing their migration, differentiation, survival, homeostasis, and morphogenesis (Paszek et al., 2005). Reorganization of its structure affects normal cell functioning; however, it can also induce severe pathological complications, including cancer development. On the other side, affecting the ECM may provide a novel target for anticancer therapies. Cells interact mechanically and molecularly with the ECM, thereby defining its properties and controlling or modulating the ECM's architecture as well. Active cell–ECM interactions result in the formation of matrices with designed composition, as it occurs during, for example, breast cancer invasion. Breast cancer cells are more deformable making them suitable for invasion (Li et al., 2008, Plodinec et al., 2012), which simultaneously requires the enhanced expression of collagen fibers stiffening the cancer microenvironment (Cox and Erler, 2011). Simultaneously, the same cancer cells can also degrade ECM that enables them to penetrate during their travel towards the bloodstream system.

The interaction of cells with the surrounding matrix is dynamic and reversible – the cells organize the matrix, and the matrix, in turn, influences cell fate (Janmey and McCulloch, 2007, Paszek et al., 2005). During cancer progression, changes in mechanical properties are observed both in the extracellular matrix and in cells. The observed stiffening of the extracellular matrix reveals the significant role of fibrosis in cancer development (Cox and Erler, 2011). Decreased deformability is a result of increased density and cross-linking levels of fibrotic fibers (Levental et al., 2009, Paszek et al., 2005). Consequently, increasing migration of cells towards higher density gradient is observed, especially for cancer cells (Clark and Vignjevic, 2015). The effects of degradation and realignment constitute a significant mechanism for ovarian cancer cells traveling through the mesothelium (Iwanicki et al., 2011). A growing tumor needs to occupy the ECM space. Thus, intuitively, growing cancer will induce tension, affecting the healthy tissue nearby. Cancer-induced tension has severe aftereffects, as it has recently been demonstrated for brain cancers (Seano et al., 2019). In the case of solid and infiltrative tumors, a significant reduction in blood vessel diameter inducing impaired oxygenation of the brain tissue, besides a smaller number of neurons was observed in the compressed, healthy tissue. On the cellular side, cells undergo specific changes in their adhesive and mechanical properties. For instance, altered deformability of cancer cells has been observed for many cancers (Lekka et al., 2012b).

In some cases, like in ovarian cancers, the increased deformability corresponds to higher invasiveness of the cells (Ketene et al., 2012b, Xu et al., 2012). Most studies highlight that the cell cytoskeleton as a primary structure is responsible for the mechanical resistance of cells and for sustaining mechanical forces. The cytoskeleton is a dynamic structure changing its organization and mechanical properties in response to altered conditions. It is a structure by which cells respond actively to mechanical forces present in their microenvironment (Janmey et al., 2009). Disseminating cancer cells sense various mechanical stresses. Initially, during the penetration through the surrounding extracellular matrix, cells sustain compressive, tensional, and shear forces, while during their translocation inside the bloodstream, mostly shear forces are present. Various mechanical conditions induce different responses in the cells, in the cytoskeleton. Although the mechanical properties of the cytoskeleton arise from a combination of mechanical properties of its structural components, in most cases, they are studied separately. The role of the actin filaments in cell mechanics seems to be highly elaborated. The mechanics and functions of the intermediate filaments and microtubular networks still remain to be elucidated.

Studies on the role of nanomechanical properties of cells, extracellular matrix, and tissues are of interest due to the potential applicability of biomechanics as a diagnostic tool. Certain diseases manifest in an altered resistance to deformation. The apparent disease is muscular dystrophy, in which the lack of dystrophin leads to a weakening of muscles (Puttini et al., 2009). On the other hand, the altered deformability of cancer cells is a manifestation of oncogenic changes. The latter can be used as a nanomechanical fingerprint of the disease, discriminating between the healthy and the cancerous state.

6.3.4 Elasticity of Cancer Cells and Tissues

Even if cancer is diagnosed for a specific organ or tissue, it is characterized by a large degree of heterogeneity. High-resolution techniques working on the nanoscale enable us to identify a group of nonspecific changes that can help the diagnosis. Emerging studies suggest that the biomechanical and biophysical properties of cells can be used as a new biomarker of changes induced by various pathologies, including cancer. Measurements carried out on cells and tissues over the past decades highlight the importance of mechanical properties in cancer progression. In the following subsections, examples of nanomechanical characterization of several cancer types are presented.

6.3.4.1 Bladder and Other Urothelial Cancers

Bladder cancer (BCa) is the ninth most common worldwide cancer, with an estimated prevalence of around 2.7 million cases and an incidence of around 350,000 new cases/year (http://www.cancer.gov/cancertopics/pdq/treatment/bladder/HealthPro fessional) occurring in southern Europe (especially, in Italy and Spain). Simultaneously, it has the highest incidence and mortality rates of BCa in the world (Antoni et al., 2017). One of the main problems in BCa treatment is the inability to efficaciously prevent high-grade nonmuscle invasive bladder cancer (NMIBC) relapse and progression, which occur in 80% and 45% of patients, respectively (Carrion and Seigne, 2002). As the above data suggest, the optimal management of these high-grade NMIBC is not entirely fulfilled by current standard protocols and poses a significant clinical challenge. Even though intravesical instillation of *Bacille Calmette-Guerin* (BCG) followed by long-term maintenance therapy results in a decreased cancer recurrence and spreading, still, the majority of patients experience relapse and progression, followed by metastatic diffusion to distant organs in the vast majority of cases.

Moreover, timely intervention to prevent patients' death is currently hampered by the lack of sensitive follow-up technologies that can detect BCa progression early. For all these reasons, proper management of BCa remains an unmet clinical need that must push the scientific community to major research commitments. As a consequence of the lack of efficient, established, prognostic, diagnostic, and predictive biomarkers of high-risk NMIBC (pT1G3 and carcinoma in situ, Cis) BCa patients undergo multiple treatments and cystoscopic assessments during their entire lifetime, with resulting poor quality of life and high healthcare costs (Anastasiadis and de Reijke, 2012, Kim, 2016). A second unmet clinical need is the management of patients with muscle-invasive BCa (MIBC). Patients with MIBC are usually advised to undergo radical cystectomy (RC) to prevent the spreading of the disease (Stein et al., 2001). At present, no urinary biomarkers or biomarker(s) in urinary cells are available, which might correlate with the progression of any form of MIBC, thus driving a personalized therapeutic decision. Furthermore, there are no prognostic clinical indicators or biomarkers for the spreading of the disease, and thus, for the stratification of MIBC patients that could benefit from more specific, systemic treatments (http://uroweb.org/guideline/bladder-cancer-muscle-invasive-and-metastatic).

The bladder is an organ, which is characterized by a large degree of mechanical flexibility that may vary from 9.6 kPa (187 mL, 8.6 mmHg) to 106.9 kPa (327 mL, 27.6 mmHg) depending on the filling volume as observed in pig bladder (Nenadic et al., 2013). This indicates large deformability of individual bladder cells that has become a driving force for the studies of the mechanical properties of these cells. Likewise, the measurement of deformability of bladder cancer cell lines by AFM approach has established an association to cancer progression (Lekka et al., 1999, Ramos et al., 2014). Deformability of living bladder cancer cells was shown to be significantly larger as compared to reference, nonmalignant cells. Almost one order magnitude difference was observed for HCV29 and Hu609 cells (nonmalignant cell cancer of the ureter), cancerous T24 and Hu456 cells (transitional cell carcinoma), and *v-ras* transfected HCV29 cells (Lekka et al., 1999). Surprisingly, there was no correlation between cancer cell deformability and invasiveness. Also, to a certain extent, larger deformability of cells is independent of cancer progression, as it has been shown in studies comparing the deformability of 5637 (urinary bladder carcinoma, grade II), HT1376 (bladder carcinoma, grade III) and T24 (transitional cell carcinoma, grade III/IV) cells with the elastic properties of nonmalignant HCV29 cells (Ramos et al., 2014). For all cancerous cells, Young's moduli were about 2–3 times lower than for nonmalignant cells. These results underline the usefulness of AFM in the detection of bladder cancer cell mechanics, which showed that regardless of the state of cancer progression, softer cells indicate a malignant phenotype. Importantly, higher deformability of bladder cancer cells has been shown to correlate with a partial lack of actin filaments and/or their depolymerization. These results demonstrated that the mechanical response of cells is dominated rather by the expression of F-actin, and not by its spatial organization.

Furthermore, HT1376 cancer cells reveal a biphasic nature of the mechanical response to altered microenvironment (Lekka et al., 2019). Mechanical properties of surrounding extracellular matrix induce threshold-dependent relations. During the initial stages of cell adhesion to soft microenvironments, cells primarily change their morphology and deformability (Figure 6.3.2).

A switch seems to be provided by a cellular deformability threshold that, in the case of nonmalignant HCV29 cells, triggers the formation of thick actin bundles accompanied by matured focal adhesions. For cancerous HT1376 cells, only a weak reorganization of actin filaments and focal adhesion formation was observed. The fact that HT1376 cells do not develop actin thick filaments confirms the essential role of actin content in maintaining the mechanical properties of bladder cancer cells. Recent studies on the role of the cell's environment on cytoskeleton remodeling have shifted the center of scientists' interest to studies of cancer invasiveness as

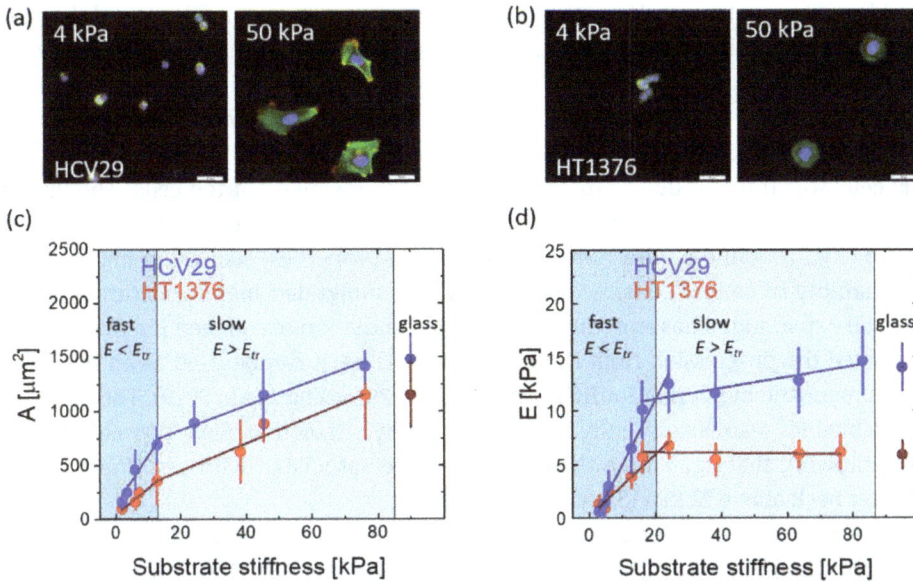

Figure 6.3.2: (a-b) Fluorescent images presenting an overlay of F-actin (green), vinculin (red), and cell nuclei (blue) recorded for bladder cancer cells cultured on substrates with various stiffnesses. (c) Spreading surface area of single cells plotted as a function of substrate stiffness. (d) The corresponding elasticity changes in cells. In both relations, a threshold-dependent character was observed (reprinted with permission from Lekka et al., 2019).

a function of how cells sense or interact with elastic gels, mimicking physiological and mechanical properties of the surrounding extracellular matrix. The mechanical properties of cells alter in response to different substrate stiffness. Thus, studies on the effect of an elastic substrate on the mechanical properties of adhering cancer cells are essential in terms of cancer cell mechanosensitivity. In exemplary experiments carried out for three bladder cancer cell lines (moderately differentiated RT112 and poorly differentiated invasive T24 and J82), cells were grown on elastic gels, with stiffness varying from 5 to 28 kPa (Abidine et al., 2018). Microrheological properties, quantified using a viscoelastic model were identified as the signature of cancer cells. Moreover, cells have been shown to adapt their properties to local conditions, especially during the contact with endothelium when they begin to reorganize actin filaments to be able to pass through.

Measurements of bladder cancer cell deformability have been applied to clinically collected voided urine cells (Shojaei-Baghini et al., 2013). Measurements were realized using micropipette aspiration, in which individual cells were soaked in a glass micropipette using pressure-controlled deformation. Results overlapped with those obtained in studies carried out on cell lines, that is, benign cells were more rigid as compared to malignant cells. These results clearly indicated that Young's modulus could be used as a biomechanical marker, in clinical practice, to enhance bladder cancer detection. More advanced approaches of biomechanics-based diagnosis for samples collected from patients who have bladder cancer have been reported for the AFM working in a ringing imaging mode combined with machine learning analysis (Sokolov et al., 2018). Surface parameters were employed to identify cancer cells. Cells were collected from urine, fixed, and imaged by AFM. Detection of cancerous cells was difficult due to the considerable heterogeneity of the cells. The use of machine learning applied directly to the sets of surface parameters derived for each image (e.g., roughness, directionality, and fractal properties) seemed to improve the detectability of cancer cells. A recent study that integrated bioinformatics analysis and cell experiments has suggested that the accumulation of collagen I in the bladder promoted the progression from NMIBC to MIBC. Thus, it can be used as an independent prognostic biomarker for BCa (Brooks et al., 2016, Zhu et al., 2019). The stiffness of the bladder neck has recently been evaluated by ultrasound shear wave elastography, showing that ages more than 45 years were associated with a stiffness of the bladder neck above 22 kPa (Sheyn et al., 2017).

Mechanical properties and accompanying alterations in actin filaments are not the only change observed during bladder cancer progression. Dynamic structure of the cell cytoskeleton also determines cell morphology. AFM has been applied to directly trace changes in bladder cancer cell morphology and elasticity during the epithelial–mesenchymal transition (EMT) induced by TGF-β1 (Wang et al., 2019). During this process, cells change their epithelial phenotype into a more migratory one (mesenchymal one). In those cells, altered cell–cell and cell–ECM interactions lead to a more extensive capability to migrate similarly as is observed during cancer progression. In experimental conditions, EMT is typically induced by transforming growth factor-β1 (TGF-β1). Based on measurements of cell topography and elasticity, T24 cancer cells change from a typical cobblestone-like shape to a more elongated spindle-shape formed after treatment with TGF-β1. In parallel, cells became more rigid because they were linked with F-actin rearrangements in these cells, especially, in the perinuclear region. This agrees with previous results (Willis et al., 2005). Another biophysical biomarker of bladder cancer progression is alteration in biochemical properties of cells, measured by combined AFM and Raman spectroscopy (Canetta et al., 2014). Instead of Raman spectroscopy, mass spectrometry can also be used in such studies (Bobrowska et al., 2019). Combined Raman/AFM techniques differentiate between normal human urothelial cells (SV-HUC-1) and bladder tumor cells (MGH-U1) with high specificity and sensitivity. Bladder cancer cells were smaller, thicker, rougher, and more deformable than normal SV-HUC-1 cells. Raman spectroscopy of cancer cells displayed typical, cancer-related changes, that is, a higher DNA, increased lipid, and decreased protein contents. Parallel change of biomechanical and biochemical properties of cells strongly indicates its potential in being used as nanobiophysical fingerprints of cancer progression.

Bladder cancer cell lines seem to be a good model not only for studies on nuclear mechanics that significantly evolve during many normal cellular processes but also in pathological conditions (Liu et al., 2014b, Worman and Courvalin, 2002). Due to a mechanical link between the ECM and nuclear lamina, it seems that the physical properties of the microenvironment surrounding the cells influence nuclear mechanics. Nuclear mechanics is mostly governed by lamina and heterochromatin that are the major mechanical components. Quantitative characterization of mechanical properties of the cell nucleus may deliver insights into mechanisms underlying cell proliferation, differentiation, as well as related diseases. Mechanical properties of cell nuclei can be measured directly or indirectly using various techniques such as micropipette aspiration (Guilak et al., 2000), substrate straining (Lammerding et al., 2005), or atomic force microscopy (Azeloglu et al., 2008). The latter technique together with the use of sharp needle tips allows penetration of the cell membrane and direct characterization of intact cell nuclei (Liu et al., 2014b). Such studies were applied to quantify the mechanical properties of the cell nucleus in two bladder cancer cell lines. A significant reorganization of the cell cytoskeleton as observed in cancer progression was a prerequisite for changes in nuclear cell mechanics. Two chosen cell lines, T24 and RT4, represent invasive and noninvasive bladder cancer cells, respectively. These cells differ mechanically: RT4 cells are more rigid than T24 cells, showing, simultaneously, that more invasive cells are more deformable (Liu et al., 2014a). By applying the needle-like tip, it has been shown that intact nuclei of noninvasive RT4 cells were more rigid than those of invasive T24 cells, that is, follow similar relations as in elasticity of whole cells. This could be correlated with the organization of actin filaments.

In summary, nanomechanical studies of bladder cancer cells show larger deformability of cancerous cells compared to normal cells, however being independent of cancer progression but simultaneously being dependent on substrate properties. Increased deformability of cells was linked to F-actin organization, but some studies showed that the actin content dominates over the spatial distribution of actin filaments. Increased deformability of cancer cells correlates with increased stiffness of cell nuclei, again demonstrated to be correlated with actin network density. Although only in the beginning, noninvasive methods for assessing areas of increased stiffness in association with collagen I deposition (i.e., fibrotic areas of the bladder) will represent a step forward in the development of technological platforms for the detection of bladder area at risk of tumor onset (i.e., relapsing regions in patients that underwent to previous transurethral bladder resection of neoplastic area) and/or progression. As a gold standard technique for measuring tissue stiffness, AFM might set up the values of the stiffness of clinical samples, such as the nonneoplastic and neoplastic bladder regions, and their fold of difference, to be used as reference values and prognostic markers for noninvasive methods. With advancing methodology of nanomechanical measurements, other biophysical properties such as morphology or biochemical composition have been shown to be parallel biomarkers of bladder cancer progression.

6.3.4.2 Primary Brain Cancers

Primary brain tumors are one of the most challenging tasks of modern medicine. Even if these tumors are rare compared to other cancers, their poor prognosis, when malignant, is devastating. Primary brain tumors are categorized according to the phylogenic origin. Tumors arising from the neuroepithelium encompass a subgroup of neoplasms collectively referred to as "gliomas." This group represents approximately 40% of primary brain tumors. These tumors are classified according to their grade (WHO classification). Tumor grade is essential when discussing treatment as well as prognosis; grade I tumors, for the most part, are well-circumscribed, noninfiltrative, and can be cured with complete surgical resection. Tumors of higher-grade infiltrate diffusely and are not amenable to cure by surgical resection alone. Despite the combination of surgery, radiotherapy, chemotherapy, and targeted therapy like bevacizumab (anti-vascular endothelial growth factor) antibody) the survival rate of grade IV (glioblastoma) is under 5% at 5 years (Gilbert et al., 2014, Omuro and DeAngelis, 2013). There is a strong need for new therapeutic approaches to treat these patients. Following the tremendous progress in understanding of physiopathological processes involved in the identification of the hallmarks of cancer (Hanahan and Weinberg, 2011), it has been increasingly recognized that the mechanical properties of the tumor microenvironment is key in cancer biology (Mierke, 2014). The mechanical properties of tissue change significantly during the progression from healthy to malignant. Due to its growth, the tumor interacts mechanically with its surroundings. It is now accepted that mechanical forces acting on cells can regulate signaling pathways responsible for cell death, division, differentiation, and migration (Katira et al., 2013). It is, then, of utmost importance to understand how the mechanical environment is regulated during tumor progression. At the molecular level, we know that complex structural changes modify the ECM during tumor initiation and progression, which would lead to mechanical modified response (Butcher et al., 2009).

6.3.4.2.1 Magnetic Resonance Elastography

Magnetic resonance elastography (MRE) belongs to a noninvasive imaging technique for quantitative measurement of the mechanical properties of biological tissues (Muthupillai et al., 1995). The technique is based on the MRE pulse sequence, which employs a magnetic-field gradient that produces changes in spin-emitted radio frequency phase signals that are proportional to spin displacement. The shift in the NMR signal phase is expressed by:

$$\varphi(\boldsymbol{r}) = \gamma \int_0^\tau G(t)u(\boldsymbol{r},t)dt \qquad (6.3.1)$$

where γ denotes the gyromagnetic ratio, G the magnetic field gradient, and u the spin displacement at position, r. Using an external actuator, it is possible to generate vibrational waves in the brain at a designed frequency, f, and to measure by MRI the wavelength, λ, of spin displacement, using local frequency estimation algorithm (Knutsson et al., 1994). The shear modulus, μ, is further determined with the hypothesis that the density of tissue, ρ, is similar to that of water with the following equation:

$$\mu = \rho \cdot \sqrt{\lambda \cdot f} \qquad (6.3.2)$$

MRE presents the great advantage of being a noninvasive technique and, thus, follows viscoelastic brain tumor properties (Kruse et al., 2008). MRE was able to demonstrate that many primary malignancies are softer than the normal brain, contrary to meningioma tumors that present a stiffer tissue (Simon et al., 2013). This preliminary work has been confirmed by a prospective study on 18 patients suffering from glioma. The stiffness (shear modulus) of tumors was compared to unaffected contralateral white matter. Gliomas were softer than healthy brain parenchyma – 2.2 kPa compared to 3.3 kPa ($p < 0.001$). Tumor stiffness has an inverse relationship with tumor grade: High-grade tumors were softer than lower grade tumors. For grades II, III, and IV, tumor stiffness was 2.7 ± 0.7 kPa, 2.2 ± 0.6 kPa, and 1.7 ± 0.5 kPa, respectively. Grade IV GBMs were significantly softer than grade II gliomas, but no statistically significant difference between grades II and III or between grades III and IV was observed. Here, the authors also looked at the impact of IDH1 mutations on stiffness. The IDH1 maker is believed to be associated with better prognosis. Tumors with an IDH1mutation were significantly stiffer than those with wild type IDH1 – 2.5 kPa versus 1.6 kPa, respectively ($p < 0.007$) (Pepin et al., 2018).

6.3.4.2.2 Ultrasonic Elastography

Measurement of local shear wave velocity could be used to quantify the stiffness of tissue. Ultrasonography can provide modalities authorizing an elastographic analysis of tissue (Chapter 6.6 on chronic liver diseases). Ultrasonic elastography is not easy for brains as the skull is a clear barrier. Despite this, a perioperative procedure could be done by a neurosurgeon, and determination of stiffness of brain tissues could be carried out. A pilot study on 63 patients diagnosed with four different brain tumor types has been conducted. Patients were suffering from low-grade and high-grade glioma (glioblastoma), meningioma and brain metastasis. The Young's moduli measured by shear wave elastography (SWE) were 23.7 ± 4.9 kPa, 11.4 ± 3.6 kPa for low grade and high grade gliomas, respectively. Meningiomas appeared stiffer with a Young modulus of 33.1 ± 5.9 kPa, and metastasis was measured at 16.7 ± 2.5 kPa. Normal brain tissue was characterized by a mean stiffness of 7.3 ± 2.1 kPa.

Moreover, low-grade glioma stiffness is different from high-grade glioma stiffness ($p = 0.01$), and normal brain stiffness is quite different from low-grade gliomas stiffness ($p < 0.01$) (Chauvet et al., 2016). These results were confirmatory of findings with MRE and quantitatively correlated, considering the following approximation between Young's modulus and shear modulus $E \approx 3 \left| G^* \right|$.

6.3.4.2.3 Atomic Force Microscopy

Several studies have revealed that many cancer cell types display a softer mechanical signature compared to normal cells (Suresh, 2007). In contrast, cancerous tissues usually appear stiffer than healthy ones. The extracellular matrix of brains is peculiar with a particularly low level of fibrous proteins and a high content of glycosamino-glycan (GAG), hyaluronan, and glycoprotein, and as a consequence, the elasticity of the brain is one of the smaller ones in our body, lower than 1 kPa. Glioblastomas are highly invasive, and the fact (through MRE or SWE) that these tumors are softer than healthy brain tissue is intriguing. An in-depth understanding of the role of mechanical cues in this disease is more than needed. AFM is one of the techniques that could give us relevant mechanical signatures in glioblastoma tissues. The AFM analysis of 15 samples from patients suffering from glioblastoma and meningioma revealed a spatial heterogeneity of elasticity in connection with tissue type. The peritumoral white matter was measured around 1 kPa, necrotic areas were softer with ∼ 0.3 kPa and non-necrotic tumor tissues were found ∼ 10 kPa. This study also confirms the stiffer mechanical signature of meningioma compared to glioblastoma (Ciasca et al., 2016). To correlate such mechanical heterogeneity with cellular events, we need to understand which are the key players of the brain elasticity. Many cell types are involved in the ECM remodeling, but in a cancerous context, innate immune cells like macrophages are the main noncancerous cellular compartment involved. As normal tissues of the human brain are impossible to obtain, the importance of immune cells in brain mechanical homeostasis could only be studied using animal samples. One tool to study dynamical and mechanical processes in brains is to use brain slices in culture from newborn rats. With this model, we can illustrate the importance of macrophages (here, resident microglial cells) by selectively depleting these cells. This could be done by clodronate-loaded liposomes that are selectively phagocytosed by macrophages. When macrophages are removed from the tissue, the elasticity of the brain is significantly increased (Figure 6.3.3).

One of the mechanisms that macrophages could use to modify the mechanical properties of brain tissues is their ability to secrete metalloproteinases. It has been shown that macrophages associated with glioblastoma use this property to facilitate glioma invasion (Hambardzumyan et al., 2015). To illustrate this process, ex-vivo brain slices could be used. Human glioblastoma cancer stem cells, derived from a patient were transduced to express GFP, and then implanted in a brain slice. Various

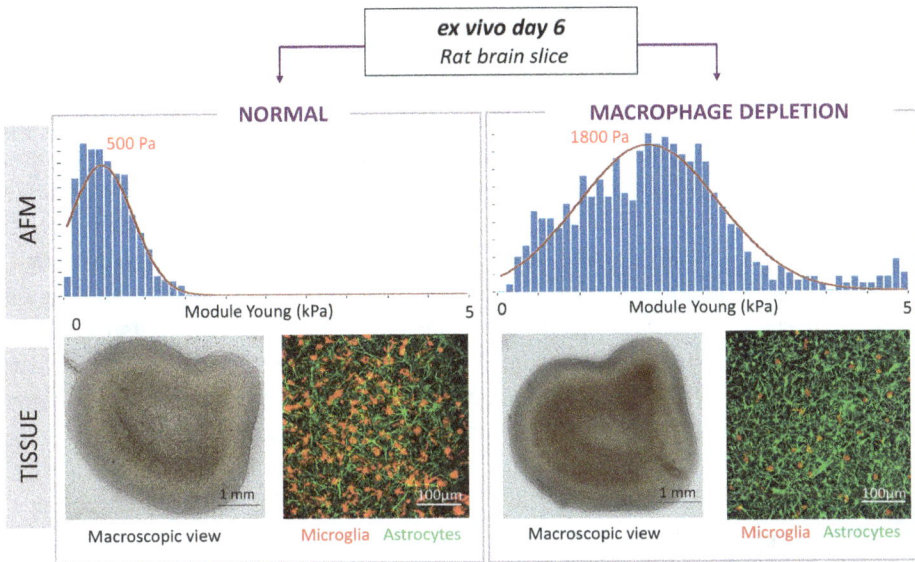

Figure 6.3.3: AFM measurement of elasticity of brain tissues when macrophages are depleted in P7 newborn rat after 6 days of ex vivo culture. IT-AFM was used with spherical borosilicate tip of 5 μm radius. Macrophages (microglia) were stained using anti-IBA1 antibody, and glial cells were stained with an anti-GFAP antibody. Images were obtained by two-photon microscopy (unpublished data, A Millet Institute for Advanced Biosciences).

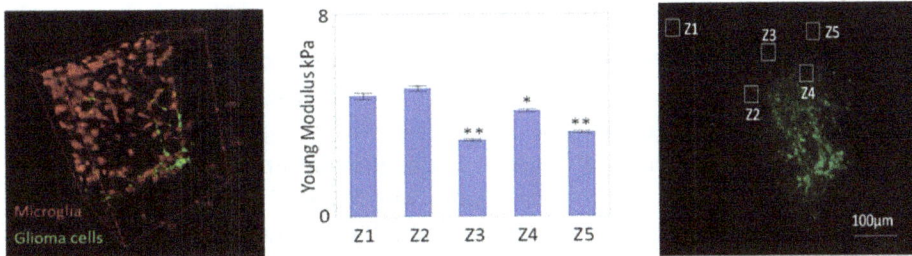

Figure 6.3.4: AFM measurement of elasticity of brain slices with implanted cancer stem cells from a patient suffering a glioblastoma. Various areas where explored (right panel). Invasion areas were associated with softer elasticity (Z3 and Z5); these were secondarily analyzed by two-photon microscopy (glioma cells transduced with GFP and microglia stained with an IBA1 antibody) showing macrophages paving the way for invasion of cancerous cells (*$p < 0.05$, **$p < 0.01$, Welch's t-test) (unpublished data, A Millet Institute for Advanced Biosciences).

areas comprising distant normal tissue, implantation area, and invading front were analyzed using AFM working in peak force mode (QNM, quantitative nanomechanical mapping). We found that invading areas are associated with a softer elasticity compared to other areas (Figure 6.3.4).

The areas where the elasticity is lower were secondarily analyzed using a two-photon microscope, and we found that macrophages in these areas pave the way for cancerous cells, illustrating the role of TAM in glioblastoma. These promising results pave the way for an AFM-based diagnostic strategy on human glioblastoma tissues. To realize it, the mechanical correlation with histological peculiarities of the brain should be addressed. The differences between cells of various types of nervous tissue might significantly contribute to the changes in elasticity of both white matter (WM) and gray matter (GM). Myelinated nerve fibers mainly occupy WM, numerous oligo-dendrocytes, and fibrous astrocytes, while GM is densely packed with unmyelinated fibers, perikaryons of nerve cells, and protoplasmic astrocytes. The content of cells' nuclei is stiffer than their processes, most likely because of the unequal local distri-bution of cell organelles (Lu et al., 2006). Myelinated nerve fibers, present in WM, are tightly bundled, resulting in strong anisotropy of the tissue, particularly when com-pared with GM (Prange et al., 2000). This property might explain why, in some studies, WM is found to be stiffer than GM (Prevost et al., 2011). Using AFM, the mechanical properties of spinal cord samples have already been reported, showing GM region to be significantly stiffer, regionally heterogeneous, and anisotropic as compared to WM region (Christ et al., 2010, Koser et al., 2015). In other studies, like the one done by Ozawa et al. (2001) no differences between WM and GM were found. This area is still requires intensive research activity.

6.3.4.2.4 Anatomical Peculiarity of Brain

Brain tumors grow in the closed environment of the almost infinitely stiff skull. Therefore, the growing mass of the tumor will induce tension that compresses nor-mal tissue surrounding the cancer. The consequences of prolonged compression have recently been presented in a mouse model of brain cancer (Seano et al., 2019). In this study, a stiff part of the mouse skull was replaced by a deformable mem-brane, to which a screw was applied. Thus, a defined compression was applied to the mouse brain to mimic prolonged compression induced by a growing tumor. In such a manner, it was possible to distinguish the mechanical effects from biochemi-cal interactions between the tumor and normal brain tissue in two forms of the tumor, nodular and infiltrative. It is somewhat obvious that prolonged stress on normal tissues was higher around nodular tumors than around the infiltrative tu-mors. Surprisingly, there was no defense mechanism present. Consequently, a re-duction of peritumoral vascular perfusion linked with impaired oxygenation of the brain and lower number of neurons in the compressed area have been observed.

6.3.4.2.5 Nanomechanics of Brain Cancer Summary

Gathered data has shown that mechanical phenotyping of brain cells can be used to classify patients according to the grade of tumors. Correlating them with precise mutation opens the doors for the new field of mechanogenomics. Moreover, nanomechanics seems to be an effective quantitative biomarker of glioblastoma also at the tissue level, highlighting the use of elastography approaches realized by MRE or ultrasonography at the macroscopic level, in clinical practice.

6.3.4.3 Breast Cancers

Breast cancer belongs to the most common cancers occurring in women. Currently, several screening techniques such as mammography, ultrasound, fine needle aspiration, or biopsy collection exist to enable identifying cancer at an early stage of progression. Early breast cancers, that is, placed in the breast or only disseminated to the axillary lymph nodes, are considered curable (usually at the level of above 70% of patients). By contrast, advanced cancers are far more difficult to cure by the currently available therapeutic options due to a large heterogeneity level of breast cancer observed at the molecular level. Thus, it is considered as a treatable disease aiming at the prolongation of survival by controlling symptoms that maintain or improve the quality of patients' life (Waks and Winer, 2019). Large developments in molecular and genetic techniques have provided us with powerful tools for the diagnosis and treatment of breast cancer patients. The standard approach includes the use of estrogen (ER) and progesterone receptor expression levels to describe biological features and endocrine responsiveness, histological grade, Ki67, and molecular signatures to evaluate proliferation and chemotherapy sensitivity, amplification status of the oncogene HER2 to stratify patients for HER2-directed treatment, and BRCA1/BRCA2 mutation status, along with other high penetrant genes for hereditary risk assessment. Treatment strategies differ according to molecular subtype of breast cancer. It includes surgery and radiation, endocrine therapy for hormone receptor-positive disease, chemotherapy, immunotherapy (e.g., anti-HER2 therapy for HER2-positive disease), and poly(ADP-ribose) polymerase inhibitors for BRCA mutation carriers. Future therapeutic concepts in breast cancer aim at personalized therapy; however, to achieve it, new biomarkers are strongly needed.

Breast cancers have been widely studied from a nanomechanical point of view, both at the cellular and tissue level, bringing results and findings in various aspects. Breast cancer cell lines have been used in studying nanomechanical properties of cells during EMT (Cascione et al., 2017), to find a correlation between cytoskeleton organization and stiffness (Calzado-Martín et al., 2016), to measure viscoelastic properties of these cells alone or in the presence of neighboring cells (Efremov et al., 2017, Schierbaum et al., 2017) and to understand the role of nanomechanics during cell invasion into collagen matrices (Staunton et al., 2016). Initially, the deformability of breast cancer cells has been evaluated using an optical stretcher (Lincoln et al., 2004). Obtained results showed that malignant MCF-7 (human breast adenocarcinoma cell line) can stretch about 5 times more as compared to nonmalignant MCF-10A breast epithelial cells. This confirmed the results obtained for bladder cancer cells (Lekka et al., 1999). Later, AFM has been applied to quantified breast cancer deformability (Li et al., 2008). This study is one of the early demonstrations that Young's moduli depend on the loading rate, showing the importance of relative measurements. Force versus indentation curves were recorded at different loading rates (from 0.03 to 1 Hz). For each dataset, the apparent Young's modulus was quantified, using Hertzian contact mechanics. It has been shown that malignant cells do not change their deformability within a whole range of loading rates. Nonmalignant breast cells were less deformable. The Young's modulus was 1.4–1.8 times larger, but in contrast to MCF-7 cells, it increases with loading rate. Together with the nanomechanical characterization of breast cells, the organization of actin filaments was investigated to find the relation between the structure of actin filaments, mechanical properties of cells, and disease. Actin filaments in malignant MCF-7 cells revealed to be disorganized in contrast to nonmalignant MCF-10A cells, in which linearly organized actin filaments were observed in the whole cell volume. The presence of the latter was associated with an increased rigidity of MCF-10A cells. The more deformable malignant cells revealed the impared organization of actin filaments. This implies that malignant cells may possess the ability to migrate more easily through the surrounding tissue matrix and small capillaries. However, AFM-based measurements of snap-frozen mammary tissues (ex vivo) showed that the malignant epithelium in situ is far stiffer than isolated breast tumor cells (Lopez et al., 2011). In further studies, the stiffness of human breast biopsies appeared to be a unique mechanical fingerprint of cancer-related changes, differentiating between normal, benign, and invasive cancers (Plodinec et al., 2012). Simultaneously, tumor progression was correlated to matrix stiffening and softening of the tumor cells that might be connected to the higher density of collagen in mammary tissues (Provenzano et al., 2008). AFM combined with confocal microscopy derived results showing that stiffening of breast cancer cells during invasion to 3D matrices composed of collagen. By affecting actomyosin contractility through Rho-associated protein kinase (ROCK) inhibitor, a significant increase in the deformability of breast adenocarcinoma cells was observed, pointing to actomyosin as a key player during the initial steps of invasion (Staunton et al., 2016).

In summary, nanomechanical studies of breast cancer cells show a larger deformability of cancerous cells, which seems to be independent of cancer progression but dependent on substrate properties. The increased deformability of cells was linked with F-actin organization, but some studies showed that the actin content dominates spatial distribution of actin filaments. The increased deformability of cancer cells correlates with increased stiffness of cell nuclei, again demonstrated to be correlated with the actin network.

6.3.4.4 Ovarian Cancers

Ovarian cancer cells, analogous to other cancer cells, are more deformable as compared to nonmalignant ovarian epithelial cells. However, studies carried out for these cells demonstrated the capability to distinguish more tumorigenic cells from less invasive types (Xu et al., 2012). A clear correlation between cell elasticity, migration, and invasion was observed between IOSE (nonmalignant immortalized ovarian surface epithelial cells) and two subpopulations of HEY (human ovarian adenocarcinoma) cells (Figure 6.3.5).

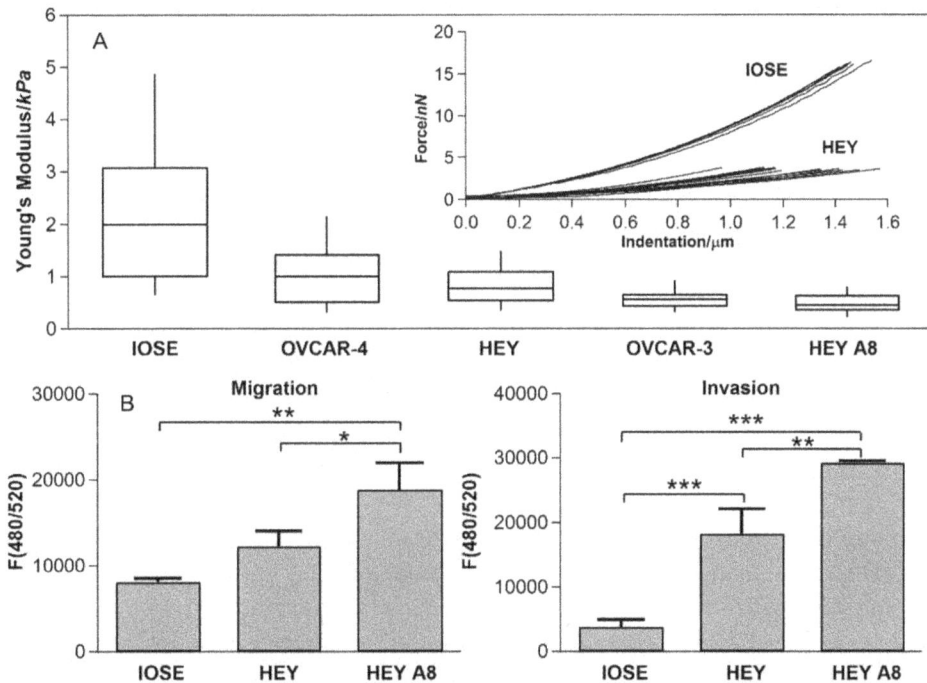

Figure 6.3.5: (A) Mechanical, (B) migratory, and (C) invasive properties of ovarian cancer cells. F(480/520) denotes the fluorescence intensity (excitation 480 nm and emission 520 nm) proportional to the number of migrating or invading cells (reprinted from Xu et al., 2012).

Similar studies were conducted on mouse ovarian surface epithelial MOSE cell lines from a primary mouse cell model for progressive ovarian cancer from C57BL/6 mice derived due to the spontaneous transformation in cell culture.

In summary, results showed that ovarian cancer cells are less deformable at benign stages of the disease progression. Deformability of cells directly increases with the advancing progression from a benign to a metastatic one. The amount of F-actin (forming actin filaments in the cytoskeleton of the cell) and its organization are directly associated with the alterations in cellular nanomechanical properties (Ketene et al., 2012b, 2012a).

6.3.4.5 Pancreatic Cancers

Pancreatic cancer is a lethal disease because most of patients are diagnosed at an advanced stage. At this point, the only treatment option is chemotherapy, which together with the inability of surgical removal, results in a noticeably short overall survival. No effective therapy is currently available for metastatic pancreatic cancers; therefore, the development of novel therapeutic options in pancreatic cancer is urgently needed.

Studies on mechanical properties of pancreatic cancers have been focused on pancreatic ductal adenocarcinoma (PDAC) due to its highly aggressive forms with extremely poor prognosis (Kim et al., 2018, Kulkarni et al., 2019, Nguyen et al., 2016, Walter et al., 2011). Several pancreatic cell lines, already measured using AFM, can be grouped as cells derived from primary cancers (Panc-1, MIA PaCa-2), and from metastasis to pleural effusion (Hs766T) and liver (PaTu8988T and PaTu8988S).

It is widely observed that cells are characterized by a wide distribution of elastic modulus, revealing, among other aspects, a large mechanical heterogeneity of cellular structures probed. Some recent study proposes "bottom-up" approach for the biophysical characterization of pancreatic cancer cell lines (Kim et al., 2018). Cells (PaTu8988T and PaTu8988S) were isolated from liver PDAC metastases. PaTu8988T presents a lower degree of differentiation in the cytoskeleton as well as faster and more disordered growth behavior. Thus, differences in their cytoskeleton structure, and consequently, in their elasticity are expected. The "bottom-up" approach relies on the way the cell cultures were – either apical (i.e., cells cultured on a glass coverslip surface) or basolateral (cells cultured on a net-shaped culture substrate). In the first approach, spherical AFM probe approaches the cells from the top, while, in the latter, from the bottom side of the cell. Obtained results showed that the basolateral approach results in a lower level of the experimental errors (a relative error: 12–17% instead of 32–33%). Simultaneously, AFM-based elasticity measurements demonstrated that PaTu8988S cells were more than 2 times more rigid as compared to the PaTu8988T cell line. These results agree with actin cytoskeleton organizations and could be correlated with predicted, larger invasiveness of the PaTu8988T cells.

The role of cell cycle in influencing the biomechanics of Panc-1 cells along with the impact of tip geometry was investigated recently (Kulkarni et al., 2019). The AFM results obtained for Panc-1 cells show that such characteristics as morphology, membrane roughness, and mechanical properties were significantly influenced by cell cycle and AFM tip geometries. The first observation is that Panc-1 cells probed with a sharp tip display larger moduli values as compared to cells measured with a blunt tip, and it agrees with previously published data for other cells (Guz et al., 2014). More interesting is that changes in mechanical properties of Panc-1 cells alter during the cell cycle in a probe-geometry-dependent manner. Cells were found to be the softest in G0/G1 phase when indented with a pyramidal probe, while they are the stiffest when probed with spherical AFM probe. A common feature is that cells arrested in S phase possess elastic moduli values that are between those of G0/G1 and G2/M phases. Although only 10–15 cells were measured in this experiment, the obtained results are valuable, showing another way of minimizing the often apparent large experimental variations by synchronizing the cell cycle phases.

In various cancers, mechanical properties of cancerous cells are related to tumor invasiveness, which has been, for instance, nicely shown for ovarian cancer (Xu et al., 2012). It is interesting whether such a relation exists for PDAC cells and to what extent mechanical properties can predict the disease aggressiveness. A study presented by Nguyen et al. (Nguyen et al., 2016) shows a relation between the invasive potential and mechanical properties of PDAC cell lines derived from human pancreatic ductal epithelium (HPDE cells), from primary tumors (Panc-1 and MIA PaCa-2) and a secondary metastatic site (Hs766T). Deformability of these cells was measured using various complementary methods such as microfiltration, single-cell microfluidic deformability cytometry, and AFM. Analogous with other cancers, noncancerous HPDE cells appeared to be the stiffest one, confirming the statement that cancerous cells are softer than the normal ones. Mechanical properties of PDAC cells were compared with their invasiveness assessed from a modified scratch wound invasion assay and a transwell migration assay. Results demonstrated that the ability of PDAC cells to migrate through pores passively is only weakly correlated with their invasive potential. However, Young's modulus of these cells reveals a strong association between cell deformability and invasive potential in PDAC cells, showing that stiffer PDAC cells are more invasive.. Furthermore, by analyzing gene expression, vimentin, actin, and lamin A were found to be the most expressed. Among them, lamin A was identified as a potential protein contributing to the variability in the mechanical properties of PDAC cells.

Some epithelial pancreatic cancer cells such as Panc-1 express unusual amounts of keratin filaments. They are a suitable system to study the role of keratin in maintaining mechanical properties of cells (Deer et al., 2010). In Panc-1 cells, keratin fibers like the other filamentous structures of the cytoskeletal scaffold constitute one part of a whole cytoskeleton; therefore, it is interesting to evaluate their impact on mechanical properties of cell cytoskeleton. One way is to compare the mechanical properties of living cells and intermediate keratin filamentous networks (Walter et al., 2011). Such studies have demonstrated that the skeletonized but structurally intact keratin network is characterized by much lower elastic modulus (~10 Pa) as compared to living cells (100–500 Pa). Such a low value may indicate negligible contribution of keratin network in mechanics of a whole cytoskeleton. In contrast, significant changes of mechanical values of living cells are observed because of a rearrangement of the keratin network. Altogether, these findings indicate that the keratin network should be studied in its intact form inside the cell as cross-linking and interlinking between various cytoskeletal components defines the dynamic structure of the cytoskeleton fully.

In summary, nanomechanical studies on pancreatic cancer cells show larger deformability of more invasive cells that could be linked with rearrangements in both the actin and the keratin networks. Apart from biomechanical characterization of these cells, their role as a model or reference cancer cell type to study various AFM-oriented methodological aspects has been shown.

6.3.4.6 Prostate Cancers

Nanomechanical measurements of prostate cancer cells have been motivated by the uncertainty of the Gleason scoring system, used to diagnose the aggressiveness of this cancer type. Until today, the question of whether the mechanical properties of cells correlate with various stages of cancer progression is not fully answered. In one of the first papers, dating back to 2008, a comparison of Young's moduli for a set of cancerous and nonmalignant cells was presented (Faria et al., 2008). In this study, the elastic properties of reference primary cells isolated from a tissue collected from a patient with benign prostate hyperplasia (BPH) were compared with well-established cancer cell lines (LNCaP and PC-3). Results show larger deformability of PC-3 and LNCaP cells. Surprisingly, the most deformable cell line was that of low metastatic potential. LNCaP cells belong to androgen-sensitive human prostate adenocarcinoma cells derived from the left supraclavicular lymph node metastasis.

Young's modulus of highly metastatic PC-3 originating from bone metastasis were comparable to that of PNT2 cells, being immortalized epithelial cells of normal adult prostate. Such results strongly indicate that the relation between the mechanical properties of cancer cells and their invasiveness or aggressiveness is overly complex. Further studies on prostate cancers confirmed the lack of such a correlation (Lekka et al., 2012a) but, simultaneously, indicated the need for a standardized protocol of AFM-based elasticity measurements, enabling a direct comparison between various laboratories. It is known that cancer cells modify their microenvironments by, for example, accumulating collagen fibers observed in optical microscope as collagen deposits in some cancer types such as breast cancer (Plodinec et al., 2012). Studies on individual properties of prostate cancer cells (for the PC-3 and LNCaP cell lines) cultured on a glass substrate coated with collagen I and fibronectin (two types of ECM proteins) showed that Young's modulus of cells can be regulated by surface properties (Docheva et al., 2010). More invasive PC-3 cells were stiffer on collagen I coated surface than on fibronectin, while LNCaP cells showed a slight tendency for the opposite relation. They seem to be more rigid when cultured on fibronectin, but Young's modulus did not exceed that of PC-3 cells. Moreover, PC-3 cells reveal higher adhesion and spreading capability to collagen I surface, which was explained by the expression of a certain repertoire of PC-3 specific surface receptors. Although there is the relation between deformability and invasiveness in prostate cancer cells, it has been reported that in one cell line (DU145 being the brain metastasis) it is possible to distinguish a population of more invasive cells (Piwowarczyk et al., 2017). Heterogeneity of cells is characteristic of cancer and seems to be essential for metastasis. A specific subpopulation of cells is preferentially observed in front of the region of prostate cancer. For DU145 prostate cancer cells, front cells express connexin-43 but, simultaneously, were also the most deformable ones. This implies that the expression of connexin-34 and the nanomechanical properties of prostate cancer cells may govern a change of prostate cells towards more invasive phenotypes. At a tissue level, in the early nineties, the so-called sonoelasticity imaging has been applied to differentiate between stiff and hard regions of some exemplary normal tissues, including prostate gland (Parker et al., 1990). Sonoelasticty imaging combines externally applied vibrations with the Doppler-based detections of abnormal regions. These early studies show no significant difference in mechanical properties of normal and benign prostate hyperplasia (BPH).

In conclusions, nanomechanics of prostate cancers measured at the single-cell level could be used to differentiate between various populations of cancer cells originating from distinct stages of cancer progression. Its use to quantify changes on the tissue level requires better-elaborated methodology as most of the prostate cancer cells are not uniform in their structure. Normal cells overlap with cancerous ones, which makes it difficult to quantify nanomechanical properties when measured in tissue.

6.3.4.7 Thyroid Cancers

Measurements of thyroid cancer demonstrated the applicability of AFM in nanomechanical assessment of primary cell lines. Cells originating from primary cell lines carry characteristics of the original tissue. Primary thyroid cells (primary cell line S748) were obtained from a surgically removed tissue surrounding a papillary carcinoma (0.7 cm diameter) of a 66-year-old female patient (Prabhune et al., 2012). The anaplastic carcinoma (primary cell line S277) was obtained from a large tumor sample of an 86-year-old female patient. Both primary cells were cultured and measured in a sequence of 3 days. Results show large variability of Young's modulus starting from 0.3 kPa. For primary cancer cells, the maximum Young's modulus is around 5 kPa, while for normal cells, it is 38 kPa.

The difference between normal and cancerous cells after one day of culture was smaller as compared to cells measured at days 2 and 3. Results show a significant increase in cell deformability. The median of the elastic modulus of normal cells was within the range of 2.2–6.9 kPa, while for cancer cells its value was smaller, that is, between 1.2 and 1.4 kPa. Cell adhesion to any surface starts with the formation of adhesive sites, followed by continuous spreading. This takes place during the first few hours of culture and reaches a steady-state phase approximately after 24 h (Hytönen and Wehrle-Haller, 2016). A threshold between continuous spreading and steady-state phase is not sharp. It explains a weak correlation between normal and cancer thyroid cells observed after 24 h of culture. Notably, the obtained difference in the mechanical properties of thyroid cells can be correlated with the organization of actin filaments. A rhodamine-phalloidin fluorescent staining shows distinct organization of the cytoskeleton in cells cultured for 3 days. A normal primary thyroid cell exhibits bundles of thick actin filaments, probably actin stress fibers. Cancer cells show less organized cytoskeletal architecture with weakly visible long actin fibers (Figure 6.3.6).

Mechanisms on how cells sense mechanical forces generated by the surrounding environment are not fully understood. In particular, there is no clear answer on how cells modulate their molecular but also biophysical properties in response to substrate stiffness. Cells cultured on hydrogels surfaces characterized by physiologically relevant stiffness reveal distinct behavior strongly dependent on the cell type, with obvious changes in the organization of actin filaments (Georges, 2005). When cells are cultured on stiff surfaces such glass coverslips or Petri dishes, in most cases, cancer cells are characterized by large deformability, linked with the remodeling of the actin cytoskeleton. When cells are cultured on compliant hydrogels, remodeling of actin cortex allows withstanding stress induced by altered microenvironment. Thus, intuitively, the overall mechanical properties of cells will be different from those cultured on a stiff surface. Indeed, thyroid cancer cells turn out to be stiffer than the normal cells (Rianna and Radmacher, 2017). Moreover, for thyroidal cells cultured on hydrogels characterized by different stiffness, various mechanical responses can be measured, that is, elasticity and dynamic viscosity of cells. Normal thyroidal cells change the mechanical properties depending on the substrate stiffness, whereas thyroidal cancer cells do not.

These results emphasize the importance of research towards the understanding of how cancer cells interact with or adapt to the soft environment to resolve mechanisms of cancer dissemination in the tissue context.

6.3.4.8 Circulating Tumor Cells (CTCs)

In recent years, we have witnessed the translation and application of information gained in preclinical studies that estimated the physical properties of tumor cells, with a growing interest in using the physical properties of cancer cells in clinical studies. Isolation of circulating tumor cells (CTCs) has been attempted with different design criteria of the devices (Aghaamoo et al., 2015, Harouaka et al., 2013). Most of the devices for isolating CTCs are based on antigen-independent approaches and target physical properties of cancer cells, such as cell size and deformability-based separation of cancer cells from the blood of patients, with one device currently being evaluated in clinical studies (Miller et al., 2018). Whereas blood cells have a size

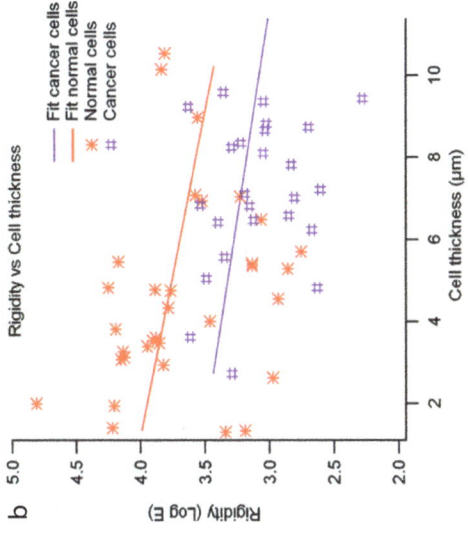

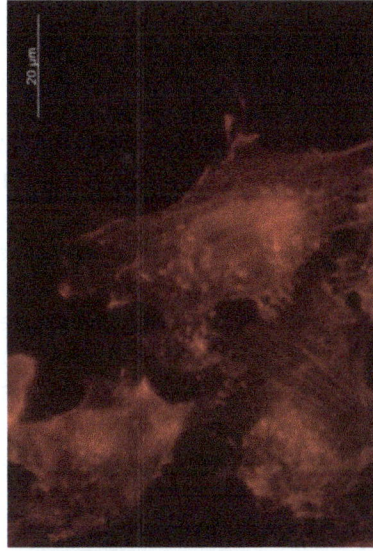

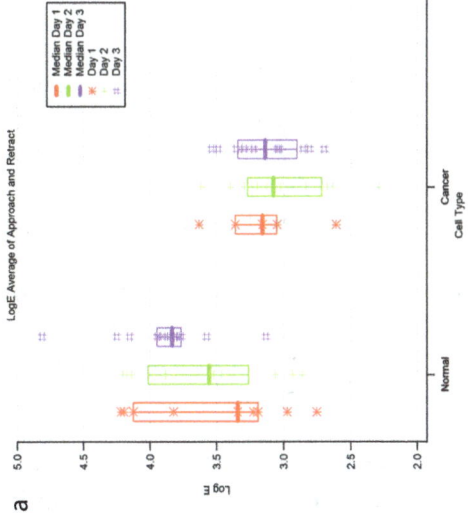

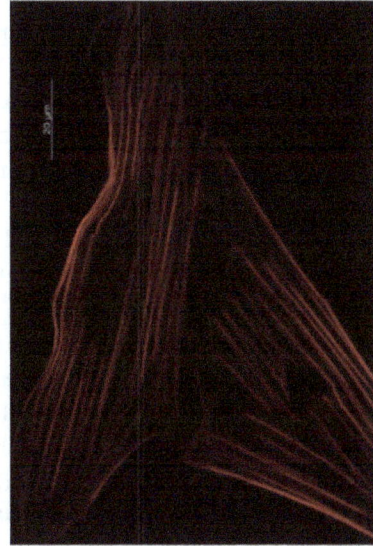

Figure 6.3.6: Variability of elastic modulus (a) in thyroid cancer in relation to cell thickness (b) and actin filament organization (c) and (d) (reprinted from Prabhune et al (2012) with modifications).

<10 μm or area <140 μm^2, CTC from several tumors are characterized by wider sized and area (Harouaka et al., 2013).

High deformability of cancerous cells from solid tumors is indicative of shorter time to extravasate and intravasate to get efficient spread on secondary sites (Lenarda et al., 2019). On the contrary, AFM indentation has confirmed that lymphocytes from chronic lymphocytic leukemia (CLL) patients have higher stiffness (i.e., lower deformability), as compared to lymphocytes in healthy samples (Zheng et al., 2015). It is still unclear why cancer cells from solid and liquid tumors have different elasticity. One of the reasons might be the ratio of nuclear to cytoplasmic volume that is different between neoplastic cells derived from solid and liquid tumors. The nuclei of CLL lymphocytes occupy almost the entire cell. Thus, it seems that the higher stiffness of the nuclei versus cytoplasm contributes to the overall decreased deformability of CLL cells. Simultaneously, these findings highlight the relevance of the physical features of the cells, and how data from preclinical studies can be translated in vivo. Indeed, the CTCs in the bloodstream represent an invaluable source of material that can be quickly and serially collected through a simple blood draw; the ability of detecting and separating CTCs is essential for early cancer detection and treatment. Thus, liquid biopsy has recently been adopted to isolate CTCs from breast, colorectal, small-cell lung and prostate tumors, and CTCs estimation and characterization have been associated with different cancer progression and survival rates (Cristofanilli et al., 2004, Vishnoi et al., 2015). For example, CTCs in bladder cancer patients have been characterized for their epithelial origin (EpCAM +, or cytokeratin + (Gazzaniga et al., 2014, 2012, Zhang et al., 2017)); moreover, a recent meta-analysis has shown the utility of estimating CTCs number in the blood of high-grade BCa patients, showing poorer overall survival for those individuals positive for CTC (hazard ratio of 3.98 and diagnostic OR for European population of 22) (Antoni et al., 2017, Azevedo et al., 2018, Zhang et al., 2017).

CTCs may represent the clonal component of the primary tumor that left the primary site, therefore being highly representative of the possible systemic metastatic spreading of the cancer. Only a few AFM-based studies have focused on their mechanical properties. Gathered results showed that CTCs isolated from prostate cancers were characterized by increased deformability following the deformability of cells they were isolated from (Chen et al., 2013, Osmulski et al., 2014).

Studies of the mechanical properties of CTCs can potentially track variations in the mechanical properties of the isolated CTCs and, simultaneously, serve as a biomarker with disease pathogenesis, progression, and metastatic potential.

6.3.5 Relativeness of Young's Modulus

The Young's modulus determined from AFM measurements is a relative value and can be used only for comparative studies in all cases when experimental conditions are conserved. Its values are calculated on the basis of the Hertz–Sneddon contact mechanics that assumes a flat surface with infinitive thickness, indented by an axi-symmetric punch (Sneddon, 1965). Living cells are not isotropic and, additionally, they reveal a viscoelastic nature. Thus, Young's moduli determined from AFM-based elasticity measurements should be treated as a relative value. Larger deformability of cancer cells has been shown for most cancers for measurements carried out on individual, isolated cells. However, the direct comparison of the results obtained in various laboratories so far is not easy.

There are multiple sources of uncertainties in AFM measurements, arising from instrumental, analytical, and biological aspects. Instrumental-related variability in elastic properties of soft samples like cells can originate from various reasons encompassing variability in the determination of deflection sensitivity, spring constant, or tip geometry. Minimization of instrumental variability has already been tackled, delivering the standardized nanomechanical AFM procedure (SNAP, Schillers et al., 2017). Using this protocol, variability in AFM-based elasticity measurements can be significantly reduced in samples with stiffnesses comparable to living cells.

Analytical sources of modulus variability encompass two aspects, that is, data acquisition and analysis. The way in which data are recorded is essential for the final Young's modulus used to compare data. Experimental settings such as time (Figure 6.3.7a) and place of poking (Figure 6.3.7b), AFM probe geometry (Figure 6.3.7c), and speed of poking (load speed, Figure 6.3.7d) have a significant effect on the determined elastic properties of cells. Other experimentally oriented factors that affect the determination of elastic modulus are maximum loading force or sample position (e.g., cell central body or periphery). Despite numerous papers showing differences between normal and cancerous cells, only a few of them have reported on the effect of probe shape on the determined Young's modulus indicating its large degree of relativeness (Chiou et al., 2013, Guz et al., 2014, Kim et al., 2013). Reported results clearly indicate that cells probed with cantilevers possessing pyramidal tips mounted at the free end reveal larger elastic moduli values (they seem to be more rigid) as compared to measurements carried out on the same cell type with spherical probes. From such research, one can impose that sharp probes can be used for AFM-based nanoindentation measurements under the conditions of keeping constant, both geometry of the probe and experimental conditions. However, it is not clear how indenting probe geometry is linked with the detection level of cancer cells. This is important for improving a diagnostic significance of AFM-based elasticity measurements, especially when absolute value of Young's modulus is difficult to be obtained.

Data analysis may also be an additional source of uncertainty. It is linked with theoretical models applied to describe nanoindentation data, especially to the choice

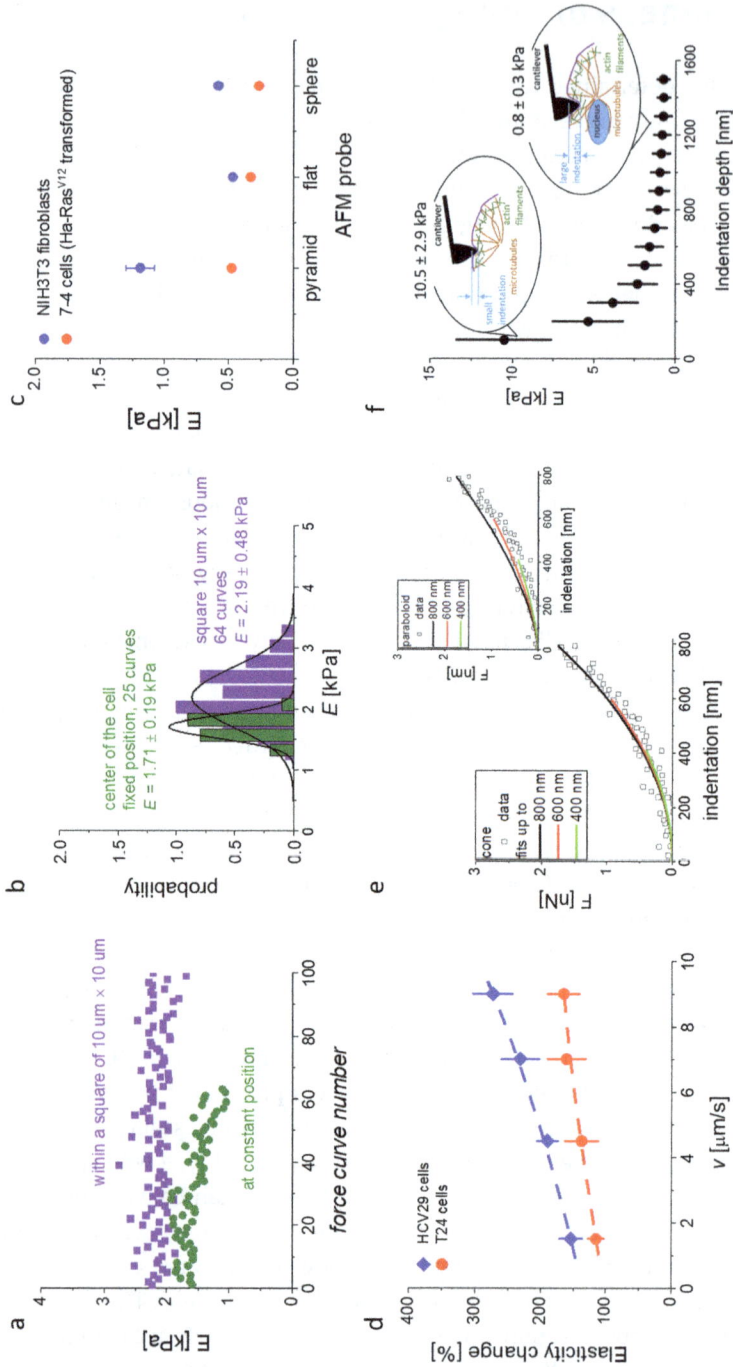

Figure 6.3.7: Experimental and analytical sources of elastic properties variability: (a) time of poking, (b) place of poking, (c) AFM probe geometry, (d) load speed, (e) choice of an approximation of the AFM probe, and (f) indentation depths (reprinted from Lekka (2016) with modifications).

of an approximation of the probing AFM tip, which is typically a pyramid (symmetric or nonsymmetric). Usually, two geometries are considered, that is, a cone or paraboloid. The approximations of the AFM tip shape define the relation between a load force and indentation depth. The cone predicts that indentation changes as $\sim x^2$, while, for the paraboloid, it is $\sim x^{3/2}$. Fitting the corresponding equation to the data may introduce some additional uncertainty (Figure 6.3.7e). Other important factors present during the data analysis are the localization of the point of contact between the indenting AFM tip and the cell's surface, range of indentation depth, or load force. The depth of indentation, due to the large structural heterogeneity of the cell interior, should also be evaluated. The results of the depth-dependent analysis show larger moduli values for small indentation depths and its decrease for large indentations (Figure 6.3.7f). The explanation can be linked to three main reasons. It, desirably, demonstrates heterogeneity of the cell interior but, simultaneously, it may reflect wrongly chosen approximation of the AFM tip shape or the presence of prestress in the sample.

Biologically related sources of uncertainty can be linked with cell culture conditions, obviously from a composition of culture medium (Nikkhah et al., 2011, Zemła et al., 2018) or modifications of surface properties by, for example, coating with poly-L-lysine (an agent increasing the number of positively charged amino groups, causing better adhesion of the cell to the glass substrate). Substrate properties influence the organization of the cell cytoskeleton and, simultaneously, cell shape (cell height, volume, and diameter). Thereby, this will affect cell mechanical properties (Docheva et al., 2010, Lekka et al., 2019, Rianna and Radmacher, 2017). For example, thyroid cancer cells appeared to be higher (thicker) than normal cells. Thus, one could expect that increased deformability of cancerous cells results from enhanced thickness of the cells, not from the cancer-related alterations. To verify it, Young's modulus was plotted against cell thickness. The relation shows a clear separation between normal and cancer cells, regardless of the cell thickness (Prabhune et al., 2012). These data demonstrate that, together with biological questions, nanomechanical measurements of thyroid cancer cells must be addressed to achieve better understanding of methodological cues influencing the determination of Young's modulus.

In summary, the relativeness of Young's modulus value seems to not affect the cancer detection capability through single-cell mechanics, as it has already been shown in many papers. It rather strongly affects the comparison of the results between various laboratories. The latter is particularly essential for diagnostic purposes where inter-laboratory cross-checking might affect the way of patient treatment. More importantly, as biomechanics may serve as a tool for understanding cell/tissue response to specific chemotherapists, understanding factors influencing the determined Young's modulus is of great importance in tailoring a personalized medical anticancer therapy.

6.3.6 Monitoring Anticancer Drugs Effect

AFM-based nanomechanical measurements, accumulated till now, clearly demonstrate that mechanical properties of pathologically altered cells and tissues are significantly different from their healthy counterparts. Although it is not common practice yet, it seems to be obvious that the nanomechanical analysis can be successfully implemented in the studies of changes induced by the action of antitumor drugs carried out for living cells (Pillet et al., 2014). In one of the early studies on the effect of actin-disrupting agents and microtubule-interfering compounds, it has been shown that disassembling actin filaments in fibroblasts resulted in a significant decrease of the Young modulus. At the same time, disaggregation of microtubules did not affect the elastic properties (Rotsch and Radmacher, 2000). The latter indicated the crucial role of the actin network in the mechanical properties of living cells. In addition, different mechanisms of disassembling actin filaments were observed: In physiological conditions, when kept constant during the experiments, cytochalasin B and D and latrunculin A treatment leads to a general softening of the cell in regions devoid of stress fibers, while, in contrast, jasplakinolide appears not to disrupt stress fibers. Such analysis can also be carried out in search of biophysical mechanism governing the effectiveness of anticancer drugs. The limitations are linked with the capability of the AFM to measure changes in mechanical properties that are mainly linked with the organization of cell cytoskeleton, in particular with actin filaments or microtubules. This directed an interest in anti-cytoskeletal agents. They can be classified according to the targeting cytoskeletal elements either as actin filaments or as microtubules. From two of these cytoskeletal networks, the microtubular system has already been an object of clinical studies and trials.

Microtubule-targeting anticancer drugs encompass three families of chemical compounds: vinca alkaloids, colchicine, and taxanes. All of them bind to tubulin dimers, leading to the diverse mechanisms of action but, regardless of the action mechanism, the final processes induced by microtubule-targeting anticancer drugs are impaired mitosis (a process of cell division) and initiated apoptosis (programmed cell death), both inhibiting cancer cell growth (Jordan and Wilson, 2004). Depending on the binding site on tubulin dimer, agents interacting with microtubules are divided into three classes. Two of them, vinca alkaloids and colchicine, upon binding to tubulin dimer, destabilize microtubule network. Taxanes, the third class of tubulin-binding agents, stabilize microtubules (Barbier et al., 2014). Determination of mechanical properties of cells treated by microtubule-targeted antitumor drug has been mostly studied for taxol-based compounds. AFM-based studies on the mechanism of taxanes interaction with microtubules showed taxol-based alterations in single protofilament conformation. Microtubules polymerize by changing a curvature of protofilaments, while taxol prevents microtubule stabilization by protecting protofilaments from curving (Elie-Caille et al., 2007). Paclitaxel is one of the antitumor drugs currently used in clinical practice. It is known that it binds to the

tubulin in tumor cells and promotes the formation of stable microtubules manifested in cell rigidity increases. In this way, it inhibits cell replication through disruption of a mitotic spindle formation and an initiation of apoptosis. In Ishikawa
and HeLa cells, apoptosis is observed after 24 h of paclitaxel treatment (Kim et al.,
2012). However, alterations in cell stiffness were dependent on the cell type. Deformability of Ishikawa cells increased in paclitaxel treated cells as compared to
control, untreated cells. In contrast, HeLa cells became more rigid after treatment
with paclitaxel for 6 and 12 h, followed by a decrease for cells treated longer than
24 h. These findings showed a correlation of cell stiffness with the paclitaxel-
mediated activation of apoptosis (Kim et al., 2012).

Nanomechanical analysis of cells' elastic properties can be applied in the detection of undesired side effects, as has been shown in chemotherapy-induced peripheral neuropathy (Au et al., 2014). Neuropathy is a major problem as it forces cancer
patients to stop their therapies, since even 50% of cancer patients who undergo chemotherapy may have sensory symptoms (Windebank, 2008). Treatment of DRG (dorsal root ganglion) neurons with vincristine and paclitaxel shows that reduced cell
elasticity in DRG neurons accompanies the development of chemotherapy-induced
peripheral neuropathy. Vincristine reduced DRG neurite formation in a dose-
dependent manner, but deformability of cells, quantified by Young's modulus,
increased upon vincristine treatment (11 kPa versus 7 kPa for untreated and vin-
cristine-treated DRG neurons, respectively). Analogously, DRG neurons treated
with paclitaxel exhibited a significant increase in the average value of Young's
modulus from 10 to 18 kPa. The two common anticancer drugs applied, for which
the mechanism of action on microtubules is opposite, showed evidence that
there is a link between cell nanomechanics and microtubule organization. Results obtained show that DRG neurons can be a model system for future development of cell-based AFM methodology for testing of upcoming antitumor agents
against neuropathy induction.

Current clinical approaches to treat cancers rely on the use of two or three well-
established anticancer drugs and profit from their synergistic effect. Mostly, taxanes
and vinca alkaloids are combined; however, recently, the interest focused on colchicine, which is also a tubulin targeted drug. Colchicine has found its main medical application in gout treatment (Dalbeth et al., 2014). Some researchers tried to use it also
for cancer treatment, despite its capability to cause severe side effects. The example of
such studies was conducted for lymphoma U937 cancer cells (Hung and Tsai, 2015).
Cells were treated with colchicine (microtubule-destabilizing agents) and taxol (micro-
tubule-stabilizing agents). Young's modulus indicated indentation depth-dependent
relation. For small indentations (less than 200 nm) the colchicine-treated cells exhibited the largest deformability, whereas the taxol-treated cells were the most rigid one.
This was possibly because taxol induces microtubule assembly and increases cell
strength, whereas colchicine induced microtubule disassembly and decreased cell
strength. For larger indentations, colchicine-treated cells were the most rigid cells as

compared to control and taxol-treated cells. This phenomenon was explained by the reduction of cell strength caused by microtubule disassembly. In another study, paclitaxel and vinorelbine have been applied alone and in combination with human lung adenocarcinoma cell line (Jung et al., 2004). Vinorelbine and paclitaxel bind differently to tubulin dimer, inducing different effects on microtubules. Paclitaxel stabilizes polymerized tubulin into nonfunctional microtubular bundles, and, in this way, it blocks the progression of mitosis. Vinorelbine blocks polymerization of microtubules impairing mitosis. Both finally trigger the apoptosis in the treated cells. A combination of these drugs acts synergistically and enhances apoptosis as a primary mechanism of cell killing. A very nice demonstrations of the use of nanomechanics as a novel determinant for screening and developing new anticancer agents have been presented for human prostate cancer cell (PC-3 cell line) treated with eight different anticancer drugs (Ren et al., 2015). These were disulfiram (DSF), paclitaxel (Taxol), tomatine, BAY 11-7082 (BAY), vaproic acid (VPA), 12-O-tetradecanoylphorbol-13-acetate (TPA), celecoxib, and MK-2206 (MK). To study the effect of each anticancer drug on deformability of PC-3 cells, nanomechanical results were compared to control untreated cells. All the drug-treated cells had a much higher Young's modulus and followed a relation: the larger the drug dose, the more rigid the cells. Together with elastic modulus quantification, two different mechanisms involved in the action of anticancer drugs were revealed. For the three drugs, DSF, MK, and taxol, the cell cytoskeleton network reconstruction may lead to stiffening or softening of the proteins' structures (e.g., filament shortening and thickening), but it may not cause changes of polymerization of actin filaments inside the cells. Although MK, taxol and DSF may reveal different mechanisms of the interactions, the similar trend of changes in Young's modulus may explain the similarity of these drugs' effects on cellular nanomechanical behavior. The other five drugs had a different effect on the cells. They changed both the elastic and viscous behavior of the PC-3 cell line. The cell cytoskeleton stiffened in response to different dynamics of actin filament polymerization, resulting in their reorganization. The commonly used MTT assay failed to show any difference in cell viability upon PC-3 cell treatment with these drugs. These findings suggest that AFM may reveal new aspects of the biological effects of anticancer drugs on cells. Microtubule-targeted anticancer drugs are not the only ones affecting the nanomechanical properties of cells. In one of the first papers, the effect of chitosan on the stiffness and glycolytic activity of bladder cancer cells was studied. Chitosan is a linear polysaccharide, derived from chitin with a potential antitumor activity linked with the inhibition of glycolytic activity of cells. Bladder cancer cells were treated with microcrystalline chitosan, with three different deacetylation degrees. While for nonmalignant HCV29 cells, the difference in the stiffness was insignificant, the results obtained for cancerous T24 cells were quite different. Stiffness of the cells increased significantly after chitosan treatment. Cell deformability changes were accompanied by an inhibition of glycolytic activity of these cells, upon chitosan treatment. Positively charged chitosan binds to the negatively charged cell membrane and induces a change in the group of glycolytic enzymes that required cytoskeleton participation. By changing stiffness, chitosan deactivated these classes of enzymes leading to the inhibition of glycolysis (Lekka et al., 2001). Other studies focused on human neutrophil peptide-1 (HNP-1), an endogenous antimicrobial peptide, which exerts a cytotoxic effect on cancerous cells (McKeown et al., 2006). After treatment with HNP-1, prostate cancer cells revealed damage to the cellular membrane accompanied by changes in the cell shape and the fragmentation of the nucleus (Gaspar et al, 2015b). Elasticity measurements showed that HNP-1-treated cells have a larger deformability related to internal damage in the cytoskeleton. Analogously, as for microtubule-targeting anticancer drugs, apoptosis was observed. In the studies on how cellular deformability changes in drug-sensitive and drug-resistance cells upon treatment with the similar anticancer drug, ovarian cancer cells (A2780 cisplatin-sensitive and A2780cis cisplatin-resistant) were treated with cisplatin (Seo et al., 2015). A2780cis cells showed about 3 times higher migratory behavior as compared to A2780 cells. Subsequently, AFM results show larger deformability of cisplatin-sensitive cells (80 ± 49 Pa) as compared to cisplatin-resistant cells (273 ± 236 Pa). Interestingly, the drug-sensitive A2780 cells showed a normal Gaussian distribution of the elastic modulus with one major peak. For cisplatin-resistant A2780cis cells, a bimodal distribution with two main peaks was obtained. More rigid cells may have obstacles in deforming themselves, but more deformable cells generate too small traction force to be able to penetrate through the matrix. Thus, A2780cis cells should be able to generate proper traction forces for invasion, possibly resulting in a better chance to escape from the drug treatment.

6.3.7 Future of AFM: From Single Cells to Tissue

Although only at the beginning, noninvasive methods for assessing areas of increased stiffness in association with collagen I deposition (i.e., fibrotic areas of the bladder) will represent a step forward in the development of technological platforms for the detection of bladder area at risk of tumor onset (i.e., relapsing regions in patients that underwent to previous transurethral bladder resection of neoplastic area) and/or progression. As a gold standard technique for measuring tissue stiffness, AFM might set up the values of the stiffness of clinical samples, such as the nonneoplastic and neoplastic bladder regions, and their fold of difference, to be used as reference values and prognostic markers for noninvasive methods.

In summary, gathered data on nanomechanical properties of cells treated with various anticancer drugs has demonstrated the functionality of AFM in investigation of cellular changes induced by them. Nanomechanical assays can be applied for screening the effectiveness of anticancer drugs, to quantify the magnitude of resistance of cells to a specific drug, or to elaborate the degree of side effects such as neuropathy. Such measurements provide an excellent tool to better understanding of how mechanical properties contribute, change, and affect cellular response to anticancer drugs. Nanomechanical assays of anticancer drug effectiveness may be beneficial for antitumor drug design that would improve cancer treatment.

References

Abidine, Y., A. Constantinescu, V. M. Laurent, V. Sundar Rajan, R. Michel, V. Laplaud, A. Duperray and C. Verdier (2018). "Mechanosensitivity of cancer cells in contact with soft substrates using AFM." Biophysical Journal 114: 1165–1175.

Aghaamoo, M., Z. Zhang, X. Chen and J. Xu (2015). "Deformability-based circulating cell separation with conical-shaped microfilters: Concept, optimization, and design criteria." Biomicrofluidics 9: 034106.

Anastasiadis, A. and T. M. de Reijke (2012). "Best practice in the treatment of nonmuscle invasive bladder cancer." Therapeutic Advances in Urology 4: 13–32.

Antoni, S., J. Ferlay, I. Soerjomataram, A. Znaor, A. Jemal and F. Bray (2017). "Bladder cancer incidence and mortality: A global overview and recent trends." European Urology 71(1): 96–108.

Au, N. P. B., Y. Fang, N. Xi, K. W. C. Lai and C. H. E. Ma (2014). "Probing for chemotherapy-induced peripheral neuropathy in live dorsal root ganglion neurons with atomic force microscopy." Nanomedicine Nanotechnology 10: 1323–1333.

Azeloglu, E. U., J. Bhattacharya and K. D. Costa (2008). "Atomic force microscope elastography reveals phenotypic differences in alveolar cell stiffness." Journal of Applied Physiology: Respiratory, Environmental and Exercise Physiology 105: 652–661.

Azevedo, R., J. Soares, A. Peixoto, S. Cotton, L. Lima, L. L. Santos and J. A. Ferreira (2018). "Circulating tumor cells in bladder cancer: Emerging technologies and clinical implications foreseeing precision oncology." Urologic Oncology 36(5): 221–236.

Barbier, P., P. O. Tsvetkov, G. Breuzard and F. Devred (2014). "Deciphering the molecular mechanisms of anti-tubulin plant derived drugs." Phytochemistry Reviews 13: 157–169.

Ben-Ze'ev, A. (1985). "The cytoskeleton in cancer cells." Biochimica et biophysica acta 780: 197–212.

Bobrowska, J., K. Awsiuk, J. Pabijan, P. Bobrowski, J. Lekki, K. M. Sowa, J. Rysz, A. Budkowski and M. Lekka (2019). "Biophysical and biochemical characteristics as complementary indicators of melanoma progression." Analytical Chemistry 91: 9885–9892.

Brooks, M., Q. Mo, R. Krasnow, P. L. Ho, Y. C. Lee, J. Xiao, A. Kurtova, S. Lerner, G. Godoy, W. Jian, P. Castro, F. Chen, D. Rowley, M. Ittmann and K. S. Chan (2016). "Positive association of collagen type I with non-muscle invasive bladder cancer progression." Oncotarget 7: 82609–82619.

Butcher, D. T., T. Alliston and V. M. Weaver (2009). "A tense situation: Forcing tumour progression." Nature Reviews Cancer 9: 108–122.

Calzado-Martín, A., M. Encinar, J. Tamayo, M. Calleja and A. San Paulo (2016). "Effect of actin organization on the stiffness of living breast cancer cells revealed by peak-force modulation atomic force microscopy." ACS Nano 10: 3365–3374.

Canetta, E., A. Riches, E. Borger, S. Herrington, K. Dholakia and A. K. Adya (2014). "Discrimination of bladder cancer cells from normal urothelial cells with high specificity and sensitivity: Combined application of atomic force microscopy and modulated Raman spectroscopy." Acta Biomater 10: 2043–2055.

Carrion, R. and J. Seigne (2002). "Surgical management of bladder carcinoma." Cancer Control : Journal of the Moffitt Cancer Center 9(4): 284–292.

Cascione, M., V. De Matteis, C. C. Toma, P. Pellegrino, S. Leporatti and R. Rinaldi (2017). "Morphomechanical and structural changes induced by ROCK inhibitor in breast cancer cells." Experimental Cell Research **360**: 303–309.

Chauvet, D., M. Imbault, L. Capelle, C. Demene, M. Mossad, C. Karachi, A. L. Boch, J. L. Gennisson and M. Tanter (2016). "In vivo measurement of brain tumor elasticity using intraoperative shear wave elastography." Ultraschall der Medizin **37**: 584–590.

Chen, C. L., D. Mahalingam, P. Osmulski, R. R. Jadhav, C. M. Wang, R. J. Leach, T. C. Chang, S. D. Weitman, A. P. Kumar, L. Sun, M. E. Gaczynska, I. M. Thompson and T. H. M. Huang (2013). "Single-cell analysis of circulating tumor cells identifies cumulative expression patterns of EMT-related genes in metastatic prostate cancer." Prostate **73**: 813–826.

Chiou, Y. W., H. K. Lin, M. J. Tang, H. H. Lin and M. L. Yeh (2013). "The influence of physical and physiological cues on atomic force microscopy-based cell stiffness assessment." PLoS One **8**: e77384.

Christ, A. F., K. Franze, H. Gautier, P. Moshayedi, J. Fawcett, R. J. M. Franklin, R. T. Karadottir and J. Guck (2010). "Mechanical difference between white and gray matter in the rat cerebellum measured by scanning force microscopy." Journal of Biomechanics **43**: 2986–2992.

Ciasca, G., T. E. Sassun, E. Minelli, M. Antonelli, M. Papi, A. Santoro, F. Giangaspero, R. Delfini and M. De Spirito (2016). "Nano-mechanical signature of brain tumours." Nanoscale **8**: 19629–19643.

Clark, A. G. and D. M. Vignjevic (2015). "Modes of cancer cell invasion and the role of the microenvironment." Current Opinion in Cell Biology **36**: 13–22.

Cox, T. R. and J. T. Erler (2011). "Remodeling and homeostasis of the extracellular matrix: Implications for fibrotic diseases and cancer." Disease Models & Mechanisms **4**(2): 165–178.

Cristofanilli, M., G. T. Budd, M. J. Ellis, A. Stopeck, J. Matera, M. C. Miller, J. M. Reuben, G. V. Doyle, W. J. Allard, L. W. M. M. Terstappen and D. F. Hayes (2004). "Circulating tumor cells, disease progression, and survival in metastatic breast cancer." The New England Journal of Medicine **351**: 781–791.

Dalbeth, N., T. J. Lauterio and H. R. Wolfe (2014). "Mechanism of action of colchicine in the treatment of gout." Clinical Therapeutics **36**(10): 1565–1479.

Deer, E. L., J. González-Hernández, J. D. Coursen, J. E. Shea, J. Ngatia, C. L. Scaife, M. A. Firpo and S. J. Mulvihill (2010). "Phenotype and genotype of pancreatic cancer cell lines." Pancreas **39**(4): 425–435.

Docheva, D., D. Padula, M. Schieker and H. Clausen-Schaumann (2010). "Effect of collagen I and fibronectin on the adhesion, elasticity and cytoskeletal organization of prostate cancer cells." Biochemical and Biophysical Research Communications **402**: 361–366.

Efremov, Y. M., W. H. Wang, S. D. Hardy, R. L. Geahlen and A. Raman (2017). "Measuring nanoscale viscoelastic parameters of cells directly from AFM force-displacement curves." Scientific Reports 7: 1541.

Elie-Caille, C., F. Severin, J. Helenius, J. Howard, D. J. Muller and A. A. Hyman (2007). "Report straight GDP-tubulin protofilaments form in the presence of taxol." Current Biology : CB **17**: 1765–1770.

Faria, E. C., N. Ma, E. Gazi, P. Gardner, M. Brown, N. W. Clarke and R. D. Snook (2008). "Measurement of elastic properties of prostate cancer cells using AFM." Analyst **133**: 1498–1500.

Fouad, Y. A. and C. Aanei (2017). "Revisiting the hallmarks of cancer." American Journal of Cancer Research 7(5): 1016–1036.

Gaspar, D., J. M. Freire, T. R. Pacheco, J. T. Barata and M. A. R. B. Castanho (2015a). "Apoptotic human neutrophil peptide-1 anti-tumor activity revealed by cellular biomechanics." Biochimica Et Biophysica Acta-Molecular Cell Research **1853**: 308–316.

Gaspar, P., M. V. Holder, B. L. Aerne, F. Janody and N. Tapon (2015b). "Zyxin antagonizes the FERM protein expanded to couple f-actin and yorkie-dependent organ growth." Current Biology : CB **25**: 679–689.

Gazzaniga, P., E. De Berardinis, C. Raimondi, A. Gradilone, G. M. Busetto, E. De Falco, C. Nicolazzo, R. Giovannone, V. Gentile, E. Cortesi and K. Pantel (2014). "Circulating tumor cells detection has independent prognostic impact in high-risk non-muscle invasive bladder cancer." International Journal of Cancer **13**: 1978–1982.

Gazzaniga, P., A. Gradilone, E. De Berardinis, G. M. Busetto, C. Raimondi, O. Gandini, C. Nicolazzo, A. Petracca, B. Vincenzi, A. Farcomeni, V. Gentile, E. Cortesi and L. Frati (2012). "Prognostic value of circulating tumor cellsin nonmuscle invasive bladder cancer: A cell search analysis." Annals of Oncology **23**: 2352–2356.

Georges, P. C. (2005). "Cell type-specific response to growth on soft materials." Journal of Applied Physiology: Respiratory, Environmental and Exercise Physiology **98**: 1547–1553.

Gilbert, M. R., J. J. Dignam, T. S. Armstrong, J. S. Wefel, D. T. Blumenthal, M. A. Vogelbaum, H. Colman, A. Chakravarti, S. Pugh, M. Won, R. Jeraj, P. D. Brown, K. A. Jaeckle, D. Schiff, V. W. Stieber, D. G. Brachman, M. Werner-Wasik, I. W. Tremont-Lukats, E. P. Sulman, K. D. Aldape, W. J. Curran and M. P. Mehta (2014). "A randomized trial of bevacizumab for newly diagnosed glioblastoma." The New England Journal of Medicine **370**: 699–708.

Guilak, F., J. R. Tedrow and R. Burgkark (2000). "Viscoelastic properties of the cell nucleus." Biochemical and Biophysical Research Communications **269**: 781–786.

Guz, N., M. Dokukin, V. Kalaparthi and I. Sokolov (2014). "If cell mechanics can be described by elastic modulus: study of different models and probes used in indentation experiments." Biophysical Journal **107**: 564–575.

Hall, A. (2009). "The cytoskeleton and cancer." Cancer Metastasis Reviews **28**: 5–14.

Hambardzumyan, D., D. H. Gutmann and H. Kettenmann (2015). "The role of microglia and macrophages in glioma maintenance and progression." Nature Neuroscience **19**(1): 20–27.

Hanahan, D. and R. A. Weinberg (2011). "Hallmarks of cancer: The next generation." Cell **144**: 646–674.

Harouaka, R. A., M. Nisic and S. Y. Zheng (2013). "Circulating tumor cell enrichment based on physical properties." Journal of Laboratory Automation **18**: 455–468.

Hung, M. and M. Tsai (2015). "Investigating the influence of anti-cancer drugs on the mechanics of cells using AFM." Bionanoscience **5**: 156–161.

Hytönen, V. P. and B. Wehrle-Haller (2016). "Mechanosensing in cell-matrix adhesions – Converting tension into chemical signals." Experimental Cell Research **343**(1): 35–41.

Iwanicki, M. P., R. A. Davidowitz, M. R. Ng, A. Besser, T. Muranen, M. Merritt, G. Danuser, T. Ince and J. S. Brugge (2011). "Ovarian cancer spheroids use myosin-generated force to clear the mesothelium." Cancer Discovery **1**: 144–157.

Janmey, P. A. and C. A. McCulloch (2007). "Cell mechanics: Integrating cell responses to mechanical stimuli." Annual Review of Biomedical Engineering **9**: 1–34.

Janmey, P. A., J. P. Winer, M. E. Murray and Q. Wen (2009). "The hard life of soft cells." Cell Motility and the Cytoskeleton **66**(8): 597–605.

Jordan, M. A. and L. Wilson (2004). "Microtubules as a target for anticancer drugs." Nature Reviews. Cancer **4**: 253–265.

Jung, M., S. Grunberg, C. Timblin, S. B. P. Vacek, D. J. Taatjes and B. T. Mossman (2004). "Paclitaxel and vinorelbine cause synergistic increases in apoptosis but not in microtubular disruption in human lung adenocarcinoma cells (A-549)." Histochemistry and Cell Biology **121**: 115–121.

Katira, P., R. T. Bonnecaze and M. H. Zaman (2013). "Modeling the mechanics of cancer: Effect of changes in cellular and extra-cellular mechanical properties." Frontiers in Oncology **3**: 145.

Ketene, A. N., P. C. Roberts, A. A. Shea, E. M. Schmelz and M. Agah (2012a). "Actin filaments play a primary role for structural integrity and viscoelastic response in cells." Integrative Biology **4**: 540–549.

Ketene, A. N., E. M. Schmelz, P. C. Roberts and M. Agah (2012b). "The effects of cancer progression on the viscoelasticity of ovarian cell cytoskeleton structures." Nanomedicine Nanotechnology **8**: 93–102.

Kim, J. H., K. Riehemann and H. Fuchs (2018). "Force spectroscopy on a cell drum: AFM measurements on the basolateral side of cells via inverted cell cultures." ACS Applied Materials & Interfaces **10**: 12485–12490.

Kim, K. S., C. H. Cho, E. K. Park, M. Jung and K. Yoon (2012). "AFM-detected apoptotic changes in morphology and biophysical property caused by paclitaxel in ishikawa and HeLa cells." PLoS One **7**(1): e30066.

Kim, W. J. (2016). "Changing landscape of diagnosis and treatment of bladder cancer." Investigative and Clinical Urology **57**: S1–S3.

Kim, Y., J. W. Hong, J. Kim and J. H. Shin (2013). "Comparative study on the differential mechanical properties of human liver cancer and normal cells." Animal Cells and Systems **17**(3): 170–178.

Knutsson, H., C. F. Westin and G. Granlund (1994). "Local multiscale frequency and bandwidth estimation." Proceedings of 1st International Conference on Image Processing **1**: 36–40.

Koser, D. E., E. Moeendarbary, J. Hanne, S. Kuerten and K. Franze (2015). "CNS cell distribution and axon orientation determine local spinal cord mechanical properties." Biophysical Journal **108**: 2137–2147.

Kruse, S. A., G. H. Rose and R. L. Ehman (2008). "Magnetic resonance elastography of the brain." Neuroimaging **39**: 231–237.

Kulkarni, T., A. Tam, D. Mukhopadhyay and S. Bhattacharya (2019). "AFM study: Cell cycle and probe geometry influences nanomechanical characterization of Panc1 cells." Biochimica Et Biophysica Acta – General Subjects **1863**: 802–812.

Lammerding, J., J. Hsiao, P. C. Schulze, S. Kozlov, C. L. Stewart and R. T. Lee (2005). "Abnormal nuclear shape and impaired mechanotransduction in emerin-deficient cells." The Journal of Cell Biology **170**: 781–791.

Lekka, M. (2016). "Discrimination between normal and cancerous cells using AFM." Bionanoscience **6**: 65–80.

Lekka, M., D. Gil, K. Pogoda, J. Dulińska-Litewka, R. Jach, J. Gostek, O. Klymenko, S. Prauzner-Bechcicki, Z. Stachura, J. Wiltowska-Zuber, K. Okoń and P. Laidler (2012a). "Cancer cell detection in tissue sections using AFM." Archives of Biochemistry and Biophysics **518**: 151–156.

Lekka, M., P. Laidler, D. Gil, J. Lekki, Z. Stachura and A. Z. Hrynkiewicz (1999). "Elasticity of normal and cancerous human bladder cells studied by scanning force microscopy." European Biophysics Journal : EBJ **28**: 312–316.

Lekka, M., P. Laidler, I. Ignacak, M. Labedz, J. Lekki, H. Struszczyk, Z. Stachura and A. Z. Hrynkiewicz (2001). "The effect of chitosan on stiffness and glycolytic activity of human bladder cells." Biochimica et biophysica acta **1540**: 127–136.

Lekka, M., J. Pabijan and B. Orzechowska (2019). "Morphological and mechanical stability of bladder cancer cells in response to substrate rigidity." Biochimica Et Biophysica Acta – General Subjects **1863**: 1006–1014.

Lekka, M., K. Pogoda, J. Gostek, O. Klymenko, S. Prauzner-Bechcicki, J. Wiltowska-Zuber, J. Jaczewska, J. Lekki and Z. Stachura (2012b). "Cancer cell recognition – Mechanical phenotype." Micron **43**: 1259–1266.

Lenarda, P., A. Coclite and P. Decuzzi (2019). "Unraveling the vascular fate of deformable circulating tumor cells via a hierarchical computational model." Cellular and Molecular Bioengineering 12: 543–558.

Levental, K. R., H. Yu, L. Kass, J. N. Lakins, M. Egeblad, J. T. Erler, S. F. T. Fong, K. Csiszar, A. Giaccia, W. Weninger, M. Yamauchi, D. L. Gasser and V. M. Weaver (2009). "Matrix crosslinking forces tumor progression by enhancing integrin signaling." Cell 139: 891–906.

Li, Q. S., G. Y. H. Lee, C. N. Ong and C. T. Lim (2008). "AFM indentation study of breast cancer cells." Biochemical and Biophysical Research Communications 374: 609–613.

Lincoln, B., H. M. Erickson, S. Schinkinger, F. Wottawah, D. Mitchell, S. Ulvick, C. Bilby and J. Guck (2004). "Deformability-based flow cytometry." Cytometry 59A: 203–209.

Liu, H., Q. Tan, W. R. Geddie, M. A. S. Jewett, N. Phillips, D. Ke, C. A. Simmons and Y. Sun (2014a). "Biophysical characterization of bladder cancer cells with different metastatic potential." Cell Biochemistry and Biophysics 68: 241–246.

Liu, H., J. Wen, Y. Xiao, J. Liu, S. Hopyan, M. Radisic, C. A. Simmons and Y. Sun (2014b). "In situ mechanical characterization of the cell nucleus by atomic force microscopy." ACS Nano 8: 3821–3828.

Lopez, J. I., I. Kang, W. K. You, D. M. McDonald and V. M. Weaver (2011). "In situ force mapping of mammary gland transformation." Integrative Biology 3: 910–921.

Lu, Y. B., K. Franze, G. Seifert, C. Steinhäuser, F. Kirchhoff, H. Wolburg, J. Guck, P. Janmey, E. Q. Wei, J. Käs and A. Reichenbach (2006). "Viscoelastic properties of individual glial cells and neurons in the CNS." Proceedings of the National Academy of Sciences of the United States of America 103: 17759–17764.

McKeown, S. T. W., F. T. Lundy, J. Nelson, D. Lockhart, C. R. Irwin, C. G. Cowan and J. J. Marley (2006). "The cytotoxic effects of human neutrophil peptide-1 (HNP1) and lactoferrin on oral squamous cell carcinoma (OSCC) in vitro." Oral Oncology 42: 685–690.

Mierke, C. T. (2014). "The fundamental role of mechanical properties in the progression of cancer disease and inflammation." Reports on Progress in Physics 77(7): 076602.

Miller, M. C., P. S. Robinson, D. Wagner and D. J. O'Shannessy (2018). "The ParsortixTM cell separation system – A versatile liquid biopsy platform." Cytometry Part A 93: 1234–1239.

Muthupillai, R., D. J. Lomas, P. J. Rossman, J. F. Greenleaf, A. Manduca and L. R. Ehman (1995). "Magnetic resonance elastography by direct visualization of propagating acoustic strain waves." Science 269: 1854–1857.

Nenadic, I. Z., B. Qiang, M. W. Urban, L. H. De Araujo Vasconcelo, A. Nabavizadeh, A. Alizad, J. F. Greenleaf and M. Fatemi (2013). "Ultrasound bladder vibrometry method for measuring viscoelasticity of the bladder wall." Physics in Medicine and Biology 58: 2675–2695.

Nguyen, A. V., K. D. Nyberg, M. B. Scott, A. M. Welsh, A. H. Nguyen, N. Wu, S. V. Hohlbauch, N. A. Geisse, E. A. Gibb, A. G. Robertson, T. C. Donahue and A. C. Rowat (2016). "Stiffness of pancreatic cancer cells is associated with increased invasive potential." Integrative Biology 8: 1232–1245.

Nikkhah, M., J. S. Strobl, E. M. Schmelz and M. Agah (2011). "Evaluation of the influence of growth medium composition on cell elasticity." Journal of Biomechanics 44: 762–766.

Omuro, A. and L. M. DeAngelis (2013). "Glioblastoma and other malignant gliomas: A clinical review." JAMA 310(17): 1842–1850.

Osmulski, P., D. Mahalingam, M. E. Gaczynska, J. Liu, S. Huang, A. M. Horning, C. M. Wang, I. M. Thompson, T. H. M. Huang and C. L. Chen (2014). "Nanomechanical biomarkers of single circulating tumor cells for detection of castration resistant prostate cancer." Prostate 74: 1297–1307.

Ozawa, H., T. Matsumoto, T. Ohashi, M. Sato and S. Sokubun (2001). "Comparison of spinal cord gray matter and white matter softness: Measurement by pipette aspiration method." Journal of Neurosurgery **95**: 221–224.

Parker, K. J., S. R. Huang, R. A. Musulin and R. M. Lerner (1990). "Tissue response to mechanical vibrations for "sonoelasticity imaging." Ultrasound in Medicine & Biology **16**: 241–246.

Paszek, M. J., N. Zahir, K. R. Johnson, J. N. Lakins, G. I. Rozenberg, A. Gefen, C. A. Reinhart-King, S. S. Margulies, D. Dembo, D. Boettiger, D. A. Hammer and V. M. Weaver (2005). "Tensional homeostasis and the malignant phenotype." Cancer Cell **8**: 241–254.

Pepin, K. M., K. P. McGee, A. Arani, D. S. Lake, K. J. Glaser, A. Manduca, I. F. Parney, R. L. Ehman and J. Huston (2018). "MR elastography analysis of glioma stiffness and IDH1-mutation status." American Journal of Neuroradiology **39**: 31–36.

Pillet, F., L. Chopinet, C. Formosa and E. Dague (2014). "Atomic force microscopy and pharmacology: From microbiology to cancerology." Biochimica et biophysica acta **1840**: 1028–1050.

Piwowarczyk, K., M. Sarna, D. Ryszawy and J. Czyz (2017). "Invasive Cx43high sub-line of human prostate DU145 cells displays increased nanomechanical deformability." Acta Biochimica Polonica **64**: 445–449.

Plodinec, M., M. Loparic, C. A. Monnier, E. C. Obermann, R. Zanetti-Dallenbach, P. Oertle, J. T. Hyotyla, U. Aebi, M. Bentires-Alj, R. Y. H. Lim and C. A. Schoenenberger (2012). "The nanomechanical signature of breast cancer." Nature Nanotechnology 7: 757–765. 10.1038/nnano.2012.167.

Prabhune, M., G. Belge, A. Dotzauer, J. Bullerdiek and M. Radmacher (2012). "Comparison of mechanical properties of normal and malignant thyroid cells." Micron **43**: 1267–1272.

Prange, M., D. F. Meaney and S. S. Margulies (2000). "Defining brain mechanical properties: effects of region, direction, and species." Stapp Car Crash Journal **44**: 205–213.

Prevost, T. P., G. Jin, M. A. De Moya, H. B. Alam, S. Suresh and S. Socrate (2011). "Dynamic mechanical response of brain tissue in indentation in vivo, in situ and in vitro." Acta Biomater 7: 4090–4101.

Provenzano, P. P., D. R. Inman, K. W. Eliceiri, J. G. Knittel, L. Yan, C. T. Rueden, J. G. White and P. J. Keely (2008). "Collagen density promotes mammary tumor initiation and progression." BMC Medicine **6**: 11.

Puttini, S., M. Lekka, O. M. Dorchies, D. Saugy, T. Incitti, U. T. Ruegg, I. Bozzoni, A. J. Kulik and N. Mermod (2009). "Gene-mediated restoration of normal myofiber elasticity in dystrophic muscles." Molecular Therapy **17**: 19–25.

Ramos, J. R., J. Pabijan, R. Garcia and M. Lekka (2014). "The softening of human bladder cancer cells happens at an early stage of the malignancy process." Beilstein Journal of Nanotechnology **5**: 447–457.

Ren, J., H. Huang, Y. Liu, X. Zheng and Q. Zou (2015). "An atomic force microscope study revealed two mechanisms in the effect of anticancer drugs on rate-dependent Young's modulus of human prostate cancer cells." PLoS One **10**: e0126107.

Rianna, C. and M. Radmacher (2017). "Influence of microenvironment topography and stiffness on the mechanics and motility of normal and cancer renal cells." Nanoscale **9**: 11222–11230.

Rotsch, C. and M. Radmacher (2000). "Drug-induced changes of cytoskeletal structure and mechanics in fibroblasts: An atomic force microscopy study." Biophysical Journal **78**: 520–535.

Satelli, A. and S. Li (2011). "Vimentin in cancer and its potential as a molecular target for cancer therapy." Cellular and Molecular Life Sciences : CMLS **68**: 3033–3046.

Schierbaum, N., J. Rheinlaender and T. E. Schäffer (2017). "Viscoelastic properties of normal and cancerous human breast cells are affected differently by contact to adjacent cells." Acta Biomater **55**: 239–248.

Schillers, H., C. Rianna, J. Schäpe, T. Luque, H. Doschke, M. Wälte, J. J. Uriarte, N. Campillo, G. P. A. Michanetzis, J. Bobrowska, A. Dumitru, E. T. Herruzo, S. Bovio, P. Parot, M. Galluzzi, A. Podestà, L. Puricelli, S. Scheuring, Y. Missirlis, R. Garcia, M. Odorico, J. M. Teulon, F. Lafont, M. Lekka, F. Rico, A. Rigato, J. L. Pellequer, H. Oberleithner, D. Navajas and M. Radmacher (2017). "Standardized nanomechanical atomic force microscopy procedure (SNAP) for measuring soft and biological samples." Scientific Reports **7**: 5117.

Seano, G., H. T. Nia, K. E. Emblem, M. Datta, J. Ren, S. Krishnan, J. Kloepper, M. C. Pinho, W. W. Ho, M. Ghosh, V. Askoxylakis, G. B. Ferraro, L. Riedemann, E. R. Gerstner, T. T. Batchelor, P. Y. Wen, N. U. Lin, A. J. Grodzinsky, D. Fukumura, P. Huang, J. W. Baish, P. K. Padera, L. L. Munn and R. K. Jain (2019). "Solid stress in brain tumours causes neuronal loss and neurological dysfunction and can be reversed by lithium." Nature Biomedical Engineering **3**: 230–245.

Seo, Y. H., Y. Jo, Y. J. Oh and S. Park (2015). "Nano-mechanical reinforcement in drug-resistant ovarian cancer cells." Biological & Pharmaceutical Bulletin **38**: 389–395.

Sheyn, D., Y. Ahmed, N. Azar, S. El-Nashar, A. Hijaz and S. Mahajan (2017). "Trans-abdominal ultrasound shear wave elastographyfor quantitative assessment of female bladder neck elasticity." International Urogynecology Journal **28**: 763–768.

Shojaei-Baghini, E., Y. Zheng, M. A. S. Jewett, W. B. Geddie and Y. Sun (2013). "Mechanical characterization of benign and malignant urothelial cells from voided urine." Applied Physics Letters **102**: 123704.

Simon, M., J. Guo, S. Papazoglou, H. Scholand-Engler, C. Erdmann, U. Melchert, M. Bonsanto, J. Braun, D. Petersen, I. Sack and J. Wuerfel (2013). "Non-invasive characterization of intracranial tumors by magnetic resonance elastography." New Journal of Physics **15**(8): 085024.

Sneddon, I. N. (1965). "The relation between load and penetration in the axisymmetric Boussinesq problem for a punch of arbitrary profile." International Journal of Engineering Science **3**: 47–57.

Sokolov, I., M. E. Dokukin, V. Kalaparthi, M. Miljkovic, A. Wang, J. D. Seigne, P. Grivas and E. Demidenko (2018). "Noninvasive diagnostic imaging using machine-learning analysis of nanoresolution images of cell surfaces: Detection of bladder cancer." Proceedings of the National Academy of Sciences of the United States of America **115**: 12920–12925.

Staunton, J. R., B. L. Doss, S. Lindsay and R. Ros (2016). "Correlating confocal microscopy and atomic force indentation reveals metastatic cancer cells stiffen during invasion into collagen in matrices." Scientific Reports **6**: 19686.

Stein, J. P., G. Lieskovsky, R. Cote, S. Groshen, A. C. Feng, S. Boyd, E. Skinner, B. Bochner, D. Thangathurai, M. Mikhail, D. Raghavan and D. G. Skinner (2001). "Radical cystectomy in the treatment of invasive bladder cancer: Long-term results in 1,054 patients." Journal of Clinical Oncology : Official Journal of the American Society of Clinical Oncology **19**: 666–675.

Suresh, S. (2007). "Biomechanics and biophysics of cancer cells." Acta Biomater **3**: 413–438.

Trivedi, D., V. Collins, E. Roberts, J. Scopetta, T. Wu, B. Crawford and Y. Nakanishi (2017). "Perforated gastric metastasis of Merkel cell carcinoma: Case report and review of the literature." Human Pathology: Case Reports **8**: 20–23.

Vishnoi, M., S. Peddibhotla, W. Yin, A. T. Scamardo, G. C. George, D. S. Hong and D. Marchetti (2015). "The isolation and characterization of CTC subsets related to breast cancer dormancy." Scientific Reports **5**: 17533.

Waks, A. G. and E. P. Winer (2019). "Breast cancer treatment: A review." JAMA **321**: 288–300.

Walter, N., T. Busch, T. Seufferlein and J. P. Spatz (2011). "Elastic moduli of living epithelial pancreatic cancer cells and their skeletonized keratin intermediate filament network." Biointerphases **6**(2): 79–85.

Wang, N., M. Zhang, Y. Chang, N. Niu, Y. Guan, M. Ye, C. Li and J. Tang (2019). "Directly observing alterations of morphology and mechanical properties of living cancer cells with atomic force microscopy." Talanta **191**: 461–468.

Willis, B. C., J. M. Liebler, K. Luby-Phelps, A. G. Nicholson, E. D. Crandall, R. M. Du Bois and Z. Borok (2005). "Induction of epithelial-mesenchymal transition in alveolar epithelial cells by transforming growth factor-β1: Potential role in idiopathic pulmonary fibrosis." The American Journal of Pathology **166**: 1321–1332.

Windebank, A. (2008). "Chemotherapy-induced neuropathy." Journal of the Peripheral Nervous System **46**: 27–46.

Worman, H. J. and J. C. Courvalin (2002). "The nuclear lamina and inherited disease." Trends in Cell Biology **12**(12): 591–598.

Xu, W., R. Mezencev, B. Kim, L. Wang, M. Donald and T. Sulchek (2012). "Cell stiffness is a biomarker of the metastatic potential of ovarian cancer cells." PLoS One **7**: e46609.

Zemła, J., J. Danilkiewicz, B. Orzechowska, J. Pabijan, S. Seweryn and M. Lekka (2018). "Atomic force microscopy as a tool for assessing the cellular elasticity and adhesiveness to identify cancer cells and tissues." Seminars in Cell and Developmental Biology **73**: 115–124.

Zhang, Z., W. Fan, Q. Deng, S. Tang, P. Wang, P. Xu, J. Wang and M. Yu (2017). "The prognostic and diagnostic value of circulating tumor cells in bladder cancer and upper tract urothelial carcinoma: A meta analysis of 30 published studies." Oncotarget **8**: 59527–59538.

Zheng, Y., J. Wen, J. Nguyen, M. A. Cachia, C. Wang and Y. Sun (2015). "Decreased deformability of lymphocytes in chronic lymphocytic leukemia." Scientific Reports **5**: 7613.

Zhu, H., H. Chen, J. Wang, L. Zhou and S. Liu (2019). "Collagen stiffness promoted non-muscle-invasive bladder cancer progression to muscle-invasive bladder cancer." OncoTargets and Therapy **12**: 3441–3457.

Marco Girasole, Simone Dinarelli

6.4 Blood Cells

6.4.1 Introduction

Blood is one of the most important fluids of our body. It consists of a buffered liquid (plasma), rich in salts, organic compounds, proteins, and three different cellular types: platelets, responsible for the coagulation; white cells (WBCs), that are the main actors of the immunity response; and the red cells (RBCs). RBCs, in particular, constitute approximately 90% of the cellular volume of the blood and represent the evolutionary-selected response to the problem of the low solubility of gases, such as oxygen and carbon dioxide, in the body fluids. This, in turn, allows cell respiration in complex organisms. Indeed, to tackle the issue of oxygen transfer in complex systems, two main approaches have developed in the multicellular species. The first is the development of large circulating proteins that are capable of capturing a huge amount of oxygen and are free-to-circulate in the body fluid; this is the case of the giant erythrocruorins diffused in many oligochaetes and crustacean (Weber and Vinogradov, 2001, Girasole et al., 2005). The second road, adopted by the vast majority of the higher organisms, consists of the development of a dedicated microenvironment that is easy to regulate and, possibly, defend: the erythrocytes.

The red blood cells contain significant concentrations (millimolar range) of hemoglobin (Hb), a very much investigated tetrameric protein (Antonini and Brunori, 1970) that contains heme-iron and is capable of cooperative and reversible binding of molecular oxygen and carbon dioxide, enabling their transport from the lungs to the peripheral tissues, and vice versa. In addition to this main physiological function, a variety of secondary roles, mostly regulative, has been suggested for Hb (Giardina et al., 1995). The presence of the cellular microenvironment is certainly very important in the protection of the structural and functional integrity of Hb as well as in the control of the protein's activity, especially with regard to the oxidation state of iron and the role of allosteric effectors. Besides this protective role, however, the main cell function of O_2/CO_2 shuttle induces several important consequences on the structural and cellular biology of the erythrocytes that makes these cells somehow unique in the body. In the next paragraph we will briefly recall a few important characteristics of the red cells that will clarify the correlation between their morphology, structure, and function, and will properly highlight the importance of studying their mechanical properties.

Acknowledgment: The authors wish to thank Dr. Giovanni Longo for the helpful discussions and the support provided.

Marco Girasole, Simone Dinarelli, Institute for the Structure of Matter, CNR, Rome, Italy

https://doi.org/10.1515/9783110989380-014

6.4.2 RBC Structure and Cellular Biology

A mature RBC has a characteristic biconcave disk shape, which is found only in this type of cells. Its typical diameter ranges between 7 and 8 μm, while its height varies from a maximum of about 1 micron at the cell edge to a minimum of a few hundred nanometers at the central concavity. The mature cell physiology is very simple, as it does not contain nucleus, mitochondria, or the entire apparata for protein production or enzymatic synthesis. It is interesting to note that, typically, the newborn erythrocytes are released into circulation as immature cells, called reticulocytes. In this transient status, the red cells still have a residual of nucleic acids and synthetic apparatus (Ney, 2011) that are lost in the very first days of maturation, but that can be very useful to quantify and understand the rate of new cell production in the body.

The lack of genetic and synthesis apparata has important consequences: for instance, the cell cannot self-replicate. The blood turnover mechanism relies on the synthesis of new cells in the bone marrow, produced by pluripotent progenitors, common to the white and red cell lines, and on the balanced removal of the senescent cell, carried out mostly by the spleen macrophages. The balance of new cells synthesis and old cells removal is clearly coupled and controlled, but can be the source of severe health problems in case of mis-regulation. This indicates why the investigation of the pattern and regulation of erythrocyte aging are important for a variety of topics. In addition, the lack of genetic and synthetic apparata has other consequences: the cell cannot regulate its own life cycle and cannot respond to environmental stimuli by adapting its protein content to the changed environmental conditions. Therefore, as a whole, the RBCs are biochemical machines that must be extremely robust and that they contain, since their release into circulation, all that they need to perform their task for their whole life.

Further unique characteristics of the RBCs are somehow linked to the significant oxygen flux from the cell body, as a consequence of their physiologic function. In this condition, the cells experience a strongly oxidative environment associated to a high level of ROS production, especially superoxide and peroxides. A consequence is that the role of the reducing power and of the other antioxidant systems is fundamental for the cell survival. Furthermore, due to the specific role of the oxygen in these cells and to the lack of mitochondria in mature cells, the production of ATP cannot be performed by means of conventional oxidative respiration. In erythrocytes, this resource is produced by following a simplified, yet robust metabolism, which is well controlled at the biochemical level. The entire amount of ATP derives from glycolysis, an ancient pathway developed when the Earth's atmosphere was still poor in oxygen. A side pathway of glycolysis, known as the shunt of Rapoport-Luebering, is responsible for the production of bis-phosphoglycerate, that is, the major allosteric effector that controls the Hb affinity for oxygen. Finally, the erythrocyte-reducing power (NADPH) arises from the pentose phosphate pathways (PPP). The NADPH is precious as it is employed in direct redox reactions as well as to maintain, in a reduced state, the glutathione, which is the primary reducing agent of the cell (and blood). To complete the picture, we cite that glycolysis and PPP are correlated, since, depending on the cell requirements, the metabolic flux can be directed towards the first or the second pathway. A role in this mechanism is played by the feedback regulation, which is triggered by the final product as well as by the control of the glucose-6-phosphate-dehydrogenase enzyme, which drives the PPP. In this particular landscape, the profound significance of cell morphology and of the relationship between the biochemical state, and the morphological and biomechanical properties of erythrocytes must be emphasized.

The cellular shape is maintained through a dense protein network present below the cell membrane and which, while performing a similar function, differs in structure, composition, and mechanical characteristics, from the usual cell cytoskeleton. In RBCs, this protein network is usually called *membrane skeleton*, to underline the tight conjugation between these two components. The proteic part consists of a robust network of coiled spectrin filaments, which is directly connected through suitable junctional complexes (containing important proteins such as actin, ankyrin, and band 4.1) to the band 3 protein, the most important and abundant integral membrane protein. This protein, in addition to playing an important functional and regulatory role in RBC's respiratory activity and glycolysis (Cluitmans et al., 2016, Kinoshita et al., 2007), guarantees a very deep contact between the membrane and the proteic skeleton. As a matter of fact, the membrane-skeleton architecture controls the cell shape and mediates the morphological response to environmental stimuli; moreover, it is responsible for the mechanical characteristics and for the remarkable elastic deformability of the erythrocytes. Figure 6.4.1 shows the intricate network of proteins, composing the RBC skeleton, for two different sample preparations, which produce a slightly stretched (panel a) and a normal (b) structure.

These exceptional mechanical characteristics are critical in ensuring a proper cell physiological function. Indeed, these cells travel along the veins and arteries that are larger than the cell size, but also through tiny capillaries that can be as narrow as a few microns, where RBCs are squeezed and experience significant mechanical solicitations. A major role in the cell shape adaptation, in response to shear stresses, is played by the cytoplasmic viscosity, which depends on the Hb concentration (Mohandas and Gallagher, 2008). Furthermore, strong osmotic stress must be endured while the cells travel across the kidneys and the spleen, and, as a result, the actual cell survival requires a skeletal architecture that is capable of enduring the stresses and to support the rapid and elastic cell squeezing. When the membrane-skeletal support fails, the cell becomes fragile and is no longer capable of working properly. Moreover, in general, membrane-skeletal failure as well as the accumulation of degraded forms of Hb (e.g., hemichromes), represent different signals that the cell is no longer able, either structurally or functionally, to carry out its physiological role. In these cases, possibly through the mediation of a complex

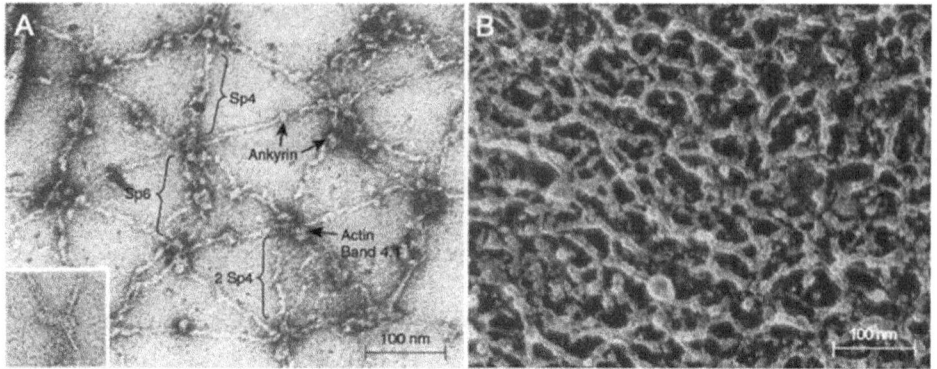

Figure 6.4.1: Electron micrograph of stretched (a) and normal (b) skeleton structure and organization. Reproduced with authorization from Lux (2016).

pattern involving both structural and functional signals (Antonelou et al., 2010, Pietraforte et al., 2007), the erythrocytes are labeled as senescent and, later, their fate will be their removal from blood circulation.

In light of the above considerations, it is clear that the mechanical properties of RBCs can be regarded as regulatory functions, whose variations due to aging or pathological alterations have consequences spanning the entire life cycle and turnover of these cells.

6.4.3 Role and Modulation of the Mechanical Properties: Cell Mechanic and Mechanotransduction

The mechanical properties of erythrocytes are very important parameters to determine the differences in their life cycle (e.g., young or senescent) or the presence of pathological situations. Such variations can be observed either very locally, in specific areas of the membrane, or throughout the whole cell surface, and in this sense, an AFM-based approach is particularly suitable, since it has the potential to reveal and measure local as well as global alterations of the cell architecture.

From the discussion of the previous paragraph it is clear that cell aging is the fundamental physiological phenomenon that governs the erythrocytes' turnover and blood homeostasis. RBC's aging is associated with biophysical and biochemical alterations of the cells, with potential consequences on blood properties, and hence different aspects of the phenomenon have been studied (Antonelou et al., 2010, Pietraforte et al., 2007, Huang et al., 2011). Moreover, the study of RBC's aging is extremely interesting as it determines progressive variations in the structural, morphological, and functional properties of the cell (Girasole et al., 2010, Dinarelli et al., 2018a), and provides a unique opportunity to understand the molecular mechanisms that modulate the bio-mechanics of erythrocytes. The variation of mechanical properties along cell aging, in particular, has practical consequences for the use of the blood for transfusion purposes. Indeed, an increased stiffness of the cell at long aging time has been observed even in blood bank conditions, and it was associated to a reduction of the transfusion efficiency (D'Alessandro et al., 2010, Huruta et al., 1998, Xu et al., 2018). Some approaches involving the use of SPM techniques are available in literature and have been focused on the variation of the nanomechanical properties as well as to their association with morphological features, both in laboratory (Girasole et al., 2012) and in blood bank conditions (Kozlova et al., 2017). The general emerging landscape shows converging evidences that, as aging increases, the observable morphological landscape is accompanied by a complex evolution of the Young's modulus and by the observation of local cytoskeletal or membrane defects, possibly related to metabolically regulated membrane detachments (Klarl et al., 2006, Borghi and Brochard-Wyart, 2007), which evolve in large-scale morphological features (e.g., vesicles or spicule). Measurements of the biomechanical data have also been proposed as a biomarker of cell quality for transfusion. Furthermore, the phenomena that accompany the morphological alteration characteristics of cell aging were recently investigated by a novel, nanoscale-sensitive approach (Ruggeri et al., 2018). The combination of AFM and IR nano-spectroscopy (see Figure 6.4.2) allowed to suggest that the biconcave-to-echinocyte transformation can be driven by oxidation, both at the membrane level and in membrane-skeletal proteins (very likely the 4.1), resulting in biomechanical alterations. These phenomena can operate in parallel to the cytoskeletal alterations, consequent to (artificial) oxidative stresses, which were observed by other groups (Sinha et al., 2015), and are also in line with previous biochemical data (Mohanty et al., 2014) that suggested oxidative stress can induce cell aging through different pathways.

As previously clarified, the cell membrane-skeleton is responsible for controlling the shape and mechanical properties of the RBC, but this is not the only task that this structure carries out. Indeed, the RBCs are capable of directly sensing environmental mechanical stimulation by means of the transmission of a mechanical deformation at the protein-skeleton interface. Such sensing (usually) results in the activation of the mechanotransduction pathway, which is devoted to transforming the mechanical stimulus into biochemical cascades.

Many of the phenomena that activate the mechanotransduction are completely physiological and are part of well consolidated biological patterns that lead to better adaptation to the environment. In other cases, the activation of the mechanotransduction can be intrinsic or even parasitic, for example, the release of ATP following cell swelling, a phenomenon that naturally occurs during the cell senescence but that can be encountered also in some pathologies or infections (Alvarez

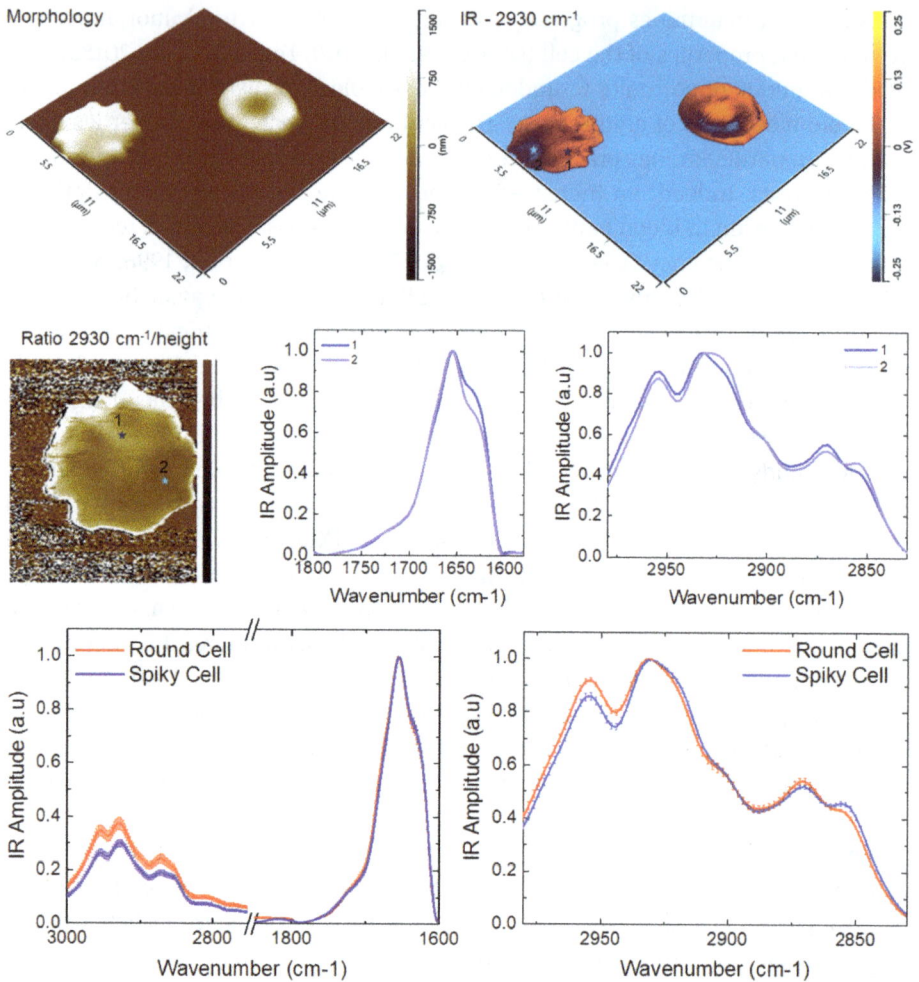

Figure 6.4.2: Chemical comparison of a biconcave and an echinocyte RBCs. 3D (a) morphology and (b) IR absorption maps of the cells. (c) Ratio images detail of the echinocyte obtained by the division of the IR signal at 2930 cm^{-1} (CH_2 asymmetric stretching, lipids) and the morphology map. Average smoothed and normalized spectra acquired in the spectroscopic (d) protein region and (e) lipid region for a bright (1) and dark (2) region of absorption within the echinocyte (each spectrum is the average of 5 independent spectra); the dark region of the cell shows sign of oxidative stress. (f,g) Averaged and normalized spectra within several dark (n = 45) and bright areas (n = 30) within the two cells in the amide band I and the lipid regions with their standard deviation. Figure reproduced with permission from Ruggeri et al. (2018).

et al., 2014, Leal Denis et al., 2016). The most interesting example of the role of mechanotransduction in the cell physiology is provided by the behavior of erythrocytes in the narrow capillaries. Indeed, at the level of the peripheral vasculature, the cells must travel along capillaries that are significantly smaller than RBCs' size.

To make the flux possible, the cell releases ATP into the extracellular medium through a mechanotransductive mechanism that is related to the sensing of the mechanical pressure exerted by the vessel walls (Forsyth et al., 2011). After release, the ATP diffuses into the endothelial cells, where it triggers the synthesis of nitric oxide (NO), the actual effector of the relaxation of the endothelium and of the dilatation of capillaries, thus permitting the cell flux. This phenomenon, as a whole, is initiated by the sensing of external mechanical stimulus, and takes advantage of the existence of a complex mechanosensing system for the translation. The release of ATP is strongly regulated, as it requires activation of a signaling pathway involving G proteins, protein kinases A, C, and one or more exit channel. Furthermore, RBCs possess several classes of mechanical receptors, such as the CFTR receptors, known for its involvement in the occurrence of cystic fibrosis, the pannexin 1, and the receptors of the Yz family (Sprague et al., 2001, Sridharan et al., 2012). As evidenced by some authors (Wan et al., 2008), environmental mechanical solicitation results in deformation of the membrane-skeleton network, which during the subsequent relaxation triggers the activation of suitable mechanosensors, and, in particular, of the pannexin I. This latter protein, whose activation requires a specific timescale for the dynamic deformation–relaxation, is the key factor determining ATP release. From the evolutionary point of view, it is interesting to note that is not unusual to discover that members of the transduction machinery play a role in other fundamental, often metabolic or regulative, cell activities.

It is interesting to conclude this discussion by citing a special case of mechanotransduction regarding the overlap between mechanobiology and gravitational biology (van Loon, 2008). Indeed, gravity or weightlessness can actually be used as a tool to understand some basic processes of life and, obviously, the mechanisms of mechanotransduction. In this sense, besides the evidence of full body adaptation to the environmental conditions (e.g., space anemia), very little information is available on RBCs exposed to simulated microgravity conditions at the cellular level (Herranz et al., 2013, Rizzo et al., 2012, Udden et al., 1995, Dinarelli et al., 2018b). Monitoring the RBCs in simulated microgravity for a long time periods allows depicting a scenario in which the cells actually sense the weightlessness conditions as a particular mechanical stimulus, and react by adapting their metabolism to this environmental condition. Such rapid metabolic adaptation, over long periods of time, is rooted in structural and morphological alterations (therefore, in modulations or alterations of functionality). Interestingly, an increase of the expelled ATP has been observed in microgravity condition, compared to static control, suggesting that the mechanotransductive machinery plays a role in the response to microgravity conditions.

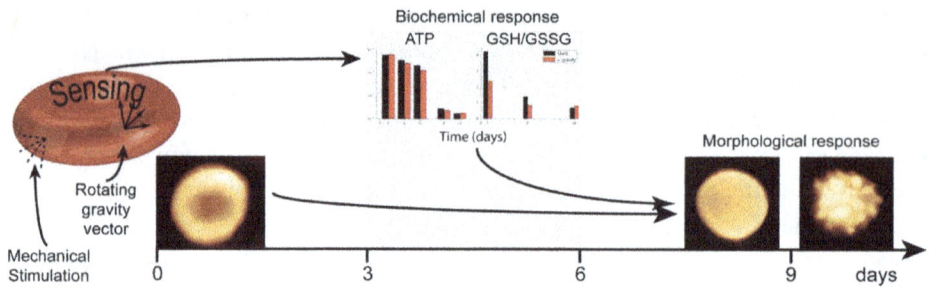

Figure 6.4.3: Sketch of the environmental sensing, signal transduction and metabolic regulation occurring in an RBC. The morphological pattern results from the integration of the mechanical and biochemical stimuli over time. Obtained with permission from Dinarelli et al. (2018b).

6.4.4 Elastic and Viscoelastic Behavior of RBCs

Blood is a complex fluid that behaves like a non-Newtonian liquid, and thus has quite complex rheological properties. On the other hand, the flow properties of blood are so important, even in a clinical perspective, that much effort has been dedicated to understanding its properties. In principle, the rheological and hemodynamical properties of blood depend on several factors: the number of cells (essentially, red cells, which are more than 90% of the total), the cell-cell interactions and, of course, the individual mechanical properties of the RBCs. Each of these critical factors has been the subject of investigation and, in the last 50 years, several studies have been dedicated to understanding the elastic and viscoelastic behavior of these cells. The first reports claiming a viscoelastic behavior in erythrocytes were available since the middle 1960s (Chien et al., 1975), from conventional studies focused on measuring the relationship between cells' stress and strain at various frequencies of external solicitation.

While the elastic properties of RBCs can be essentially traced back to the structure and architecture of the cell skeleton, their viscoelastic behavior is more complex to describe and understand, but it is fundamental to comprehend the capacity of the cells to adapt to the stresses experienced in the circulation, and to modulate their shape accordingly. Viscoelasticity is usually associated to the behavior of the cell membrane, which introduces viscosity, bending elasticity, resistance to area increase, and shear stress (Fischer, 2004, Kloppel and Wall, 2011).

Since the measurement of viscoelastic properties of RBCs is dependent on the measurement method and on the investigated biosystem, the data collected by means of different techniques, including micropipette aspiration, optical tweezers, rheometry, or flicker spectroscopy, should be compared with some caution and, sometimes, it suffers from limitation due to the low throughput (Ito et al., 2017). In this landscape, AFM stands out as particularly apt single-cell tool for such characterizations, as it provides a very accurate control and evaluation of the applied loads, together with an extremely precise measurement of the cell deformation, which is critical to describe qualitatively different behaviors of the cells. Indeed, AFM investigations are available since the pioneering work of Radmacher et al. (1993), which described the use of the microscope in force modulation mode, to evidence and map the viscoelastic properties of cells.

In a typical AFM investigation, conventional force curves collected with low surface sampling frequencies (1–10 Hz or less) can be employed to quantify the elastic properties of materials, since their frequency allows time for the surface relaxation. On the other hand, acquisition of force curves at higher frequencies can provide a glimpse to the viscoelastic behavior of the cells, since the resulting response is characterized by the frequency dependence of stress–deformation relationships. This behavior has been theoretically described by the Attard model (Attard, 2001). The model, by introducing two time-dependent elastic limits and a characteristic relaxation time, predicts that for slow interactions, the material's modulus contains only the elastic part, while for high frequency probing also, the viscous contribution is relevant.

A detailed description of the effects observable from a force–volume measurement on RBCs is reported by Bremmell et al. (2006). These authors used a colloidal probe, AFM employing a silica glass sphere, to get a detailed measurement of the deformation of the immobilized RBCs on a surface. The authors measured the hysteresis in the forward vs backward force curves, which increases as a function of the load and the probe speed, in a way identified as a typical viscoelastic behavior. Furthermore, using the Attard model, they successfully detected anelastic modulus at low deformations, and several, time-dependent viscoelastic moduli, observed at higher level of cell deformations (i.e., by fitting different part of the force curves). Related results have been obtained recently by another group, which observed, through low frequency AFM measurement, a nonlinear behavior of the mechanical properties of RBCs at high deformation levels (Kozlova et al., 2018). The authors tested the effects of stressors (glutaraldehyde and hemin) that induced a stiffening of the cells, and determined the indentation depth after which the cell mechanics stops following the classical Hertz law. The response of the cells at high indentation seems likely to be related to the different viscoelastic properties of the different bilayers that compose the systems.

In the same years, an interesting experimental approach was developed to measure the elasticity (storage) and the loss components of the complex shear modulus for RBCs (Puig-de-Morales-Marinkovic et al., 2007). In this work, ferromagnetic beads were attached to RBCs in order to directly apply torsional forces using a periodic magnetic field in the frequency range of 0.1–100 Hz, which allowed exploring the cell responses from the purely elastic regime to a significantly viscoelastic behavior. Remarkably, the measurement was at the single cell level. In this case, the authors found that the elastic component of the complex shear modulus was frequency-independent and dominated

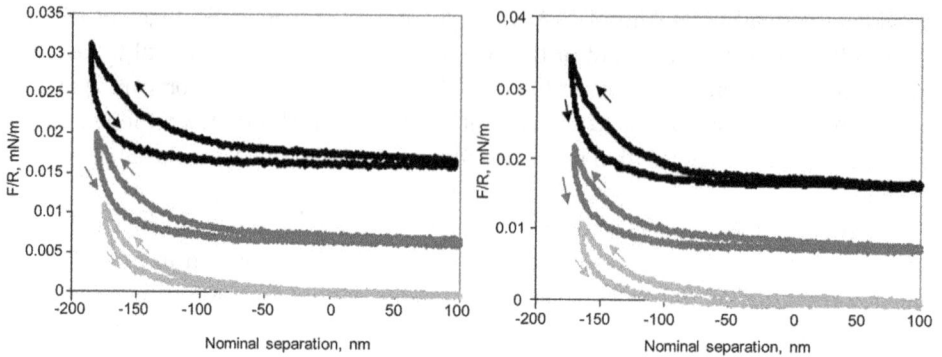

Figure 6.4.4: Left: Interaction force between a silica probe and a RBC in PBS solution as a function of nominal separation and at different approach rates: 0.6 m/s (bottom), 1.2 m/s (middle, vertically shifted) and 2.8 m/s (top, vertically shifted). Right: Interaction force between a silica probe and an RBC in PBS solution at increasing maximum loading: 0.011 mN/m (bottom), 0.0135 mN/m (middle, vertically shifted) and 0.017 mN/m (top, vertically shifted). Figure obtained with permission from Bremmell et al. (2006).

the response at the low and medium frequencies. On the other hand, the loss modulus increased with the frequency, becoming dominant at higher frequency values, with a power law that was not previously predicted. Interestingly, the trend and values observed for the RBCs were greatly different from analogous values measured with the same technique for other cells such as fibroblasts or epithelial cells (Laudadio et al., 2005, Puig-de-Morales et al., 2004).

It is interesting to note that several reports are available on the relationship between the cell properties and the characteristics of their motion in vivo (Forsyth et al., 2011, 2010, Yoon et al., 2008). Indeed, cell shape strongly influences the flow behavior of whole blood. In recent years, several authors attempted to predict the resting cell shape and the shape adaptation mechanisms that take place under flow, mostly by coupling experimental data to the development of simulation or model approaches (Kloppel and Wall, 2011, Barns et al., 2017, Sen et al., 2005). In some cases, the influence of physiological or pathological effectors on the blood flow and hemorheological properties has also been evaluated. For instance, by measuring cell biomechanics and the adhesion forces between erythrocytes in the presence of fibrinogen, both in normal cells and in patients affected by arterial hypertension, Guedes and coworkers recently suggested a role for this protein in producing hemorheological alteration. The enhanced cell-cell adhesion, which was larger in patients with arterial hypertension showing higher fibrinogen level, was suggested to be mediated by a mechanism due to a protein-dependent bridging of two erythrocytes (Guedes et al., 2017, 2019).

Overall, the available data highlight the complex motion and dynamics of the RBCs shape occurring in microchannels, and evidence the dependence of these parameters on the flow condition, shear stress, fluid viscosity, and elastic force of the membrane (Takeishi et al., 2019), although environmental biochemical factor might also play a role.

Recent systematic experimental and simulation approaches (Reichel et al., 2019), however, have evidenced that for a fixed microchannel dimension, the flow condition does not select a single cell shape, and that a distribution of shapes can be possible. Remarkably, the authors suggest that such a distribution can result from the specific nanomechanical conditions of single cells. In this way, a potentially very interesting correlation can be obtained between the cellular (aging-dependent) nano-mechanical properties, the cell shape, and the flow dynamics. In this three-actor relationship, a major role is expected to be played by the viscoelastic behavior of RBCs.

Among the various biophysical techniques to characterize cellular mechanics, optical tweezers (OT) allow studying single RBC mechanics under well-controlled experimental conditions. The first report of direct tensile stretching of RBCs was proposed by Hénon et al. (1999), who attached two silica beads to the opposite ends of a single erythrocyte, and imposed tensile elastic deformation by moving the trapped beads in opposite directions. The obtained stiffness values of 2.5 +/− 0.4 μN/m were lower than the typical shear modulus values (4–10 μN/m) obtained from experiments carried out with micropipette aspiration techniques (Boal, 2002, Discher et al., 1998, Evans and Skalak, 1979).

In the same period, Sleep et al. (1999) measured the elastic properties of human RBCs with a different OT configuration, in which one bead was held fixed and the other was moved in order to induce uniaxial tensile deformation in the cell. The resulting stiffness was nearly two orders of magnitude larger than the one obtained by Henon. Such differences, probably arose from the severe assumptions on the model chosen for the data analysis and from the viscoelastic contribution that was not considered in the study. Yoon et al. (2008) investigated the stress relaxation, following a fast deformation and the effect of varying the strain rate. They found that the stress values follow a power-law decay, which reaches a plateau; that the cell's elasticity follows a power-law increase as a function of strain rate; and that the cell stiffness shows a 3-fold larger increase as the cell is deformed at higher strain rates. Interestingly, since these exponential functions violate the linear superimposition principle, RBCs' response cannot be explained within the framework of linear viscoelasticity, but this nonlinearity is absent at small strain rates.

In the research on the elastic and viscoelastic properties of RBCs and their consequences on the hemorheology, a primary role must be assigned to the theoretical approaches. Indeed, it must be mentioned that the classic Attard theory is probably the simplest one, but is not the only approach employed to calculate or model the elasticity and viscoelasticity of RBCs. On the contrary, on these specific subjects, the production of new experimental data goes parallel to the development of theoretical or numerical methods, which are often preferred in practice to predict the cell shape and its dynamic in-flow conditions.

The most diffused simulation approaches to the study of RBCs, besides the usual ab initio calculations, are based on (a) finite element analysis, (b) coarse-grained molecular dynamics, (c) dissipative particle dynamics, or (d) coarse-grained particle methods. Even though all the methods have been widely used, they have inherent advantages or disadvantages, depending on the case under study. For instance, the finite element analysis is very accurate but its use is complex when the liquid phase must be considered; the coarse-grained molecular dynamics is very precise, but requires an enormous amount of calculations, and, as a consequence, it is mostly focused on patches of the membrane rather than on the entire cell. The other particle-based methods (dissipative particle dynamics and coarse-grained particle) have lesser resolution on the membrane and can include a liquid phase, which makes them preferable when the behavior of the entire cell must be considered (e.g., for shape dynamic or hemorheology).

Several works can be readily found and reviewed in the literature (Barns et al., 2017), which have been developed using approaches based on the finite element analysis (Kloppel and Wall, 2011, Feng and Klug, 2006), the first principle molecular dynamics (Li et al., 2005), and on the other numerical, computational or analytical methods (Pozrikidis, 2003, Prado et al., 2015). Often simulations have been used to model the specific experimental results; thus, the calculation strategies have common frameworks. For instance, they attribute stretch resistance to the cell skeleton, but they differ in the treatment of the membrane that is sketched in different ways. In the most accurate case, the membrane is modeled as a bilayer film with bending resistance, areal incompressibility, and pressed by an incompressible fluid on both sides. In addition, different physical acting forces have been taken into account, depending on the specific task and on the specific simulation method employed. For instance, Kloppel and Wall (2011) described a modeling approach based on finite element calculations that works properly in validating optical tweezers data. They modeled the nonlinear elastic and viscoelastic behavior of cells, and emphasized the importance of membrane viscosity and cytoplasm for dynamic cellular motion.

Overall, the numerical calculations and the theoretical approaches employed were proven very useful in describing specific aspects of the erythrocyte's mechanical properties in static as well as in dynamic conditions. Such theoretical approaches certainly constitute a precious support to the experiments, despite a certain degree of model specificity in the choice of simulation parameters and a certain fragmentation in the results, often a consequence of the characteristics of the calculations. Yet, in spite of the many improvements obtained in recent years and the general good agreement obtained with respect to specific experimental data, a solution to the complex problems related to the viscoelastic behavior of erythrocytes and to their influence on the hydrodynamic motion of the blood is still under way.

6.4.5 Pathological Cases: Overview and Discussion

The biconcave shape and the corresponding mechanical properties of erythrocytes are important indicators of cellular health. Indeed, since the stiffness of RBCs affects clinically relevant elements, such as the blood viscosity (Forsyth et al., 2010, Diez-Silva et al., 2010), the reversible deformability of the cells influences the whole blood properties. Certainly, the fact that the erythrocytes' mechanical properties can be used as a biomarker, stimulates research, especially from a clinical standpoint. Furthermore, since it is likely that the alterations of the mechanical properties of erythrocytes are connected to or mediated by cellular biochemical states, their variations are expected to be a consequence of the cell's network of interactions and of the intracellular signaling pathways. Many evidences support this idea, such as the fact that repeated shear stresses that are capable of inducing specific biochemical patterns cause measurable alterations to the spectrin structure (Johnson et al., 2007). Furthermore, there is evidence that phosphorylation of skeletal proteins, mediated by PKC or of band III proteins, can be associated to cytoskeletal stiffening (Picas et al., 2013). Finally, it has been reported that the calcium influx (which is well known for its ability to activate or regulate intracellular cascades), either mediated by mechanical stimuli or by specific effectors (e.g., A23187), is associated with a reduction in cellular deformability (Kim et al., 2015, Muravyov and Tikhomirova, 2013). This latter finding has been also been validated by direct AFM measurements that showed calcium-dependent stiffening of the RBC's skeleton (Liu et al., 2005).

Obviously, the determination of a relationship between the biochemical status and the nanomechanical properties of the cells has important consequences for the monitoring or detection of pathological conditions. Indeed, it is generally accepted that pathological states produce biochemical or metabolic alterations within the cells and differences in communication between cells. In the case of blood, this scenario is particularly interesting as the blood cells come in direct contact (possibly through the mediation of plasma proteins, such as albumin or fibrinogen) with the effectors or toxins released into circulation from all the systemic diseases. In this sense, the range of the pathologies that could be potentially monitored by blood analyses is very large, yet, to date, the number of diseases that have been correlated to the analysis of the biomechanical status of the red cells is still relatively small (Tomaiuolo, 2014).

These studies have been mostly focused on the cases in which the pathological agent is directly associated to the cell and the most relevant examples can be roughly grouped in:

(i)   parasitosis, the most important of which is, by far, malaria, whose etiology is due to the colonization of the red cells by the parasite *Plasmodium falciparum*;

(ii) genetic defects associated to red cell structural disorders – which can be due to direct skeletal impairment, such as hereditary spherocytosis or hereditary elliptocytosis, or to inherited hemoglobin instability, such as for sickle cell anemia or thalassemia (Dulińska et al., 2006);

(iii) defects of the mechanotransduction apparatus;

(iv) systemic pathologies, in which a direct or indirect effect on the cellular structure or function can be hypothesized.

Following this scheme, we will briefly review the main results obtained from the investigation of these pathologies, with special emphasis on the role of biomechanical properties.

(i) According to the WHO, malaria is a life-threatening disease caused by parasites that are transmitted to humans through the bites of infected female Anopheles mosquitoes. In 2017, there have been approximately 220 million cases of malaria (more than 90% in Africa), with more than 400.000 deaths (more than 90% in Africa). A key feature in the pathogenesis of malaria is the parasite's ability to change the biomechanical properties of the erythrocytes, as the cells infected with mature stages of *P. falciparum* have lesser deformability and higher adhesivity to the endothelium. These alterations cause the infected RBCs to accumulate in the spleen, leading to life-threatening complications (Glenister et al., 2009) that have also been investigated theoretically (Pivkin et al., 2016). While these works confirm the occurrence of cell stiffening and lesser deformability, they found that the most important parameter influencing the RBC retention in the spleen is the surface area loss, which, in these cells, can be rather large. On the other hand, cell stiffening and adhesivity to the endothelium (CD32) is enhanced by a large *Plasmodium falciparum* protein, translocated from the parasite to the RBC membrane after the infection (Glenister et al., 2009). At the same time, the RESA (ring-infected erythrocyte surface antigen) parasite protein complex migrates to the cell membrane, where it interacts with the spectrin, leading to reduced cell deformability (Diez-Silva et al., 2010).

The biomechanics of the RBCs have been also investigated through optical tweezers. In a very well-designed study, Suresh (2006) evaluated the effects of *P. falciparum* on the elastic behavior of infected erythrocytes at different developmental stage of the parasite. Optical images and the elastic response of the RBC, as a function of the intracellular developmental stage f the parasite, are illustrated in Figure 6.4.5. The results indicate that during the course of a 48 h period after the invasion of the RBC, the effective stiffness of the cell increases by more than a factor of 10.

As for the AFM-based studies, they were very important to detect the stiffening of the infected RBCs, also observed in hepatic cells (Eaton et al., 2012). AFM was employed to investigate many specific aspects of the infections and of the parasite development in RBCs. For instance, the high affinity of the parasite-infected cell for the heparin, and its potential pathological consequences for the higher adhesion of infected cell, have been reported and quantitatively evaluated by monitoring the nanomechanical properties of RBCs (Subramani et al., 2015, Valle-Delgado et al., 2013). The skeletal structure and the distribution of knobs formed by *P. falciparum* proteins (see Figure 6.4.5) on the erythrocyte membrane-skeleton were deeply

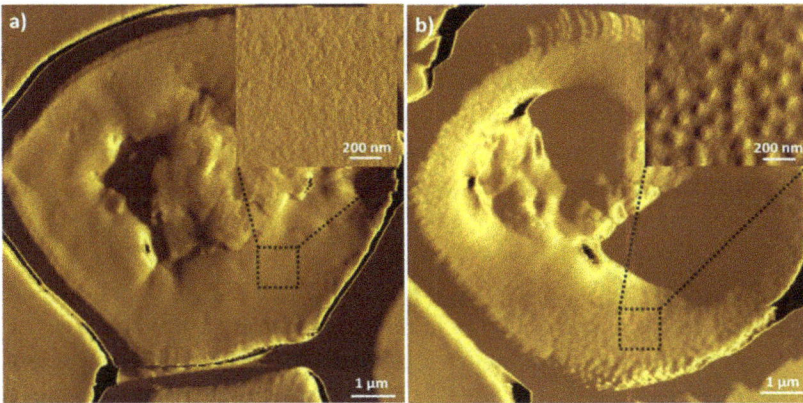

Figure 6.4.5: AFM images of an RBC with (a) or without (b) the membrane knobs produced by the *P. falciparum* infection. Obtained with permission from Subramani et al. (2015).

investigated as a function of the parasite's development stage (Nagao et al., 2000, Shi et al., 2009, 2013). The presence of the parasite and some of its subcellular structure were also investigated by the AFM-IR spectroscopy, which has demonstrated the capability to monitor the parasite during its development (Perez-Guaita et al., 2018). Furthermore, a cell–cell communication strategy of the parasite, based on the release of cargo vesicles has been evidenced and characterized (Regev-Rudzki et al., 2013). Moreover, Sisquella et al. (2017) have recently found evidence of a mechanism to connect the infection with the biomechanical alterations of the cells: the *P. falciparum,* by binding to the RBC's glycophorin protein, directly activates a signaling pathway that stimulates second messengers and a phosphorylation cascade, affecting the skeletal proteins and inducing a dramatic cell deformation. The pathway results in a modification of the viscoelastic properties of the host membrane, a circumstance that the authors claim to favor the parasite invasion.

As a whole, we can conclude that the potential of the AFM-based technique in the study of malaria has proven very interesting, as its application allowed deciphering many aspects of the infection and of the alterations induced on the cell physiology, which contribute to a thorough understanding of this pathology and could be important for its medical treatment.

(ii) Among the pathologies correlated to the observation of morphological anomalies of the cells, several disorders of RBCs are caused by genetic defects of membrane-skeletal components. These anomalies are the consequence of mutations in RBC skeletal proteins, such as spectrin, ankyrin, Band 3, and proteins 4.1 and 4.2, whose impairment produces defects in the elastic and shear moduli of the membrane, as suggested by the micropipette aspiration experiments (Tomaiuolo, 2014, Waugh, 1987).

The most common and characterized among these pathologies are hereditary spherocytosis (HS) and hereditary elliptocytosis (HE). Several investigations on these ailments have been carried out using AFM's direct measurement of the Young's modulus and also by theoretical simulations. These revealed that, in diseased cells, the difference in the cytoskeletal organization is a measurable stiffening, compared to healthy cells (Dulińska et al., 2006, Dumitru et al., 2018). More in detail, it appears that both the membrane and the skeletal component of the pathological cells have properties that are different from those of healthy RBCs.

Sickle cell anemia is another special case of genetic defect due, in this case, to hemoglobin instability. It is believed that sickle cell anemia has diffused as a form of resistance to some strains of malaria, as sickle cell patients develop RBCs alterations that complicate the development of malaria's parasite in the cell. Sickle cell anemia is due to a recessive point defect in the DNA, which determines a single amino-acid substitution in Hb. The consequences, however, are quite dramatic, as the mutated proteins become unstable in low oxygen conditions (e.g., peripheral circulation) and stick together, forming the aggregates that induce the "sickle" cell deformation (Rees et al., 2010). The biomechanical characterization of this pathology has been investigated in several works (Maciaszek et al., 2011, Maciaszek and Lykotrafitis, 2011, Zhang et al., 2017), revealing that the Young's modulus of RBCs from patients with sickle cell trait (i.e., only one mutated allele) is much larger than in normal cells, and that these cells have a stronger adhesion force to epithelial cells. The experiments, interpreted in terms of Hertz theory by comparing two different tip shapes, allowed to conclude that the Hb polymerization has an effect on the cell skeleton organization, but also that therapeutic treatments can significantly impact this pathological alteration. On this topic, optical tweezer investigations are also reported in literature. For instance, Brandao et al. (2003) evaluated the elasticity of RBCs from patients affected by sickle cell anemia with HbSS or HbAS mutation as well as with HbSS mutation, treated with hydroxyurea (HU, a common medical treatment for the disease). They found that RBCs from homozygous and heterozygous patients were significantly stiffer than those of healthy subjects, and that the RBCs from patients treated with HU presented stiffness values comparable to those of healthy erythrocytes.

(iii) There are several examples of pathological consequences related to misfunctioning or to genetic disorders affecting elements of the mechanotransductive pathways. In addition to the well-known case of the cystic fibrosis, which is related to defects of the CFTR receptors (Lasalvia et al., 2016), there is the case of the hereditary xerocytosis, a blood disorder associated to the dehydration of RBCs which can be traced back to a mutation in the gene coding of PIEZO I protein (Zarychanski et al., 2012). This is a transmembrane, pore-forming protein involved in mechanosensing and transduction that has been recently proven to be involved in the conservation of the volume of RBCs, and whose alteration can result in an imbalance of the cell osmolarity and metabolism. These considerations reinforce the idea that mechanotransduction is a very important pathway, deeply carved in the structure and functioning of this class of cells.

(iv) Some cases of systemic pathologies, such as diabetes mellitus (DM), especially type 2 (i.e., the insulin resistant), systemic lupus erythematosus (SLE) (Ebner et al., 2011), neurodegenerative disorders, such as Alzheimer's disease (AD), and rheumatoid arthritis (RA), have been investigated using AFM.

In the case of RA, the AFM approach has demonstrated reduced cell elasticity, which is directly correlated to the presence of anomalous cell morphology and to defects of the band 3 distribution across the membrane. Interestingly, AFM allowed the correlation of these alterations to the continuous exposure of the blood cells to inflammatory molecules and free radicals that are characteristics of RA (Olumuyiwa-Akeredolu et al., 2019). Such association is even more stimulating if interpreted from the perspective of representing a model for several pathologies, including neurodegenerative diseases and cancer, which are associated with the spread of inflammation or oxidative stress at the systemic level. In particular, the environmental injury observed in RA resembles the case of favism, a metabolic disorder of the cell metabolism due to the impairment of the critical enzyme that produces reducing power, the G6PD. This exposes the RBC to environmental oxidation and corresponds to an increase in cell stiffness, which has been measured using AFM (Dulińska et al., 2006).

Other cases of diffused pathological syndromes, for instance, coronary diseases and hypertension (Lekka et al., 2005), have been analyzed and the blood cells were compared to RBCs from smokers. Despite the fact that the latter are the most affected, some effects were also detectable in coronary disease and hypertension. The most prominent was a larger spread of Young's modulus values, which reflected the intrinsic variance of the cell behavior arising from the different clinical spectra of the considered pathologies.

On the other hand, relatively few studies have focused on the investigation of the morphological and nanomechanical differences between RBCs in neurodegenerative pathologies, such as AD. For instance, the investigation of the interaction between RBCs and Aβ peptide, which is known to be involved in the development of AD, would highlight the undoubtable interaction that occurs at the blood level, both in the pathological case and, very likely, also in pre-pathological situations. In these cases, some AFM-based approaches are already available and have provided evidence of morphological effects and biochemical regulation of cell metabolism due to the presence of Aβ peptide. Furthermore, it was found that erythrocytes from AD patients, which have a high ferritin level, also exhibit higher stiffness, compared to cells from AD patients with normal ferritin level or compared to a healthy subject, thus suggesting a role of ferritin in modulating the cell stiffness (Carelli-Alinovi et al., 2016, 2019a, Pretorius and Kell, 2014).

Among the systemic pathologies investigated with AFM, DM is probably the most studied disease. This is due to the large diffusion of the pathology that affects, with different degrees of severity, more than 400 million people worldwide (AlSalhi et al., 2018). DM is a widely diffused metabolic disorder characterized by high level of glucose in the blood for a prolonged period of time, and in the vast majority of the cases, by insulin resistance (type 2 DM). Hemorheological abnormalities, such as higher blood viscosity, can be observed in DM patients and correlated to impaired cellular deformability (Fornal et al., 2006). As an example, the nanomechanical properties of healthy and DM erythrocytes were compared through high resolution mapping. The viscoelastic cell behavior was identified and evaluated by considering the hysteresis of the forward and backward force curves (Ciasca et al., 2015). A general heterogeneous distribution of the Young's modulus across the cells was evidenced, with higher values in the center. Moreover, the existence of a relevant contribution of the viscous forces in determining the cell mechanics was claimed. A similar system was also characterized through theoretical modeling by Chang et al. (2017), where a particle-based, two-component cell model was used to simulate the behavior of whole cells in terms of their biomechanical properties and of the blood hemorheological properties. The most relevant cell parameters involved in the description of static cell deformations were determined, indicating a dominant role for membrane viscosity. Such simulations also showed that in shear flow conditions, RBCs undergo complex shape dynamics that can be described by taking into account membrane elasticity, cell geometry, and viscous dissipation, and the model can reproduce the increased viscosity of DM blood, compared to healthy subjects.

Several authors investigated the shape alterations, mechanical properties, and adhesivity of erythrocytes from diabetic patients, reporting converging evidence of higher stiffness, up to three times the control values (Lekka et al., 2005), and of a larger number of shape anomalies compared to healthy controls. The morphological anomalies ranged from large-scale shape alterations, such as higher number of spherocytic and crenated cells, to larger size of the membrane vesicles, leading to a stronger loss of surface area, which, as previously recalled, can be a major determinant for the cell survival in the circulation. In addition, the characterization of the membrane surface can be used to evidence additional defects, such as signs of mild corrosion related to the interaction with higher concentration of glucose over time (AlSalhi et al., 2018, Jin et al., 2010). Finally, a higher adhesivity of cells from diabetic patients has been observed, and this fact is especially important in the contact with endothelial cells (Jin et al., 2010, Zhang et al., 2015), and might be considered a determinant for the complication at the circulatory level.

From what is discussed above, it appears that a common trait in the vast majority of the pathologies that affect (or influence) RBCs is the induction of an increased stiffness, and correspondingly, a reduction of cellular elasticity and deformability. This type of alteration correlates with a worsening of cellular functionality, in terms of O_2/CO_2 shuttle and, overall, in a reduction of the cell's life time due to the increased retention of the impaired erythrocytes in the spleen (Pivkin et al., 2016). Typical consequences can range from alterations in cell turnover to the development of anemic syndromes. Moreover, these modifications of erythrocyte biomechanics, while altering the elastic properties of individual cells, are also associated to alterations of cell-cell interactions, with consequences on the viscosity and hemodynamics of blood that can cause stress to the cardiovascular system.

From a molecular point of view, it is very difficult to provide a unifying view for pathologies of very different etiology. However, it is interesting to note that some common traits can be observed in the behavior of environmental transduction systems, which are very effective in these cells, both for the chemical and mechanical cascades, and in the behavior of secondary messengers. Typical examples are Ca^{+2} and protein kinases that act as decisive regulators of the structural characteristics of the skeleton, and of its conjugations with the plasma membrane, triggering, among other things, important phenomena such as cellular vesiculation (Huisjes et al., 2018, Carelli-Alinovi et al., 2019b).

6.4.6 Other Blood Cells: White Blood Cells

White blood cells (WBC), also called leukocytes, are the principal regulators of the human immune system. Similar to RBCs, these are circulating cells that, during their life, undergo extensive morphological changes, endure shear flow while adhered to surfaces, and continuously interact with other cells. Their cytoskeleton mediates several signal transduction pathways that can rapidly transduce an environmental physical signal into biochemical changes that lead to cytoskeletal rearrangement. Any impairment of these mechanosensing capabilities leads to serious issues, such as vessel blockage (i.e., leukostasis), and to an immune response that can be damaged at several levels. Indeed, cell deformability is thought to play a key role in leukostasis, as stiffer cells are more prone to mechanically obstruct a vessel (Lichtman and Rowe, 1982, Porcu et al., 2002, Lichtman et al., 1987). A recent review has clarified this point, describing the role of mechanical forces in the regulation of cell migration, cell-surface receptor activation, intracellular signaling, and intercellular communication (Huse, 2017), highlighting the biological ramifications of these effects in various immune cells types.

Recently, several new technical innovations in the biophysical field have been employed to evaluate the mechanical forces in immune cells. In general, however, the field remains in its infancy. Our understanding of the immunoreceptor mechanotransduction, for example, is limited to antigen receptors and to a small number of well-characterized cell adhesion molecules. To be more specific, the major issue in mapping the elastic properties of WBCs by AFM arises from their nonadhesive behavior on standard substrates, and conflict with the cell's capability to respond to substrate functionalization and chemical composition. In order to perform an AFM mapping on passive (not-activated) leukocytes, two major problems must be

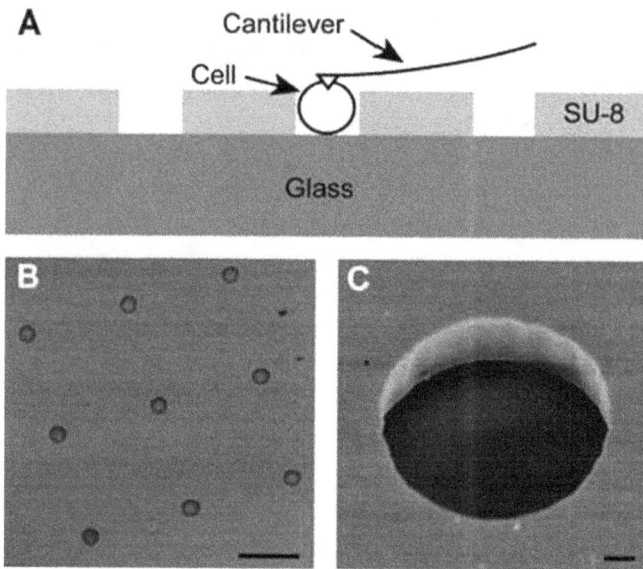

Figure 6.4.6: Microfabricated wells for force microscopy measurements. Panel (A): schematic diagram of the experimental setup that shows how the cell were mechanically immobilized in the holes. Panels (B) and (C): SEM micrograph of the microwells, scale bar 50 μm and 2 μm, respectively. Figure reproduced with permission from Rosenbluth et al. (2006).

addressed: the first is that most of the cells tend to slip under the tip pressure, while the remaining cells are leukocytes with altered adhesive characteristics; the second problem is the use of functionalized surfaces, which can activate the cells, leading to an extensive alteration of morphologies and mechanical properties (Frank, 1990). The complexity in performing AFM and OT characterizations has been compensated by the extensive use of other techniques, such as micropipette aspiration, to determine the mechanical properties of fully differentiated leukocytes (Dong et al., 1988, Needham and Hochmuth, 1990, Tsai et al., 1993). Among the very few studies involving AFM characterization of leukocytes, it is worth citing Rosenbluth et al. (2006) who have developed an interesting AFM experimental setup to measure the deformability of human myeloid and lymphoid leukemia cells, and neutrophils al low deformation rates by mechanically immobilizing the cells in micro-fabricated wells. The authors, as shown in Figure 6.4.7, found that myeloid (HL60) cells were 18-fold stiffer than lymphoid (Jurkat) cells and 6-fold times stiffer than human neutrophils, on average (855 +/− 670, 48 +/− 35 and 156 +/− 87 Pa, respectively), and evaluated the viscosity contribution on the mechanical properties by varying the z-piezo extension rates from 24 to 8,643 nm/s. In HL60, the apparent stiffness remains relatively constant up to 415 nm/s, but increases monotonically at higher piezo extension rates. These cells show no increase in the apparent stiffness when piezo extension rate is increased from 24 to 415 nm/s, indicating that within

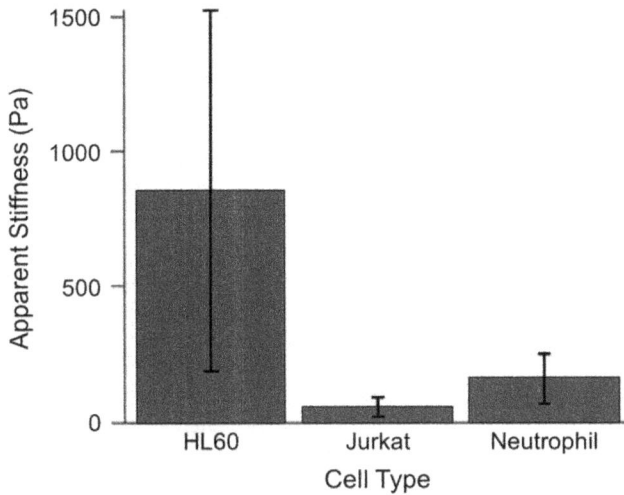

Figure 6.4.7: Measured stiffness at low z-piezo extension rates (415 nm/s), HL60 cells have an average stiffness of 855 +/−670 Pa ($n = 60$), whereas Jurkat cells are significantly softer: 48 +/−35 Pa ($n = 37$). Neutrophils have an average stiffness of 156 +/−87 Pa ($n = 26$), significantly softer than HL60 and significantly stiffer than Jurkat cells. All values are reported as mean +/− standard deviation, Student's t-test was used to determine statistically significant differences ($p < 0.001$). Figure reproduced with permission from Rosenbluth et al. (2006).

this range the measurements are not significantly influenced by viscosity. This same viscoelastic response was observed in Jurkat cells and neutrophils.

References

AlSalhi, M. S., S. Devanesan, K. E. AlZahrani, M. AlShebly, F. Al-Qahtani, K. Farhat K, et al. (2018). "Impact of diabetes mellitus on human erythrocytes: Atomic force microscopy and spectral investigations." International Journal of Environmental Research and Public Health **15**: 2368.

Alvarez, C. L., J. Schachter, A. A. de Sá Pinheiro, L. S. Silva, S. V. Verstraeten, P. M. Persechini, et al. (2014). "Regulation of extracellular ATP in human erythrocytes infected with plasmodium falciparum." PLOS ONE **9**: e96216.

Antonelou, M. H., A. G. Kriebardis and I. S. Papassideri (2010). "Aging and death signalling in mature red cells: From basic science to transfusion practice." Blood Transfus **8**: s39.

Antonini, E. and M. Brunori (1970). "Hemoglobin." Annual Review of Biochemistry **39**: 977–1042.

Attard, P. (2001). "Interaction and deformation of viscoelastic particles: Nonadhesive particles." Physical Review E **63**: 061604.

Barns, S., M. A. Balanant, E. Sauret, R. Flower, S. Saha and Y. Gu (2017). "Investigation of red blood cell mechanical properties using AFM indentation and coarse-grained particle method." Biomedical Engineering Online **16**: 140.

Boal, D. (2002). Mechanics of the cell. New York, Cambridge, UK, Cambridge University Press.

Borghi, N. and F. Brochard-Wyart (2007). "Tether extrusion from red blood cells: Integral proteins unbinding from cytoskeleton." Biophysical Journal **93**: 1369–1379.

Brandao, M., A. Fontes, M. Barjas-Castro, L. Barbosa, F. Costa, C. Cesar, et al. (2003). "Optical tweezers for measuring red blood cell elasticity: Application to the study of drug response in sickle cell disease." European Journal of Haematology **70**: 207–211.

Bremmell, K. E., A. Evans and C. A. Prestidge (2006). "Deformation and nano-rheology of red blood cells: An AFM investigation." Colloids and Surfaces B **50**: 43–48.

Carelli-Alinovi, C., S. Dinarelli, M. Girasole and F. Misiti (2016). "Vascular dysfunction-associated with Alzheimer's disease." Clinical Hemorheology and Microcirculation **64**: 679–687.

Carelli-Alinovi, C., S. Dinarelli, B. Sampaolese, F. Misiti and M. Girasole (2019a). "Morphological changes induced in erythrocyte by amyloid beta peptide and glucose depletion: A combined atomic force microscopy and biochemical study." Biochimica et Biophysica Acta – Biomembranes **1861**: 236–244.

Carelli-Alinovi, C., S. Dinarelli, B. Sampaolese, F. Misiti and M. Girasole (2019b). "Morphological changes induced in erythrocyte by amyloid beta peptide and glucose depletion: A combined atomic force microscopy and biochemical study." Biochimica et Biophysica Acta – Biomembranes **1861**: 236–244.

Chang, H.-Y., X. Li and G. E. Karniadakis (2017). "Modeling of biomechanics and biorheology of red blood cells in type 2 diabetes mellitus." Biophysical Journal **113**: 481–490.

Chien, S., R. G. King, R. Skalak, S. Usami and A. L. Copley (1975). "Viscoelastic properties of human blood and red cell suspensions." Biorheology **12**: 341–346.

Ciasca, G., M. Papi, S. Di Claudio, M. Chiarpotto, V. Palmieri, G. Maulucci, et al. (2015). "Mapping viscoelastic properties of healthy and pathological red blood cells at the nanoscale level." Nanoscale **7**: 17030–17037.

Cluitmans, J. C., F. Gevi, A. Siciliano, A. Matte, J. K. Leal, L. De Franceschi, et al. (2016). "Red blood cell homeostasis: Pharmacological interventions to explore biochemical, morphological and mechanical properties." Frontiers in Molecular Biosciences **3**: 10.

D'Alessandro, A., G. Liumbruno, G. Grazzini and L. Zolla (2010). "Red blood cell storage: The story so far." Blood Transfus **8**: 82–88.

Diez-Silva, M., M. Dao, J. Han, C. T. Lim and S. Suresh (2010). "Shape and biomechanical characteristics of human red blood cells in health and disease." MRS Bulletin / Materials Research Society **35**: 382–388.

Dinarelli, S., G. Longo, S. Krumova, S. Todinova, A. Danailova, S. G. Taneva, et al. (2018a). "Insights into the morphological pattern of erythrocytes' aging: Coupling quantitative AFM data to microcalorimetry and Raman spectroscopy." Journal of Molecular Recognition **31**: e2732.

Dinarelli, S., G. Longo, G. Dietler, A. Francioso, L. Mosca, G. Pannitteri, et al. (2018b). "Erythrocyte's aging in microgravity highlights how environmental stimuli shape metabolism and morphology." Scientific Reports **8**: 5277.

Discher, D. E., D. H. Boal and S. K. Boey (1998). "Simulations of the erythrocyte cytoskeleton at large deformation. II. Micropipette aspiration." Biophysical Journal **75**: 1584–1597.

Dong, C., R. Skalak, K. L. Sung, G. W. Schmid-Schonbein and S. Chien (1988). "Passive deformation analysis of human leukocytes." Journal of Biomechanical Engineering **110**: 27–36.

Dulińska, I., M. Targosz, W. Strojny, M. Lekka, P. Czuba, W. Balwierz, et al. (2006). "Stiffness of normal and pathological erythrocytes studied by means of atomic force microscopy." Journal of Biochemical and Biophysical Methods **66**: 1–11.

Dumitru, A. C., M. A. Poncin, L. Conrard, Y. F. Dufrêne, D. Tyteca and D. Alsteens (2018). "Nanoscale membrane architecture of healthy and pathological red blood cells." Nanoscale Horizons **3**: 293–304.

Eaton, P., V. Zuzarte-Luis, M. M. Mota, N. C. Santos and M. Prudêncio (2012). "Infection by Plasmodium changes shape and stiffness of hepatic cells." Nanomedicine: NBM **8**: 17–19.

Ebner, A., H. Schillers and P. Hinterdorfer (2011). "Normal and pathological erythrocytes studied by atomic force microscopy." Methods in Molecular Biology (Clifton, N.J.) **736**: 223–241.

Evans, E. A. and R. Skalak (1979). "Mechanics and thermodynamics of biomembranes: Part 1." CRC Critical Reviews in Biomedical Engineering **3**: 181–330.

Feng, F. and W. S. Klug (2006). "Finite element modeling of lipid bilayer membranes." Journal of Computational Physics **220**: 394–408.

Fischer, T. M. (2004). "Shape memory of human red blood cells." Biophysical Journal **86**: 3304–3313.

Fornal, M., M. Lekka, G. Pyka-Fościak, K. Lebed, T. Grodzicki, B. Wizner, et al. (2006). "Erythrocyte stiffness in diabetes mellitus studied with atomic force microscope." Clinical Hemorheology and Microcirculation **35**: 273–276.

Forsyth, A. M., J. Wan, P. D. Owrutsky, M. Abkarian and H. A. Stone (2011). "Multiscale approach to link red blood cell dynamics, shear viscosity, and ATP release." Proceedings of the National Academy of Sciences of the USA **108**: 10986–10991.

Forsyth, A. M., J. Wan, W. D. Ristenpart and H. A. Stone (2010). "The dynamic behavior of chemically "stiffened" red blood cells in microchannel flows." Microvascular Research **80**: 37–43.

Frank, R. S. (1990). "Time-dependent alterations in the deformability of human neutrophils in response to chemotactic activation." Blood **76**: 2606–2612.

Giardina, B., I. Messana, R. Scatena and M. Castagnola (1995). "The multiple functions of hemoglobin." Critical Reviews in Biochemistry and Molecular Biology **30**: 165–196.

Girasole, M., A. Arcovito, A. Marconi, C. Davoli, A. Congiu-Castellano, A. Bellelli, et al. (2005). "Control of the active site structure of giant bilayer hemoglobin from the Annelid Eisenia foetida using hierarchic assemblies." Applied Physics Letters **87**: 87–90.

Girasole, M., S. Dinarelli and G. Boumis (2012). "Structural, morphological and nanomechanical characterisation of intermediate states in the ageing of erythrocytes." Journal of Molecular Recognition: JMR **25**: 285–291.

Girasole, M., G. Pompeo, A. Cricenti, G. Longo, G. Boumis, A. Bellelli, et al. (2010). "The how, when, and why of the aging signals appearing on the human erythrocyte membrane: An atomic force microscopy study of surface roughness." Nanomedicine: NBM **6**: 760–768.

Glenister, F. K., K. M. Fernandez, L. M. Kats, E. Hanssen, N. Mohandas, R. L. Coppel, et al. (2009). "Functional alteration of red blood cells by a megadalton protein of Plasmodium falciparum." Blood **113**: 919–928.

Guedes, A. F., F. A. Carvalho, C. Moreira, J. B. Nogueira and N. C. Santos (2017). "Essential arterial hypertension patients present higher cell adhesion forces, contributing to fibrinogen-dependent cardiovascular risk." Nanoscale **9**: 14897–14906.

Guedes, A. F., C. Moreira, J. B. Nogueira, N. C. Santos and F. A. Carvalho (2019). "Fibrinogen–erythrocyte binding and hemorheology measurements in the assessment of essential arterial hypertension patients." Nanoscale **11**: 2757–2766.

Hénon, S., G. Lenormand, A. Richert and F. Gallet (1999). "A new determination of the shear modulus of the human erythrocyte membrane using optical tweezers." Biophysical Journal **76**: 1145–1151.

Herranz, R., R. Anken, J. Boonstra, M. Braun, P. C. Christianen, M. de Geest M, et al. (2013). "Ground-based facilities for simulation of microgravity: Organism-specific recommendations for their use, and recommended terminology." Astrobiology **13**: 1–17.

Huang, Y. X., Z. J. Wu, J. Mehrishi, B. T. Huang, X. Y. Chen, X. J. Zheng, et al. (2011). "Human red blood cell aging: Correlative changes in surface charge and cell properties." Journal of Cellular and Molecular Medicine 15: 2634–2642.

Huisjes, R., A. Bogdanova, W. W. van Solinge, R. M. Schiffelers, L. Kaestner and R. Van Wijk (2018). "Squeezing for Life–Properties of red blood cell deformability." Frontiers in Physiology 8(9): 656.

Huruta, R. R., M. L. Barjas-Castro, S. T. Saad, F. F. Costa, A. Fontes, L. C. Barbosa, et al. (1998). "Mechanical properties of stored red blood cells using optical tweezers." Blood 92: 2975–2977.

Huse, M. (2017). "Mechanical forces in the immune system." Nature Reviews. Immunology 17: 679–690.

Ito, H., R. Murakami, S. Sakuma, C. H. D. Tsai, T. Gutsmann, K. Brandenburg, et al. (2017). "Mechanical diagnosis of human erythrocytes by ultra-high speed manipulation unraveled critical time window for global cytoskeletal remodeling." Scientific Reports 7: 43134.

Jin, H., X. Xing, H. Zhao, Y. Chen, X. Huang, S. Ma, et al. (2010). "Detection of erythrocytes influenced by aging and type 2 diabetes using atomic force microscope." Biochemical and Biophysical Research Communications 391: 1698–1702.

Johnson, C. P., H. Y. Tang, C. Carag, D. W. Speicher and D. E. Discher (2007). "Forced unfolding of proteins within cells." Science 317: 663–666.

Kim, J., H. Lee and S. Shin (2015). "Advances in the measurement of red blood cell deformability: A brief review." Journal of Cellular Biotechnology 1: 63–79.

Kinoshita, A., K. Tsukada, T. Soga, T. Hishiki, Y. Ueno, Y. Nakayama, et al. (2007). "Roles of hemoglobin allostery in hypoxia-induced metabolic alterations in erythrocytes: Simulation and its verification by metabolome analysis." The Journal of Biological Chemistry 282: 10731–10741.

Klarl, B. A., P. A. Lang, D. S. Kempe, O. M. Niemoeller, A. Akel, M. Sobiesiak, et al. (2006). "Protein kinase C mediates erythrocyte 'programmed cell death' following glucose depletion." American Journal of Physiology. Cell Physiology 290: C244–53.

Kloppel, T. and W. A. Wall (2011). "A novel two-layer, coupled finite element approach for modeling the nonlinear elastic and viscoelastic behavior of human erythrocytes." Biomechanics and Modeling in Mechanobiology 10: 445–459.

Kozlova, E., A. Chernysh, V. Moroz, V. Sergunova, O. Gudkova and E. Manchenko (2017). "Morphology, membrane nanostructure and stiffness for quality assessment of packed red blood cells." Scientific Reports 7: 7846.

Kozlova, E., A. Chernysh, E. Manchenko, V. Sergunova and V. Moroz (2018). "Nonlinear biomechanical characteristics of deep deformation of native RBC membranes in normal state and under modifier action." Scanning 2018: 1–13.

Lasalvia, M., S. Castellani, P. D'Antonio, G. Perna, A. Carbone, A. L. Colia, et al. (2016). "Human airway epithelial cells investigated by atomic force microscopy: A hint to cystic fibrosis epithelial pathology." Experimental Cell Research 348: 46–55.

Laudadio, R. E., E. J. Millet, B. Fabry, S. S. An, J. P. Butler and J. J. Fredberg (2005). "Rat airway smooth muscle cell during actin modulation: Rheology and glassy dynamics." American Journal of Physiology. Cell Physiology 289: C1388-95.

Leal Denis, M. F., H. A. Alvarez, N. Lauri, C. L. Alvarez, O. Chara and P. J. Schwarzbaum (2016). "Dynamic regulation of cell volume and extracellular ATP of human erythrocytes." PLOS ONE 11: e0158305.

Lekka, M., M. Fornal, G. Pyka-Fosciak, K. Lebed, B. Wizner, T. Grodzicki, et al. (2005). "Erythrocyte stiffness probed using atomic force microscope." Biorheology 42: 307–317.

Li, J., M. Dao, C. T. Lim and S. Suresh (2005). "Spectrin-level modeling of the cytoskeleton and optical tweezers stretching of the erythrocyte." Biophysical Journal **88**: 3707–3719.

Lichtman, M. A. and J. M. Rowe (1982). "Hyperleukocytic leukemias: Rheological, clinical, and therapeutic considerations." Blood **60**: 279–283.

Lichtman, M. A., J. Heal and J. M. Rowe (1987). "Hyperleukocytic leukaemia: Rheological and clinical features and management." Baillière's Clinical Haematology. 1: 725–746.

Liu, F., H. Mizukami, S. Sarnaik and A. Ostafin (2005). "Calcium-dependent human erythrocyte cytoskeleton stability analysis through atomic force microscopy." Journal of Structural Biology **150**: 200–210.

Lux, S. E. (2016). "Anatomy of the red cell membrane skeleton: Unanswered questions." Blood **127**: 187–199.

Maciaszek, J. L., B. Andemariam and G. Lykotrafitis (2011). "Microelasticity of red blood cells in sickle cell disease." Journal of Strain Analysis for Engineering Design **46**: 368–379.

Maciaszek, J. L. and G. Lykotrafitis (2011). "Sickle cell trait human erythrocytes are significantly stiffer than normal." Journal of Biomechanics **44**: 657–661.

Mohanty, J., E. Nagababu and J. M. Rifkind (2014). "Red blood cell oxidative stress impairs oxygen delivery and induces red blood cell aging." Frontiers in Physiology **5**: 84.

Mohandas, N. and P. G. Gallagher (2008). "Red cell membrane: Past, present, and future." Blood **112**: 3939–3948.

Muravyov, A. V. and I. A. Tikhomirova (2013). "Role molecular signaling pathways in changes of red blood cell deformability." Clinical Hemorheology and Microcirculation **53**: 45–59.

Nagao, E., O. Kaneko and J. A. Dvorak (2000). "Plasmodium falciparum-infected erythrocytes: Qualitative and quantitative analyses of parasite-induced knobs by atomic force microscopy." Journal of Structural Biology **130**: 34–44.

Needham, D. and R. M. Hochmuth (1990). "Rapid flow of passive neutrophils into a 4 microns pipet and measurement of cytoplasmic viscosity." Journal of Biomechanical Engineering **112**: 269–276.

Ney, P. A. (2011). "Normal and disordered reticulocyte maturation." Current Opinion in Hematology **18**: 152–157.

Olumuyiwa-Akeredolu, O., M. J. Page, P. Soma and E. Pretorius (2019). "Platelets: Emerging facilitators of cellular crosstalk in rheumatoid arthritis." Nature Reviews. Rheumatology **15**: 237–248.

Perez-Guaita, D., K. Kochan, M. Batty, C. Doerig, J. Garcia-Bustos, S. Espinoza, et al. (2018). "Multispectral atomic force microscopy-infrared nano-imaging of malaria infected red blood cells." Anal. Chem **90**: 3140–3148.

Picas, L., F. Rico, M. Deforet and S. Scheuring (2013). "Structural and mechanical heterogeneity of the erythrocyte membrane reveals hallmarks of membrane stability." ACS nano **7**: 1054–1063.

Pietraforte, D., P. Matarrese, E. Straface, L. Gambardella, A. Metere, G. Scorza, et al. (2007). "Two different pathways are involved in peroxynitrite-induced senescence and apoptosis of human erythrocytes." Free Radical Biology & Medicine **42**: 202–214.

Pivkin, I. V., Z. Peng, G. E. Karniadakis, P. A. Buffet, M. Dao and S. Suresh (2016). "Biomechanics of red blood cells in human spleen and consequences for physiology and disease." Proc. Natl. Acad. Sci. **113**: 7804–7809.

Prado, G., A. Farutin, C. Misbah and L. Bureau (2015). "Viscoelastic transient of confined red blood cells." Biophysical Journal **108**: 2126–2136.

Pretorius, E. and D. B. Kell (2014). "Diagnostic morphology: Biophysical indicators for iron-driven inflammatory diseases." Integrative Biology **6**: 486–510.

Porcu, P., S. Farag, G. Marcucci, S. R. Cataland, M. S. Kennedy and M. Bissell (2002). "Leukocytoreduction for acute leukemia." Therapeutic Apheresis **6**: 15–23.

Pozrikidis, C. (2003). "Numerical simulation of the flow-induced deformation of red blood cells." Annals of Biomedical Engineering **31**: 1194–1205.

Puig-de-Morales-Marinkovic, M., K. T. Turner, J. P. Butler, J. J. Fredberg and S. Suresh (2007). "Viscoelasticity of the human red blood cell." American Journal of Physiology. Cell Physiology **293**: C597–605.

Puig-de-Morales, M., E. Millet, B. Fabry B, D. Navajas, N. Wang, J. P. Butler, et al. (2004). "Cytoskeletal mechanics in adherent human airway smooth muscle cells: Probe specificity and scaling of protein-protein dynamics." American Journal of Physiology. Cell Physiology **287**: C643–54.

Radmacher, M., R. W. Tillmann and H. E. Gaub (1993). "Imaging viscoelasticity by force modulation with the atomic force microscope." Biophysical Journal **64**: 735–742.

Rees, D. C., T. N. Williams and M. T. Gladwin (2010). "Sickle-cell disease." Lancet **376**: 2018–2031.

Regev-Rudzki, N., D. W. Wilson, T. G. Carvalho, X. Sisquella, B. M. Coleman, M. Rug, et al. (2013). "Cell-cell communication between malaria-infected red blood cells via exosome-like vesicles." Cell **153**: 1120–1133.

Reichel, F., J. Mauer, A. A. Nawaz, G. Gompper, J. R. Guck and D. Fedosov (2019). "High throughput microfluidic characterization of erythrocyte shapes and mechanical variability." Biophysical Journal **116**: 123a–4a.

Rizzo, A. M., P. A. Corsetto, G. Montorfano, S. Milani, S. Zava, S. Tavella, et al. (2012). "Effects of long-term space flight on erythrocytes and oxidative stress of rodents." PLOS ONE **7**: e32361.

Rosenbluth, M. J., W. A. Lam and D. A. Fletcher (2006). "Force microscopy of nonadherent cells: A comparison of leukemia cell deformability." Biophysical Journal **90**: 2994–3003.

Ruggeri, F., C. Marcott, S. Dinarelli, G. Longo, M. Girasole, G. Dietler, et al. (2018). "Identification of oxidative stress in red blood cells with nanoscale chemical resolution by infrared nanospectroscopy." International Journal of Molecular Sciences **19**: 2582.

Sen, S., S. Subramanian and D. E. Discher (2005). "Indentation and adhesive probing of a cell membrane with AFM: Theoretical model and experiments." Biophysical Journal **89**: 3203–3213.

Shi, H., A. Li, J. Yin, K. Tan and C. Lim (2009). "AFM study of the cytoskeletal structures of malaria infected erythrocytes." 13th International Conference on Biomedical Engineering: Springer 1965–1968.

Shi, H., Z. Liu, A. Li, J. Yin, A. G. Chong, K. S. Tan, et al. (2013). "Life cycle-dependent cytoskeletal modifications in Plasmodium falciparum infected erythrocytes." PLOS ONE **8**: e61170.

Sinha, A., T. T. Chu, M. Dao and R. Chandramohanadas (2015). "Single-cell evaluation of red blood cell bio-mechanical and nano-structural alterations upon chemically induced oxidative stress." Scientific Reports **5**: 9768.

Sisquella, X., T. Nebl, J. K. Thompson, L. Whitehead, B. M. Malpede, N. D. Salinas, et al. (2017). "Plasmodium falciparum ligand binding to erythrocytes induce alterations in deformability essential for invasion." Elife **6**: e21083.

Sleep, J., D. Wilson, R. Simmons and W. Gratzer (1999). "Elasticity of the red cell membrane and its relation to hemolytic disorders: An optical tweezers study." Biophysical Journal **77**: 3085–3095.

Sprague, R. S., M. L. Ellsworth, A. H. Stephenson and A. J. Lonigro (2001). "Participation of cAMP in a signal-transduction pathway relating erythrocyte deformation to ATP release." American Journal of Physiology. Cell Physiology **281**: C1158–64.

Sridharan, M., E. A. Bowles, J. P. Richards, M. Krantic, K. L. Davis, K. A. Dietrich, et al. (2012). "Prostacyclin receptor-mediated ATP release from erythrocytes requires the voltage-dependent anion channel." American Journal of Physiology Heart and Circulatory Physiology **302**: H553–9.

Subramani, R., K. Quadt, A. Jeppesen, C. Hempel, J. Petersen, T. Hassenkam, et al. (2015). "Plasmodium falciparum-Infected Erythrocyte Knob Density Is Linked to the PfEMP1 Variant Expressed." mBio **6**: e01456–15.

Suresh, S. (2006). "Mechanical response of human red blood cells in health and disease: Some structure-property-function relationships." Journal of Materials Research **21**: 1871–1877.

Takeishi, N., H. Ito, M. Kaneko and S. Wada (2019). "Deformation of a red blood cell in a narrow rectangular microchannel." Micromachines **10**: 199.

Tomaiuolo, G. (2014). "Biomechanical properties of red blood cells in health and disease towards microfluidics." Biomicrofluidics **8**: 051501.

Tsai, M. A., R. S. Frank and R. E. Waugh (1993). "Passive mechanical behavior of human neutrophils: Power-law fluid." Biophysical Journal **65**: 2078–2088.

Udden, M. M., T. B. Driscoll, M. H. Pickett, C. S. Leach-Huntoon and C. P. Alfrey CP (1995). "Decreased production of red blood cells in human subjects exposed to microgravity." Journal of Laboratory and Clinical Medicine **125**: 442–449.

van Loon, J. J. W. A. (2008). "Mechanomics and physicomics in gravisensing." Microgravity Science and Technology **21**: 159.

Valle-Delgado, J. J., P. Urbán and X. Fernandez-Busquets (2013). "Demonstration of specific binding of heparin to Plasmodium falciparum-infected vs. non-infected red blood cells by single-molecule force spectroscopy." Nanoscale **5**: 3673–3680.

Wan, J., W. D. Ristenpart and H. A. Stone (2008). "Dynamics of shear-induced ATP release from red blood cells." Proceedings of the National Academy of Science **105**: 16432–16437.

Waugh, R. E. (1987). "Effects of inherited membrane abnormalities on the viscoelastic properties of erythrocyte membrane." Biophysical Journal **51**: 363–369.

Weber, R. E. and S. N. Vinogradov (2001). "Nonvertebrate hemoglobins: Functions and molecular adaptations." Physiological Reviews **81**: 569–628.

Xu, Z., Y. Zheng, X. Wang, N. Shehata, C. Wang and Y. Sun (2018). "Stiffness increase of red blood cells during storage." Microsystems and Nanoengineering **4**: 17103.

Yoon, Y.-Z., J. Kotar, G. Yoon and P. Cicuta (2008). "The nonlinear mechanical response of the red blood cell." Physical Biology **5**: 036007.

Zarychanski, R., V. P. Schulz, B. L. Houston, Y. Maksimova, D. S. Houston, B. Smith, et al. (2012). "Mutations in the mechanotransduction protein PIEZO1 are associated with hereditary xerocytosis." Blood **120**: 1908–1915.

Zhang, J., K. Abiraman, S.-M. Jones, G. Lykotrafitis and B. Andemariam (2017). "Regulation of active ICAM-4 on normal and sickle cell disease RBCs via AKAPs is revealed by AFM." Biophysical Journal **112**: 143–152.

Zhang, S., H. Bai and P. Yang (2015). "Real-time monitoring of mechanical changes during dynamic adhesion of erythrocytes to endothelial cells by QCM-D." ChemComm **51**: 11449–11451.

Marta Targosz-Korecka

6.5 Diabetes and Endothelial Nanomechanics

6.5.1 Introduction

Diabetes and the related hyperglycemia are severe health problems, globally. The elevated glucose level causes metabolic disorders of many cell types (glucose toxicity) and, in consequence, leads to cellular dysfunction and various types of complications such as heart disease, stroke, kidney damage, and nerve damage (Vinik and Vinik, 2003, Fowler, 2008, Yang et al., 2011, Michaelson et al., 2014). The vascular endothelium is particularly vulnerable to the hyperglycemia effect. Due to its location, the endothelium is a direct "victim" of hyperglycemia and is complicit in the development of vascular diseases in diabetes (Duffy et al., 2006). Moreover, a prolonged elevated glucose level is a very promoting environment for carcinogenesis. Epidemiological studies indicate the relation between chronic hyperglycemia and an increased risk of several types of cancer development (Pothiwala et al., 2009, Vigneri et al., 2009, Zhou et al., 2010, Li et al., 2019).

The biological and medical context of the influence of hyperglycemia on cells motivates us to check the relationship between the cells' nanomechanics and the biological aspects of dysfunction development. In this chapter, the relationship between hyperglycemia and the changes in the endothelial nanomechanics in both in vitro and ex vivo observations will be discussed.

6.5.2 Hyperglycemia-Induced Endothelial Cells Stiffening

In most cases, the starting points of diabetes complication development are vascular injury or complications directly related to vascular dysfunction (Kaiser et al., 1993, Tabit et al., 2010). Endothelium is one of the main effector organs that regulate the physiology of the vascular system. This single layer of endothelial cells (ECs) controls the vascular tone by production of vasoprotective factors like nitric oxide (NO). It regulates the blood flow in vessels and controls coagulation and fibrinolysis (Kemeny

Acknowledgment: This research was supported by the 1.1.2 PO IG EU project POMOST FNP: "Elasticity parameter and strength of cell to cell interaction as a new marker of endothelial cell dysfunction in hyperglycemia/hypoglycemia."

Marta Targosz-Korecka, Center for Nanometer-Scale Science and Advanced Materials, NANOSAM, Faculty of Physics, Astronomy and Applied Computer Science, Jagiellonian University

https://doi.org/10.1515/9783110989380-015

et al., 2013). Hyperglycemia has an impact on endothelium metabolism and leads to ECs dysfunction, which is, in particular, manifested by the deficiency of the bioavailable NO, reduced endothelium mediated vasorelaxation, overproduction of growth factors, and increased expression of adhesion molecules and inflammatory genes (Altannavch et al., 2004, Tabit et al., 2010). In normoglycemic ECs, there are three metabolic pathways: glycolysis process, mitochondria ROS signaling, and NO production by eNOS, which cooperate and control the EC's metabolism and function. Hyperglycemia disrupts this cooperation by modifying the glycolysis pathway, enhancing mitochondrial biogenesis, and inhibiting NADPH oxidase activity that results in an increased ROS production.

In consequence, the production of NO by eNOS is diminished, and eNOS are transformed to uncoupled form, which, in turn, enhances the ROS production. These changes cause the overproduction of ROS in ECs (Duffy et al., 2006, Quintero et al., 2006, Busik et al., 2008, Yao and Brownlee, 2010, Tang et al., 2014). The excess of ROS induces a cascade of events, leading to a failure of the function of some proteins (nonenzymatic glycation process), and even to irreversible changes with epigenetic character (metabolic memory) (Ceriello, 2012). The effect of these changes is the progressive long-term endothelial dysfunction, which is reflected in the limited bioavailability of NO and other vascular endothelial mediators. Consequently, the functioning of the vascular system is disturbed, which leads to serious diseases such as hypertension or atherosclerosis. The biological and physiological aspects of the hyperglycemia-induced endothelial dysfunction are well known. In this context, the interesting question is how hyperglycemia influences the nanomechanical properties of ECs.

Hyperglycemia, as have been shown in Chen et al. (2013), increases the stiffness of ECs, which is associated with actin polymerization and a decreased NO production by ECs. In this study, human umbilical vein ECs were exposed to constant hyperglycemia (25 mM) and variable hyperglycemia (25/5 mM) conditions for 7 days. Authors have proven that both constant and fluctuating hyperglycemia harm the ECs; however, the difference in the changes of the stiffness and NO production between the cells exposed to constant and fluctuating high glucose conditions were not significant. In general, stiffening of ECs and the related decreased NO production are hallmarks of endothelial dysfunction (Fels et al., 2012). The disruption in NO availability leads to disorders of vasodilatation and hypertension, and it is one of the reasons for the vascular complication.

The effect of chronic hyperglycemia on the nanomechanical properties of ECs was studied by Targosz-Korecka et al. (2013). The ECs of the EA.hy926 line were exposed to chronic long-term hyperglycemia for almost 80 days. In the experiments, EA.hy926 ECs were cultured under hyperglycemic conditions (glucose concentration 25 mM), from the first passage up to the 26–28 passage. This continuous, long-term experiment showed that within the first 20 passages, a gradual increase of the EC stiffness could be observed (Figure 6.5.1).

After the twentieth passage, there was a decrease in EC stiffness. At the same time, a decrease in cell population and an increase in the number of apoptotic cells

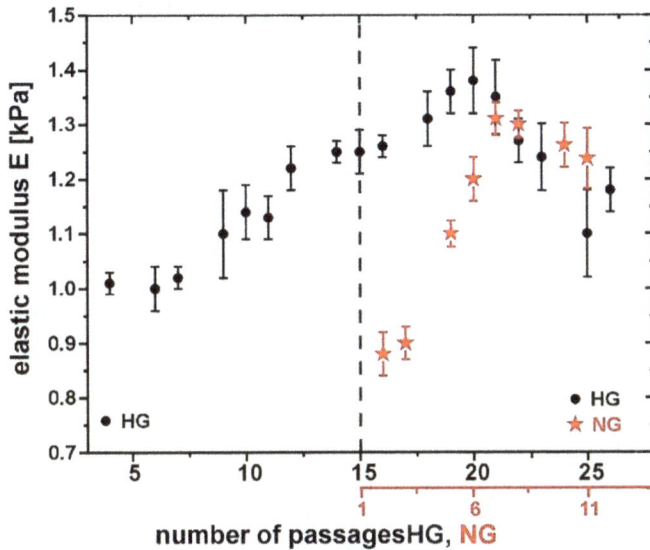

Figure 6.5.1: Changes in the modulus of elasticity of endothelial cells subjected to chronic hyperglycemia (black spots) and after normalization of glucose level (red points). The return of the elastic modulus values to the level corresponding to the cells cultured in hyperglycemia was referred to as "stiffness memory," by an analogy to the metabolic memory effect. The entire experiment (26 passages) lasted about three months. HG – high glucose, NG – low glucose (*reprinted with permission from* Targosz-Korecka et al., 2013).

was observed. Such a result showed that metabolic changes, which occur in ECs as a result of chronic hyperglycemia are reflected in the process of a gradual increase in stiffness of these cells.

In the same publication, it has been shown that the changes in cell stiffness caused by chronic hyperglycemia are irreversible. Normalization of the glucose level after the fourteenth passage (horizontal dotted line in Figure 6.5.1) did not cause permanent normalization of EC stiffness. Although, directly after the change of the glucose concentration in the culture medium from 25 mM (HG) to 5 mM (NG), the elastic modulus decreased, after subsequent passages, the stiffness of the cells began to increase again, to the level corresponding to the value obtained for the cell permanently grown in hyperglycemic condition. The observed effect of permanent changes in the stiffness of ECs under hyperglycemic conditions has been referred to as "stiffness memory." The obtained results could be related to clinical observations (Ceriello, 2012), which showed that short-term normalizations of glucose levels in people with chronic hyperglycemia might be not sufficient to reduce the risk of complications in the cardiovascular system. The effect of "stiffness memory" is a coherent element of metabolic memory, that is, the phenomenon of irreversible epigenetic changes at the cellular level, induced by chronic hyperglycemia.

6.5.3 Modification of the Endothelial Glycocalyx in Hyperglycemia

Glycocalyx plays an essential role in the nanomechanics of ECs. A brush-like polymer sugar-rich layer covers the cell and plays an important role in cell mechanobiology, as mechanosensor and an interface between the extracellular matrix and the cell (Reitsma et al., 2007). In pathological environments such as hyperglycemia, reduction of the glycocalyx thickness and density as well as the collapse of its internal structure significantly change the mechanical properties of this external cell overlay, leading to impairment of the cell functionality (Tarbell and Pahakis, 2006, Eskens et al., 2013).

In recent years, the nanoindentation method with AFM probe has been successfully applied to detect and characterize the mechanical properties of the glycocalyx layer (Wang and Hascall, 2004, Wang et al., 2011). In those experiments, a spherical AFM probe mounted on the soft spring cantilever was used, because of a larger interaction area between probe and sample. If the cell surface is covered by the glycocalyx, at first, the probe squeezes the glycocalyx layer and then indents the cell body (Figure 6.5.2A) The deflection of the cantilever during approach and squeezing is recorded as a function of the relative vertical sample/scanner movement, in the form of force–distance curve or force–indentation curve. On the curve shown in Figure 6.5.2B, one can recognize two distinct regions that correspond to the deformation of two layers with different properties. The first range, for a small indentation depth, corresponds to the squeezing of the glycocalyx layer, whereas

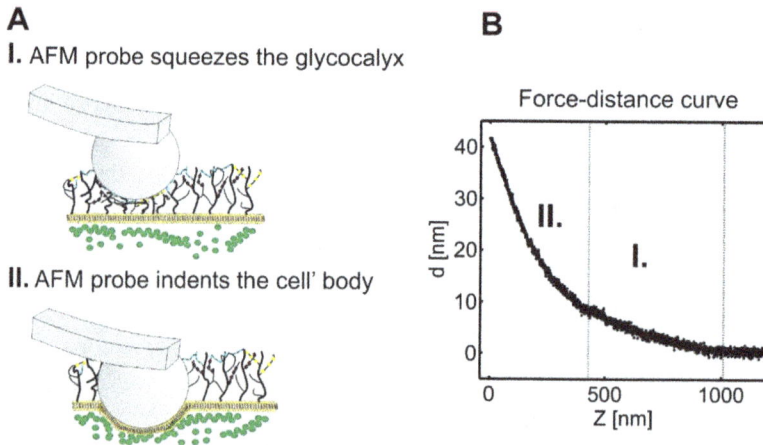

Figure 6.5.2: Principle of nanomechanical detection and characterization of endothelial glycocalyx with an AFM probe. (A) Schematics of nanoindentation process. At shallow depths, the probe senses the glycocalyx layer (region I) and, at larger indentation depth, the cell body (region II). (B) A typical nanoindentation curve recorded for an endothelial cell.

the second one, for larger indentation, corresponds to mechanical response of the cell body (Singleton, 2014).

Measurements of changes in the glycocalyx structure under hyperglycemic condition were carried out for pulmonary human aortal EC by Malek-Zietek et al. (2017). The cells were incubated in hyperglycemic condition in various time regimes – "short-term" (below 2 h) and "prolonged" (up to 24 h). For incubation times below 2 h, an increase in the length of the glycocalyx-forming chains was observed, compared to normoglycemia (Figure 6.5.3). This rather surprising effect of the increase in the length of glycocalyx chains in hyperglycemia could be, however, explained based on the observations in (Wang and Hascall, 2004, Wang et al., 2011) and the increase in the synthesis of hyaluronan (one component of glycocalyx) in hyperglycemia.

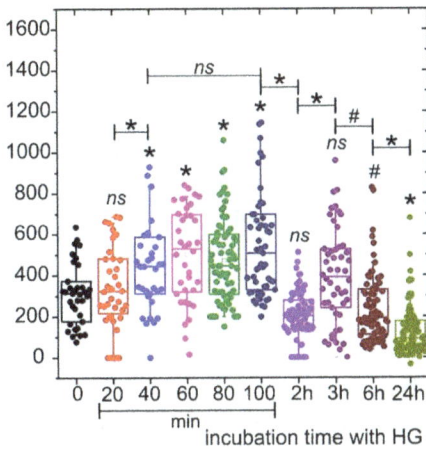

Figure 6.5.3: Hyperglycemia-induced two-step alteration of glycocalyx length. For short-term hyperglycemia (0–100 min), the synthesis of hyaluronic molecules can be observed, which may increase the glycocalyx length. In prolonged hyperglycemia (2–24 h), a significant decrease of the pulmonary human aortal endothelial cell glycocalyx length was observed (reprinted with permission from Malek-Zietek et al. (2017)).

An AFM experiment with the enzymatic removal of hyaluronan by hyaluronidase validated this hypothesis. For a prolonged time of hyperglycemia, the reduction of the glycocalyx layer was noted by decreasing the glycocalyx length and density. This experiment has shown that changes of glycocalyx occurred dynamically, and synthesis or removal of glycocalyx molecules occur in a response to biological stimulation such as hyperglycemia.

6.5.4 Nanomechanical Properties of Vascular Endothelium in Diabetes: An Ex Vivo Experiment

Changes of the endothelial stiffness in progression of diabetes were also studied in an ex vivo experiment on the murine model of type II diabetes. In in vitro measurements of ECs, we deal only with hyperglycemia. In the development of diabetes, in addition to hyperglycemia, other factors such as increased cholesterol and triglyceride levels, or protein glycation products might be important, as well. Therefore, ex vivo measurements performed on the endothelium from mice aorta revealed a more complete picture of the endothelial nanomechanical changes in diabetes. Experiments described by Targosz-Korecka et al. (2017) were performed for C57BLKsJ-db/db mice (homozygotes with a leptin gene mutation), in which type II diabetes develops with age. The C57BLKsJ-db / + mouse (heterozygotes, with the normal leptin gene) was used as a reference. The descending aorta was collected from both groups of mice at ages 11, 12, 16, and 19 weeks. In the cited research, the nanoindentation experiments focused on the characterization of the nanomechanical properties of the glycocalyx layer and endothelium. The layered structure of the aorta as well as the surface heterogeneity surface required a new method for data classification. Therefore, the analysis and tailored data classification methods were applied in order to distinguish the area of endothelium covered with glycocalyx and the area without the glycocalyx layer (Figure 6.5.4A).

For diabetic mice, a statistically significant increase in the stiffness of the endothelial layer was observed. The increase in endothelial stiffness obtained for ex vivo measurements correlated with the reduction of the effective surface coverage with glycocalyx (Figure 6.5.4B and C). However, the length of the glycocalyx chains was gradually reduced, with the progression of diabetes. This means that in the development of diabetes, the degradation of glycocalyx manifests first as a reduction of the glycocalyx coverage. The shortening of the glycocalyx brush takes place at a later stage. Moreover, the elasticity modulus values obtained in the experiment indicated an increase in endothelial stiffness in the progression of diabetes, which is a prognostic factor for the development of hypertension and other pathological vascular changes in patients with diabetes.

The nanomechanical method applied to the study of endothelial *"mechanical dysfunction"* that occurs in diabetes provided important complementary information about the reduction of the glycocalyx coverage and gradual degradation of the glycocalyx length. From the medical point of view, the correlation between endothelial stiffness and the reduction of glycocalyx coverage is interesting. Both the mechanical properties of the endothelium and the structure of glycocalyx play a significant role in the process of regulating the arterial balance between vasoconstriction and vasoconstriction. Hence, the dysfunctional changes of both these parameters broaden knowledge about the development of vascular diseases, which, in diabetes, are dynamic and lead to many complications. The applied method of nanoindentation with the AFM probe combined with the developed data classification method provided complementary information on the cells' elasticity and the glycocalyx characterization.

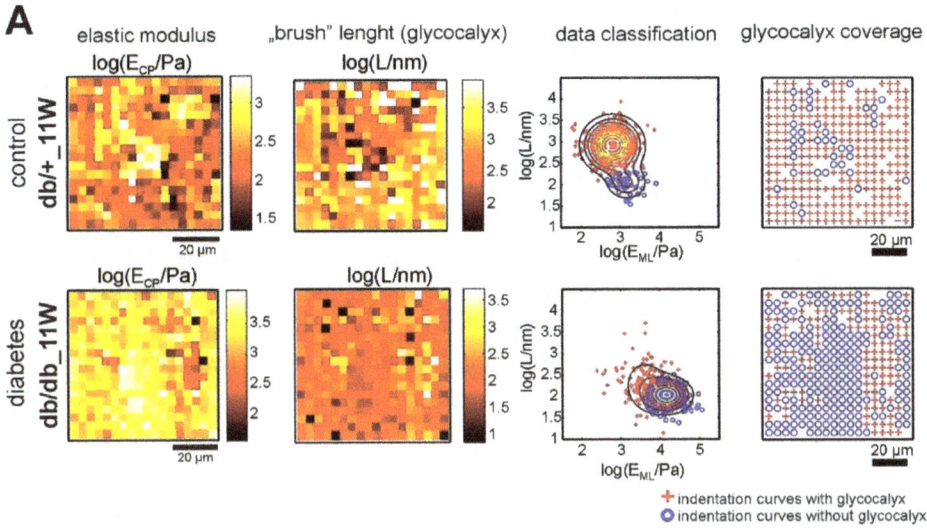

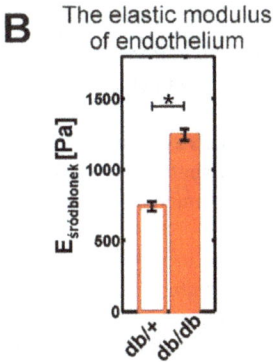

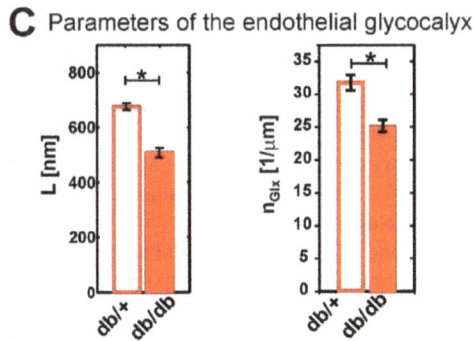

Figure 6.5.4: Diabetes impairs the glycocalyx layer and causes endothelial stiffening: results from an ex vivo experiment on diabetic mice. (A) Examples of spatial maps (from left to right) of the elastic modulus, brush length (glycocalyx), data classification and determined spatial coverage of the glycocalyx. Top row: db/+ control mouse. Bottom row: db/db diabetic mouse. As a result of the data analysis and classification procedure, a significant reduction glycocalyx coverage was demonstrated in diabetic mice at an early stage of diabetes (B) and (C) Aggregate result of an ex vivo experiment carried out on type II diabetes mellitus mice model. Endothelial elasticity modulus (B) and glycocalyx parameters (C), that is, length and coverage of the endothelial surface with glycocalyx (reprinted with permission from Targosz-Korecka et al. (2017)).

6.5.5 Summary

This chapter focused on studies of endothelial nanomechanics in the context of hyperglycemia and diabetes. Endothelium, as an organ regulating the vascular system, plays a crucial role in the formation of vascular complications in diabetes. The nanoindentation method is a unique method that provides complementary information on the cell elasticity as well as the glycocalyx structure. The main effect of the hyperglycemia on the ECs is the cell stiffening and degradation of the glycocalyx layer.

It should be emphasized that hyperglycemia has a harmful effect on a variety of cells and organs. For example, hyperglycemia also increases the stiffness of the insulinoma cell line that is related to the ion channel activation (Yang et al., 2011). Moreover, in clinical observations, one of the most severe problems of diabetes patients is the increase in the risk of cancer development and cancer metastasis. As was shown, hyperglycemia significantly increases the adhesive interaction between cancer and ECs, which promotes attaching the cancer cells to ECs and crossing the vessel walls (Malek-Zietek et al., 2017).

References

Altannavch, T., K. Roubalova, P. Kucera and M. Andel (2004). "Effect of high glucose concentrations on expression of ELAM-1, VCAM-1 and ICAM-1 in HUVEC with and without cytokine activation." Physiological Research **53**(1): 77–82.

Busik, J. V., S. Mohr and M. B. Grant (2008). "Hyperglycemia-induced reactive oxygen species toxicity to endothelial cells is dependent on paracrine mediators." Diabetes **57**(7): 1952–1965.

Ceriello, A. (2012). "The emerging challenge in diabetes: The "metabolic memory"." Vascular Pharmacology **57**(5–6): 133–138.

Chen, X., L. Feng and H. Jin (2013). "Constant or fluctuating hyperglycemias increases cytomembrane stiffness of human umbilical vein endothelial cells in culture: Roles of cytoskeletal rearrangement and nitric oxide synthesis." BMC Cell Biology **14**(1): 22.

Duffy, A., A. Liew, J. O'Sullivan, G. Avalos, A. Samali and T. O'Brien (2006). "Distinct effects of high-glucose conditions on endothelial cells of macrovascular and microvascular origins." Endothelium **13**(1): 9–16.

Eskens, B. J., C. J. Zuurbier, J. van Haare, H. Vink and J. W. van Teeffelen (2013). "Effects of two weeks of metformin treatment on whole-body glycocalyx barrier properties in db/db mice." Cardiovascular Diabetology **12**(1): 175.

Fels, J., P. Jeggle, K. Kusche-Vihrog and H. Oberleithner (2012). "Cortical actin nanodynamics determines nitric oxide release in vascular endothelium." PLoS One **7**(7): e41520.

Fowler, M. J. (2008). "Microvascular and macrovascular complications of diabetes." Clinical Diabetes **26**(2): 77–82.

Kaiser, N., S. Sasson, E. P. Feener, N. Boukobza-Vardi, S. Higashi, D. E. Moller, S. Davidheiser, R. J. Przybylski and G. L. King (1993). "Differential regulation of glucose transport and transporters by glucose in vascular endothelial and smooth muscle cells." Diabetes **42**(1): 80–89.

Kemeny, S. F., D. S. Figueroa and A. M. Clyne (2013). "Hypo-and hyperglycemia impair endothelial cell actin alignment and nitric oxide synthase activation in response to shear stress." PloS One **8**(6): e66176.

Li, W., X. Zhang, H. Sang, Y. Zhou, C. Shang, Y. Wang and H. Zhu (2019). "Effects of hyperglycemia on the progression of tumor diseases." Journal of Experimental & Clinical Cancer Research **38** (1): 327.

Malek-Zietek, K. E., M. Targosz-Korecka and M. Szymonski (2017). "The impact of hyperglycemia on adhesion between endothelial and cancer cells revealed by single-cell force spectroscopy." Journal of Molecular Recognition **30**(9): e2628.

Michaelson, J., V. Hariharan and H. Huang (2014). "Hyperglycemic and hyperlipidemic conditions alter cardiac cell biomechanical properties." Biophysical Journal **106**(11): 2322–2329.

Pothiwala, P., S. K. Jain and S. Yaturu (2009). "Metabolic syndrome and cancer." Metabolic Syndrome and Related Disorders **7**(4): 279–288.

Quintero, M., S. L. Colombo, A. Godfrey and S. Moncada (2006). "Mitochondria as signaling organelles in the vascular endothelium." Proceedings of the National Academy of Sciences **103**(14): 5379–5384.

Reitsma, S., D. W. Slaaf, H. Vink, M. A. Van Zandvoort and M. G. A. Oude Egbrink (2007). "The endothelial glycocalyx: Composition, functions, and visualization." Pflügers Archiv-European Journal of Physiology **454**(3): 345–359.

Singleton, P. A. (2014). "Hyaluronan regulation of endothelial barrier function in cancer." Advances in Cancer Research **123**: 191–209.

Tabit, C. E., W. B. Chung, N. M. Hamburg and J. A. Vita (2010). "Endothelial dysfunction in diabetes mellitus: Molecular mechanisms and clinical implications." Reviews in Endocrine and Metabolic Disorders **11**(1): 61–74.

Tang, X., Y.-X. Luo, H.-Z. Chen and D.-P. Liu (2014). "Mitochondria, endothelial cell function, and vascular diseases." Frontiers in Physiology **5**: 175.

Tarbell, J. M. and M. Pahakis (2006). "Mechanotransduction and the glycocalyx." Journal of Internal Medicine **259**(4): 339–350.

Targosz-Korecka, M., G. D. Brzezinka, K. E. Malek, E. Stępień and M. Szymonski (2013). "Stiffness memory of EA. hy926 endothelial cells in response to chronic hyperglycemia." Cardiovascular Diabetology **12**(1): 96.

Targosz-Korecka, M., M. Jaglarz, K. E. Malek-Zietek, A. Gregorius, A. Zakrzewska, B. Sitek, Z. Rajfur, S. Chlopicki and M. Szymonski (2017). "AFM-based detection of glycocalyx degradation and endothelial stiffening in the db/db mouse model of diabetes." Scientific Reports **7**(1): 1–15.

Vigneri, P., F. Frasca, L. Sciacca, G. Pandini and R. Vigneri (2009). "Diabetes and cancer." Endocrine-Related Cancer **16**(4): 1103–1123.

Vinik, A. I. and E. Vinik (2003). "Prevention of the complications of diabetes." American Journal of Managed Care **9**(3): S63–S80.

Wang, A., C. de la Motte, M. Lauer and V. Hascall (2011). "Hyaluronan matrices in pathobiological processes." The FEBS Journal **278**(9): 1412–1418.

Wang, A. and V. C. Hascall (2004). "Hyaluronan structures synthesized by rat mesangial cells in response to hyperglycemia induce monocyte adhesion." Journal of Biological Chemistry **279** (11): 10279–10285.

Yang, R.-G., N. Xi, K. W.-C. Lai, B.-H. Zhong, C. K.-M. Fung, C.-G. Qu and D. H. Wang (2011). "Nanomechanical analysis of insulinoma cells after glucose and capsaicin stimulation using atomic force microscopy." Acta Pharmacologica Sinica **32**(6): 853–860.

Yao, D. and M. Brownlee (2010). "Hyperglycemia-induced reactive oxygen species increase expression of the receptor for advanced glycation end products (RAGE) and RAGE ligands." Diabetes **59**(1): 249–255.

Zhou, X., Q. Qiao, B. Zethelius, K. Pyörälä, S. Söderberg, A. Pajak, C. Stehouwer, R. Heine, P. Jousilahti and G. Ruotolo (2010). "Diabetes, prediabetes and cancer mortality." Diabetologia **53**(9): 1867–1876.

Arnaud Millet, Thomas Decaens

6.6 Chronic Liver Diseases

Chronic liver diseases are an increasing cause of death worldwide. Even if the causes leading to liver dysfunction is broad, a common physiopathological step is the appearance of fibrosis leading to cirrhosis. This clinical stage is associated with the development of hepatocellular carcinoma (HCC), the main type of primary cancer of the liver, which present a devastating prognosis. From a physiopathological point of view, the appearance of cirrhosis is related to a mechanical modification of liver tissues offering a clue to diagnose the disease and propose new prognostic markers. In this chapter, we will review the various methodologies used so far to address this question.

6.6.1 Cirrhosis: A Common Feature of Chronic Liver Diseases

Cirrhosis is an increasing cause of morbidity and mortality in developed countries. This condition results from various liver injuries that lead to a diffuse nodular regeneration surrounded by dense fibrotic septa. It is the 14th most common cause of death worldwide corresponding to 1 million deaths per year (Tsochatzis et al., 2014). Cirrhosis is the main indication for liver transplant in Europe (Blachier et al., 2013). The main causes depend on the geographical origin of patients. In developed countries, hepatitis C virus, alcohol misuse, and nonalcoholic liver diseases are the predominant cause, and this contrasts to the situation found in southeast Asia or sub-Saharan Africa where the leading cause is hepatitis B virus. The prevalence of cirrhosis is not easily determined; nevertheless, a French screening program has found a prevalence of 0.3% (Blachier et al., 2013).

A striking finding of the histopathology of liver is the fact that, irrespective to the aetiology of the liver disease (toxic, genetic, auto-immune, or infectious), the liver tissue will ultimately be characterized by an excess of deposition of extracellular matrix leading to fibrosis. Pathologists use the METAVIR classification to stage fibrosis: F0 corresponds to the absence of fibrosis, F1 is associated with portal fibrosis without

Arnaud Millet, Institute for Advanced Biosciences, Grenoble-Alpes University, Inserm U1209–CNRS UMR 5309 and Research Department University Hospital of Grenoble Alpes, Grenoble, France
Thomas Decaens, Univ. Grenoble Alpes; Institute for Advanced Biosciences, Grenoble-Alpes University, Inserm U1209–CNRS UMR 5309 and Hepatology Department University Hospital of Grenoble Alpes, Grenoble, France

https://doi.org/10.1515/9783110989380-016

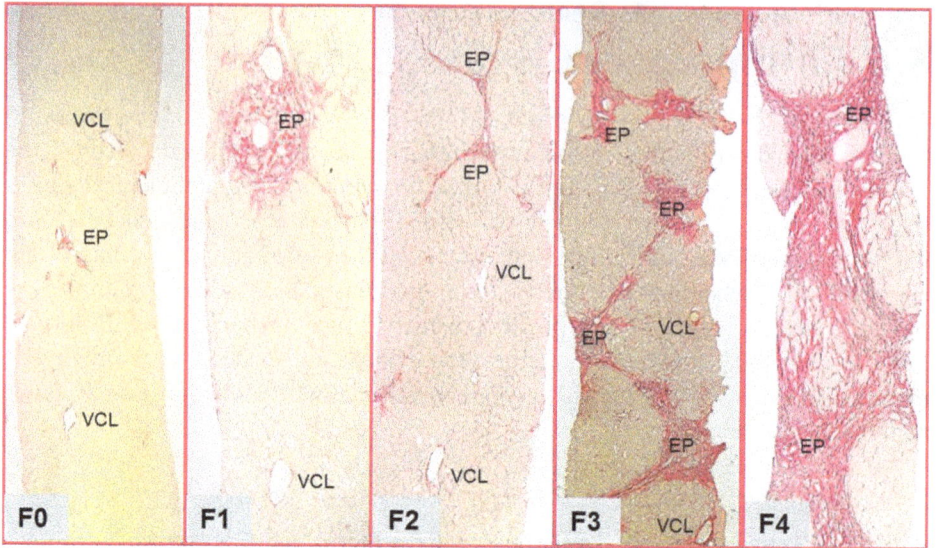

Figure 6.6.1: Sirius red staining of liver tissues illustrating the various stage of the METAVIR classification (EP, portal area; VCL = centrolobular vein) (images from Pr Nathalie Surm, Pathology Department, University Hospital of Grenoble Alpes).

septa, in F2 rare septa are found, numerous septa characterize the F3 stage, and cirrhosis is described as F4 (Bedossa and Poynard, 1996). These different stages are illustrated in Figure 6.6.1.

Patients suffering from cirrhosis do have a high mortality rate and usually present various decompensation episodes. Cirrhosis is associated with a hepatocellular insufficiency, impacting the ability to synthesize proteins like albumin. The progression of fibrosis will also increase the vascular resistance leading to a portal hypertension as well as an increase of the porto-hepatic venous pressure gradient. From the clinical point of view, cirrhosis is associated with ascites (liquid collection in the peritoneum) which could be spontaneously infected leading to peritonitis; esophageal varices could bleed; liver failure and HCC are the ending stage of the disease. When compensated, the diagnosis of cirrhosis could be challenging. Liver biopsy is the gold standard for assessing the stage of liver fibrosis, but it is an invasive procedure. Classical imaging techniques like ultrasound echography, computed tomography (CT), or magnetic resonance imaging (MRI) lack specificity and sensitivity to detect the first stages of the disease. Serum biomarkers, even if useful, do not correlate perfectly with the fibrosis stage. Benefiting from clinical knowledge that cirrhotic livers are stiffer than normal ones, the last decade has shown an increased interest in the mechanical characterization of tissues in the context of chronic liver diseases.

6.6.2 Liver Stiffness: A New Clinical Parameter

The way to quantitatively assess the stiffness of a material is advantageously done by measuring its Young's modulus or elasticity. The physical content of this parameter could be easily grasped as followed: If we apply a force F on a surface S of a body, the elastic response will correspond to a compression in the direction of the force. The shrinkage of the body could be quantified by the difference between the resulting size l and the steady-state size l_0. The relation that links these physical quantities is called the Hooke's law:

$$\frac{F}{S} = E\frac{l_0 - l}{l_0} \quad (*)$$

The proportionality parameter E is the Young's modulus; its unit is in Pa (the same as a pressure).

The intuitive meaning of this equation is that in order to deform a body by a certain percentage of its size, one must apply a force that will be higher for stiffer material compared to softer ones, and the Young's modulus is a quantification of this property. Biological tissues like liver have a Young's modulus typically of few kilopascals. Before describing the various techniques used to measure the stiffness of livers, we need to understand what are the various parameters that could modify it. The first factor is our target: the extracellular matrix. But external and internal pressure of the organ (blood outflow, respiratory dysfunction, inflammation, etc.) could sensibly increase the measured stiffness. Another point that should be raised is the fact that from a physical point of view, the Young's modulus is a static parameter. In practice, it is not what we directly measure as we always obtained a viscoelastic response related to the finite characteristic time our techniques. This is a concern as stiffness usually increases with the frequency of mechanical stimulus.

Various excellent reviews have been published describing the different techniques used to assess liver's stiffness. We have summarized the different methods used in Table 6.6.1, adapted from Mueller and Sandrin (2010).

Many studies have addressed the clinical validation of liver stiffness as a marker of fibrosis. Despite some variability, it has been possible to propose cutoff values for stiffness of stage F3 and for F4. F1 and F2 fibrosis stages present mild increase stiffness and are not discriminated by shear wave methods. In a recent meta-analysis, transient elastography measurement of liver stiffness in apparently healthy individuals has been addressed. The mean stiffness in truly healthy nonobese individuals was 4.68 kPa (95% CI, 4.64–4.73). Statistically significant modifiers of stiffness included diabetes, dyslipidemia, waist circumference, serum transaminases (AST), and systolic blood pressure (Bazerbachi et al., 2019). It also appears clear that the proposed cutoff values for fibrosis stages should be adapted when inflammation and cholestasis are present. In the following table, we have summarized the different studies validating the connection between stiffness (measured by

Table 6.6.1: A summary of various methods used to quantify mechanical properties of the liver.

	Method	Vibration mode/ source	Frequency	Advantages	Limitations
Static elastography	Quasi-static compression	None	NA	Widely available	Qualitative
Magnetic resonance elastography	Shear wave	Continuous mechanical actuator	50–60 Hz	2D/3D stiffness mapping	Expansive, pacemaker
Acoustic radiation force impulse (ARFI)	Shear wave	Transient radiation force		Ascites	No clinical validation
Vibration-controlled transient elastography (Fibroscan®)	Shear wave	Transient radiation force	50 Hz	Clinical validation	Obesity, ascites, limited range of stiffness values (1.5–75 kPa)

shear wave method) and fibrosis stages in various liver diseases, in an updated version of a Table 6.6.2 published in Mueller and Sandrin (2010).

In addition to this clinical validation of the correlation between liver stiffness and cirrhosis, a met-analysis included 32 studies for the use of transient elastography to detect the presence of esophageal varices. The pooled sensitivity, specificity, and diagnostic odds ratio were 0.8 (95% CI, 0.78–0.86), 0.68 (95% CI, 0.62–0.74), and 10 (95% CI, 7–14) for any EV; 0.81 (95% CI, 0.77–0.85), 0.72 (95% CI, 0.66–0.77), and 11 (95% CI, 8–15) for substantial EV; and 0.92 (95% CI, 0.83–0.96), 0.78 (95% CI, 0.70–0.85), and 40 (95% CI, 15–107) for large EV (Cheng et al., 2018). Transient elastography appeared, also, as a relevant clinical tool for assessment of complications of cirrhosis in a non-invasive manner.

The clinical validation performed in these studies were associated with a selection of patients for which the transient elastography could be performed. Indeed, various factors have been identified that could increase liver stiffness:
- Hepatitis (inflammation)
- Obstructive cholestasis
- Liver congestion
- Cellular infiltrations
- Amyloidosis

Steatosis, even if a major provider of chronic liver disease, is usually associated with a decrease of liver stiffness. In order to cope with these confounding factors, the assessment of fibrosis by transient elastography in a clinical setting could be described as follows (Figure 6.6.2).

Table 6.6.2: A correlation between stiffness and fibrosis stage in various liver diseases.

Patients	Number of patients	Correlation fibrosis-stiffness	AUC F3	AUC F4	Cutoff F3 (kPa)	Cutoff F4 (kPa)	References
HCV	193	ND	0.9	0.95	9.5	12.5	Castéra et al. (2005)
HCV	935	ND	0.89	0.91	ND	ND	Kettaneh et al. (2007)
HCV/HIV	72	0.48 ($p < 0.0001$)	0.91	0.97	ND	11.9	de Lédinghen et al. (2006)
HBV	202	0.65 ($p < 0.001$)	0.93	0.93	ND	11	Marcellin et al. (2009)
ALD	103	0.72 ($p < 0.014$)	0.9	0.92	11	19.5	Nguyen-Khac et al. (2008)
ALD	101	0.72 ($p < 0.001$)	0.91	0.92	8	11.5	Mueller et al. (2010)
NAFLD	246	ND	0.92	0.95	7.9		Wong et al. (2010)
PBC/PSC	101	0.84 ($p < 0.0001$)	0.95	0.96	9.8	17.3	Corpechot et al. (2006)
PBC	80	ND		0.96	ND	ND	Gómez-Dominguez et al. (2008)
Prospective study for liver biopsy	148	0.77 ($p < 0.0001$)	0.96	0.99	11.5	18.1	Guibal et al. (2016)
HBV/NAFLD	94	ND	0.95	0.96	8.7	10.9	Zhang et al. (2019)

Abbreviations: HVC, hepatitis C virus; HBV, hepatitis B virus; ALD, alcoholic liver disease; NAFLD, nonalcoholic fatty liver disease; PBC, primary biliary cirrhosis; PSC, primary sclerosing cirrhosis; AUC, area under the curve; ND, not determined.

6.6.3 Is There a Place for Atomic Force Microscopy in Liver Stiffness Assessment?

Atomic force microscopy (AFM) is a technique coming from physics that has been used to assess mechanical signature in cell biology for more than a decade and recently its use on human tissues has gain interest (Plodinec et al., 2012). AFM presents several advantages, such as the possibility of combining topographical and mechanical (rheological) imaging with high spatial and force resolution, as well as adding specific mapping abilities using functionalized probes. This surface technique is only applicable on tissue samples from surgical procedure or biopsies. Fresh samples are released from internal or external pressure, and the elasticity of

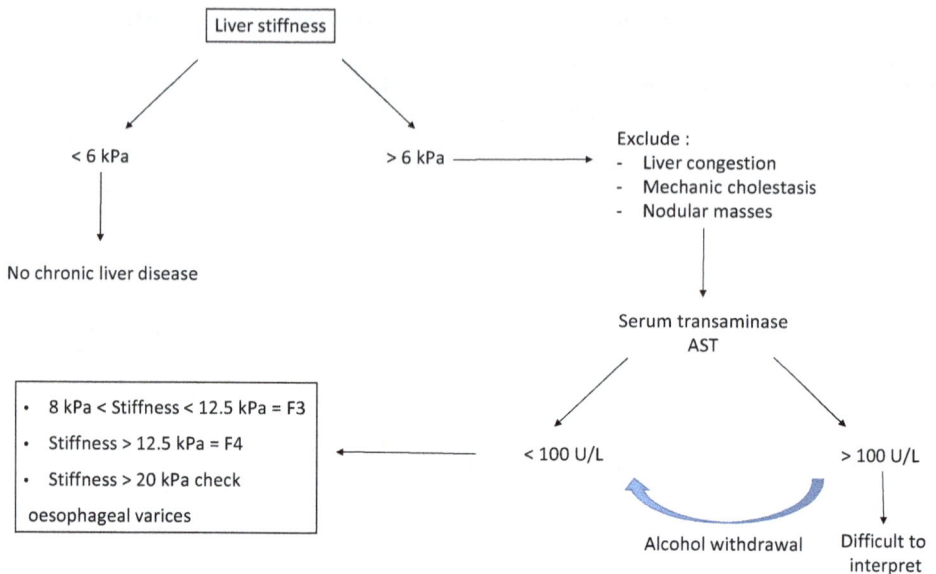

Figure 6.6.2: Assessment of liver stiffness according to various clinical conditions (Mueller and Sandrin, 2010).

the tissue measured by AFM will be directly related to extracellular matrix deposition and cellular infiltration. Few studies have used AFM to study liver tissues. One study directly addressed the question of the mechanical signature during hepatocarcinogenesis using the DEN (diethylnitrosamine) rat model. In this model, animals develop HCC on a cirrhotic liver. Using a V-shaped probe, the authors were able to follow the increase of rat liver stiffness during the course of the disease. Control rats without exposition to DEN were found to have a liver stiffness of 0.18 ± 0.06 MPa. Rats exposed to 8 weeks of DEN were found to have an increased stiffness at 0.25 ± 0.06 MPa and cirrhotic rats after 12 weeks of DEN presented a stiffness of 0.39 ± 0.06 MPa. HCC appeared after 16 weeks of DEN without differences from the cirrhotic state. The stiffness was measured at 0.42 ± 0.07 MPa (Gang et al., 2009). A recent study specifically addressed the mechanical signature associated with the HCC apparition. Using a pyramidal probe, they find a bimodal distribution of elasticity in samples from human livers. They also reported that HCC was associated with a downward shift of the lowest elasticity peak and could be used as a mechanical fingerprint of malignancy (Tian et al., 2015). These promising results need to be confirmed, but AFM appears as a new tool for diagnosis and prognosis in chronic liver diseases.

References

Bazerbachi, F., S. Haffar, Z. Wang, J. Cabezas, M. T. Arias-Loste, J. Crespo, S. Darwish-Murad, M. A. Ikram, J. K. Olynyk, E. Gan, et al. (2019). "Range of normal liver stiffness and factors associated with increased stiffness measurements in apparently healthy individuals." Clinical Gastroenterology and Hepatology: The Official Clinical Practice Journal of the American Gastroenterological Association 17: 54–64.e1.

Bedossa, P. and T. Poynard (1996). "An algorithm for the grading of activity in chronic hepatitis C. The METAVIR cooperative study group." Hepatology Baltimore, MD 24: 289–293.

Blachier, M., H. Leleu, M. Peck-Radosavljevic, D.-C. Valla and F. Roudot-Thoraval (2013). "The burden of liver disease in Europe: A review of available epidemiological data." Journal of Hepatology 58: 593–608.

Castéra, L., J. Vergniol, J. Foucher, B. Le Bail, E. Chanteloup, M. Haaser, M. Darriet, P. Couzigou and V. De Lédinghen (2005). "Prospective comparison of transient elastography, Fibrotest, APRI, and liver biopsy for the assessment of fibrosis in chronic hepatitis C." Gastroenterology 128: 343–350.

Cheng, F., H. Cao, J. Liu, L. Jiang, H. Han, Y. Zhang and D. Guo (2018). "Meta-analysis of the accuracy of transient elastography in measuring liver stiffness to diagnose esophageal varices in cirrhosis." Medicine (Baltimore) 97: e11368.

Corpechot, C., A. El Naggar, A. Poujol-Robert, M. Ziol, D. Wendum, O. Chazouillères, V. de Lédinghen, D. Dhumeaux, P. Marcellin, M. Beaugrand, et al. (2006). "Assessment of biliary fibrosis by transient elastography in patients with PBC and PSC." Hepatology Baltimore, MD 4: 1118–1124.

Gang, Z., Q. Qi, C. Jing and C. Wang (2009). "Measuring microenvironment mechanical stress of rat liver during diethylnitrosamine induced hepatocarcinogenesis by atomic force microscope." Microscopy Research and Technique 72: 672–678.

Gómez-Dominguez, E., J. Mendoza, L. García-Buey, M. Trapero, J. P. Gisbert, E. A. Jones and R. Moreno-Otero (2008). "Transient elastography to assess hepatic fibrosis in primary biliary cirrhosis." Alimentary Pharmacology & Therapeutics 27: 441–447.

Guibal, A., G. Renosi, A. Rode, J. Y. Scoazec, O. Guillaud, L. Chardon, M. Munteanu, J. Dumortier, F. Collin and T. Lefort (2016). "Shear wave elastography: An accurate technique to stage liver fibrosis in chronic liver diseases." Diagnostic and Interventional Imaging 97: 91–99.

Kettaneh, A., P. Marcellin, C. Douvin, R. Poupon, M. Ziol, M. Beaugrand and V. de Lédinghen (2007). "Features associated with success rate and performance of FibroScan measurements for the diagnosis of cirrhosis in HCV patients: A prospective study of 935 patients." Journal of Hepatology 46: 628–634.

de Lédinghen, V., C. Douvin, A. Kettaneh, M. Ziol, D. Roulot, P. Marcellin, D. Dhumeaux and M. Beaugrand (2006). "Diagnosis of hepatic fibrosis and cirrhosis by transient elastography in HIV/hepatitis C virus-coinfected patients." Journal of Acquired Immune Deficiency Syndromes 41: 175–179.

Marcellin, P., M. Ziol, P. Bedossa, C. Douvin, R. Poupon, V. de Lédinghen and M. Beaugrand (2009). "Non-invasive assessment of liver fibrosis by stiffness measurement in patients with chronic hepatitis B." Liver International: Official Journal of the International Association for the Study of the Liver 29: 242–247.

Mueller, S. and L. Sandrin (2010). "Liver stiffness: A novel parameter for the diagnosis of liver disease." Hepatic Medicine: Evidence and Research 2: 49–67.

Mueller, S., G. Millonig, L. Sarovska, S. Friedrich, F. M. Reimann, M. Pritsch, S. Eisele, F. Stickel, T. Longerich, P. Schirmacher, et al. (2010). "Increased liver stiffness in alcoholic liver disease: Differentiating fibrosis from steatohepatitis." World Journal of Gastroenterology 16: 966–972.

Nguyen-Khac, E., D. Chatelain, B. Tramier, C. Decrombecque, B. Robert, J.-P. Joly, M. Brevet, P. Grignon, S. Lion, L. Le Page, et al. (2008). "Assessment of asymptomatic liver fibrosis in alcoholic patients using fibroscan: Prospective comparison with seven non-invasive laboratory tests." Alimentary Pharmacology & Therapeutics **28**: 1188–1198.

Plodinec, M., M. Loparic, C. A. Monnier, E. C. Obermann, R. Zanetti-Dallenbach, P. Oertle, J. T. Hyotyla, U. Aebi, M. Bentires-Alj, R. Y. H. Lim, et al. (2012). "The nanomechanical signature of breast cancer." Nature Nanotechnology **7**: 757–765.

Tian, M., Y. Li, W. Liu, L. Jin, X. Jiang, X. Wang, Z. Ding, Y. Peng, J. Zhou, J. Fan, et al. (2015). "The nanomechanical signature of liver cancer tissues and its molecular origin." Nanoscale **7**: 12998–13010.

Tsochatzis, E. A., J. Bosch and A. K. Burroughs (2014). "Liver cirrhosis." Lancet (London, England) **383**: 1749–1761.

Wong, V. W.-S., J. Vergniol, G. L.-H. Wong, J. Foucher, H. L.-Y. Chan, B. Le Bail, P. C.-L. Choi, M. Kowo, A. W.-H. Chan, W. Merrouche, et al. (2010). "Diagnosis of fibrosis and cirrhosis using liver stiffness measurement in nonalcoholic fatty liver disease." Hepatology Baltimore, MD **51**: 454–462.

Zhang, G.-L., Q.-Y. Zhao, C.-S. Lin, Z.-X. Hu, T. Zhang and Z.-L. Gao (2019). "Transient elastography and ultrasonography: Optimal evaluation of liver fibrosis and cirrhosis in patients with chronic hepatitis B concurrent with nonalcoholic fatty liver disease." BioMed Research International **2019**: 3951574.

Prem Kumar Viji Babu, Manfred Radmacher

6.7 Mechanics of Dupuytren's Disease

Dupuytren's disease is a connective tissue disorder of the hand, which leads to, and to a larger extent causes, alteration in the pathological tissue mechanical properties. The primary function of connective tissue in interplay with tendons and ligaments is to support various cellular types by providing a suitable microenvironment mainly defined by macromolecular proteins found in the constituting cells and the surrounding extracellular matrix (ECM). Investigating the mechanical behavior of tissues started in the late 1950s, where the mechanical response and structure of the connective tissues were characterized on the macro level. This load-bearing biological tissue withstands the mechanical tension created within the body and thus restores the overall tissue and organ-level orientation to its original configuration. Most of the connective tissues are abundantly enriched with collagen and elastin, which makes them the predominant mechanome. Large-scale mechanical studies were conducted using the so-called Instron testing machine on these collagenous and elastinous tissues and their varying proportions in particular tissues directly affect its mechanical response (Greenman et al, 2012). Stress-relaxation tests were conducted on tissues extracted from different parts of the human body including lung, aorta, tendon, dura mater, pericardium, or skin from the thoracic or abdominal regions (Dunn et al, 1983). Nevertheless, there is still a need to conduct micro- or nanoscale mechanics on connective tissues using nanoindenters or atomic force microscopy (AFM). This could help to understand cell and ECM architecture in tissues and also help to follow their dynamic interplay at the tissue level. Basically, ECM components such as collagen, elastin, glycoproteins, proteoglycans, and glycosaminoglycans are synthesized by the cells of connective tissue, namely fibroblasts. Connective-tissue-linked diseases are mostly originating by the disordered biochemical or biophysical cues that are invoked in these fibroblasts and its surrounding matrices. The pathological aspects of connective tissue are mainly investigated from the viewpoint of protein expression level such as its secretion, orientation, degradation, and dynamics. But, further exploration in diseased tissue mechanics leads to the basic understanding of the pathological tissue elasticity, thus comparing it to the healthy tissue phenotype. This can be related to the identification of many connective-tissue-related diseases such as fibrosis. The formation of abundant connective tissue filaments creates a dense and hardened environment which leads to organ-specific fibrosis, namely in the lung, kidney, liver, heart, and skin (Jun et al., 2018). For example, the stiffening in human lung fibrosis compared to healthy lung in both native and decellularized states was observed by AFM (Booth et al., 2012).

Prem Kumar Viji Babu, Manfred Radmacher, Institute of Biophysics, University of Bremen, Bremen, Germany

https://doi.org/10.1515/9783110989380-017

Dupuytren's disease is a fibrocontractive disease that affects the cords and nodular region of the palmar aponeurosis, which leads to the permanent contracture of the finger joints (Figure 6.7.1A). Distortion of tissue architecture by the excessive

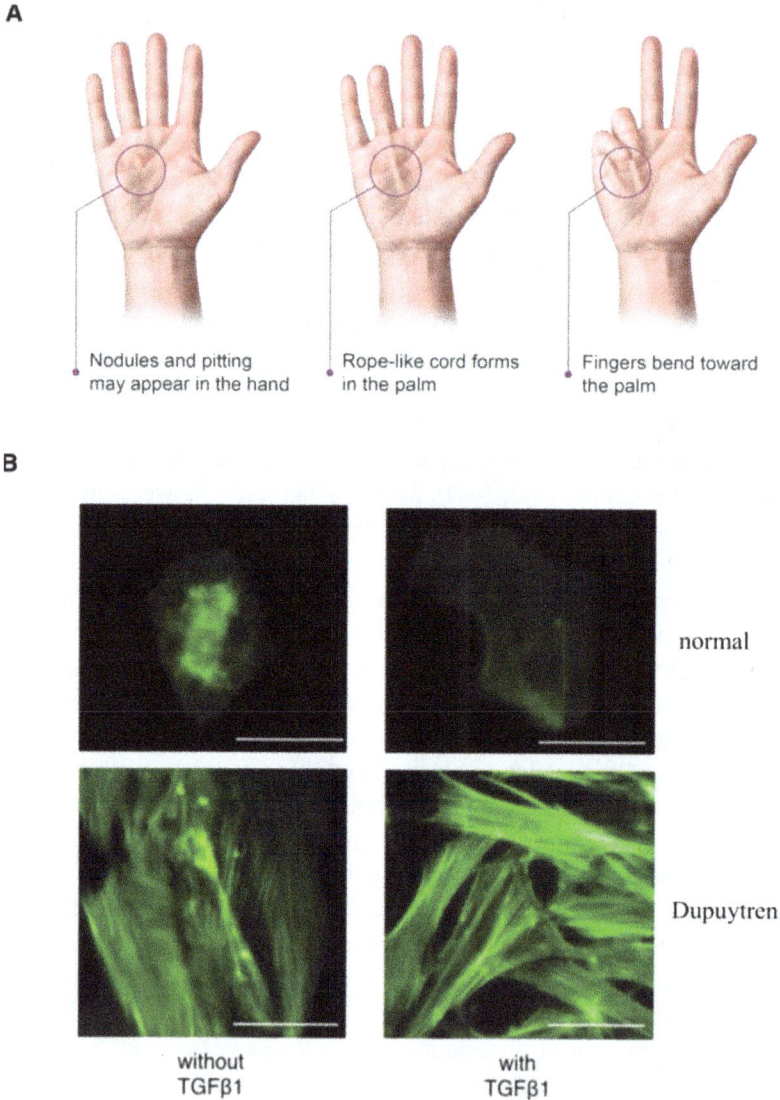

A

Nodules and pitting
may appear in the hand

Rope-like cord forms
in the palm

Fingers bend toward
the palm

B

normal

Dupuytren

without
TGFβ1

with
TGFβ1

Figure 6.7.1: (A) Schematic drawing of the stepwise cords and nodules stiffening, which leads to finger contracture (reproduced without permission from https://cliffordcraig.org.au/dupuytrens-disease-research-update/). (B) Normal fibroblast from normal healthy dermal region of the skin and Dupuytren's fibroblast from palmar fascia of the same Dupuytren's diseased patients shows distinguishable phenotypes: large and no/less stress fiber formation in Dupuytren's and normal fibroblasts, respectively. Scale bar: 50 μm.

deposition of ECM and uncontrollable exertion of contractile forces on ECM by myofibroblasts leads to the fibrotic environment. From the mesenchyme family, fibroblasts synthesize, degrade, and maintain the ECM, thus providing a meshwork for
other cell types to coordinate at the tissue and subtissue level organization and
function. Fibroblasts, by acquiring myofibroblast phenotype, actively participate in

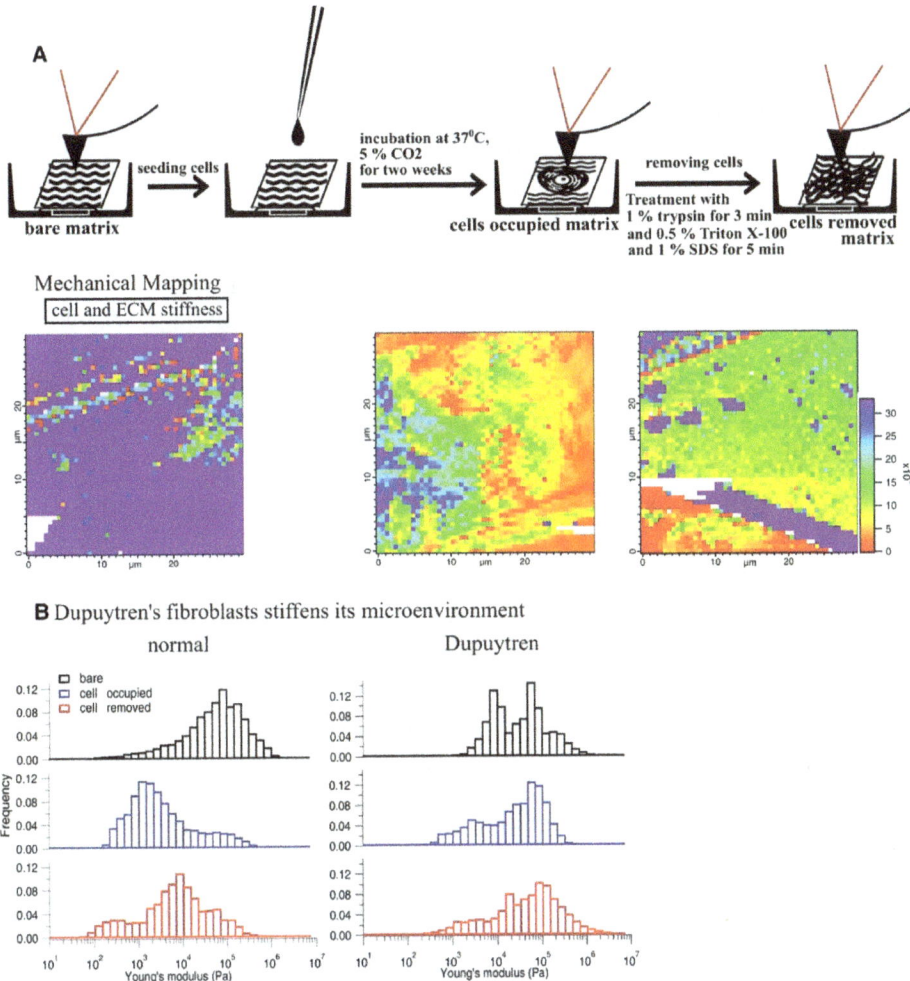

Figure 6.7.2: (A) Schematic drawing of the experimental setup to follow the fibroblast and its
microenvironment mechanics by recording mechanical maps of high resolution. Maps were
obtained on the decellularized/synthesized matrices at each step: bare, cell occupied, and cell
removed in order to follow the mechanical changes. (B) Histogram representation of Young's
modulus shows that normal fibroblast softens the matrices on the larger degree whereas
Dupuytren's fibroblast stiffens its microenvironment (Reprinted with permission from Viji Babu
et al. (2019)).

wound healing. However, loss of their apoptosis or reversion to a fibroblast phenotype after the wound has healed results in various pathological reactions related to connective tissue disorder. Viji Babu et al. have studied the mechanical and migratory properties of fibroblasts and myofibroblasts extracted from the same Dupuytren's patient (Figure 6.7.1B). They observed that normal fibroblasts are softer and migrate faster than Dupuytren's fibroblasts, which exhibit a myofibroblast-like phenotype (Viji Babu et al., 2018). This can be inferred from the expression of alpha-smooth muscle actin and the formation of long stress fibers in Dupuytren's fibroblasts. In this study, it has been found that Dupuytren's fibroblast mechanics depends on substrate stiffness, showing the interplay between cell mechanics and ECM stiffness and vice versa.

In another study from the same research group, AFM force spectroscopic experiment was designed to mimic their subtissue level dynamics of cell and ECM by seeding normal and Dupuytren's fibroblasts on decellularized matrices (Figure 6.7.2A).

These decellularized matrices, differing in their biochemical and biophysical cues, are used to study the interdependent interplay between fibroblasts and their microenvironment. Normal fibroblasts tend to soften its surroundings, whereas Dupuytren's fibroblasts stiffen the decellularized matrix, MatriDerm (Viji Babu et al., 2019) (Figure 6.7.2B). These findings could be related in the future to the tissue-level mechanics in such connective tissue disorders, including Dupuytren's contracture explored using AFM.

References

Booth, A. J., R. Hadley, A. M. Cornett, A. A. Dreffs, S. A. Matthes, J. L. Tsui, K. Weiss, J. C. Horowitz, V. F. Fiore, T. H. Barker and B. B. Moore (2012). "A cellular normal and fibrotic human lung matrices as a culture system for in vitro investigation." American Journal of Respiratory and Critical Care Medicine 186(9): 866–876.

Dunn, M. G. and F. H. Silver (1983). "Viscoelastic behaviour of human connective tissues: Relative contribution of viscous and elastic components." Connective Tissue Research 12(1): 59–70.

Greenman, P. E. (Ed.) (2012). Concepts and mechanisms of neuromuscular functions: An international conference on concepts and mechanisms of neuromuscular functions. Springer Science & Business Media.

Jun, J. I. and L. F. Lau (2018). "Resolution of organ fibrosis." The Journal of Clinical Investigation 128(1): 97–107.

Viji Babu, P. K., C. Rianna, G. Belge, U. Mirastschijski and M. Radmacher (2018). "Mechanical and migratory properties of normal, scar, and Dupuytren's fibroblasts." Journal of Molecular Recognition 31(9): e2719.

Viji Babu, P. K., C. Rianna, U. Mirastschijski and M. Radmacher (2019). "Nano-mechanical mapping of interdependent cell and ECM mechanics by AFM force spectroscopy." Scientific Reports 9: 12317.

Wolfgang H. Goldmann

6.8 Intermediate Filaments

Abstract: Intermediate filaments (IFs) are one of the three types of cytoskeletal polymers that resist tensile and compressive forces in cells. They cross-link with each other as well as with actin filaments and microtubules by means of proteins such as desmin, lamin (A/C), plectin, and filamin C. Mutations in these proteins can lead to a wide range of pathologies, some of which exhibit mechanical failure of the skin, skeletal, or heart muscle.

Keywords: cellular mechanics, desmin, filamin C, intermediate filaments, lamin (A/C), plectin

6.8.1 Introduction

Intermediate filaments (IFs) are found in many cell types and are part of the actin filament and microtubule cytoskeleton (Figure 6.8.1). They extend throughout the cytoplasm connecting the nuclear and cell membrane and are responsible for cell morphology and mechanics (Capetanaki et al., 2007, Fletcher and Mullins, 2010, Etienne-Manneville, 2018). While extra-sarcomeric IFs constitute a filamentous network through a number of cross-linking and regulatory proteins in cells that connect membrane-anchored structures with Z-disks, sarcomeric IF proteins integrate the cytoskeleton with organelles such as mitochondria and nuclei. Various IF protein types have been described in many cell types, whose staggered assembly into protofilaments impart high tensile strength, thus enhancing their resistance to compression, stretching, and bending forces (Herrmann et al., 2009, Goldman et al., 2011, Köster et al., 2015, Herrmann and Aebi, 2016, Brennich et al., 2019).

In the following, the four prominent proteins (desmin, lamin (A/C), plectin, and filamin C) from an IF network will be described in terms of their biological, disease, and mechanical function in living cells. Such IF proteins have been reported as important contributors to cellular contractility and prestress and serve as molecular "guy wires" that facilitate the transfer of mechanical loads between the cell surface

Acknowledgments and funding: The author thanks Drs. Ben Fabry, Rolf Schröder, Harald Herrmann, and Oliver Friedrich for their advice and support and Mr. Paul Gahman (MA) for proofreading the manuscript. This book chapter was funded by the Marie Sklodowska-Curie Action, Innovative Training Networks (ITN) 2018, EU grant agreement no. 812772.

Wolfgang H. Goldmann, PhD. Department of Physics, Biophysics Group, Friedrich-Alexander-University Erlangen-Nuremberg, Erlangen, Germany

https://doi.org/10.1515/9783110989380-018

and the nucleus, and thereby stabilize microtubules and actin filaments (Winter and Goldmann, 2015). A complete list of all IF proteins and their characteristics in various cell lines is provided in Cooper (2000).

6.8.2 Desmin and Diseases

Desmin is the most commonly studied disease entities in human myofibrillar myopathies. It belongs to the group of IFs that form 3D extra-sarcomeric filamentous networks in cells and is responsible for a number of functions, including maintenance of sarcomeres, specific positioning of organelles, and cell signaling. Desmin-deficient cells compromise the general organization of muscle fibers in that they misalign and mislocate myofibrils and mislocate nuclei and mitochondria. In certain neurodegenerative diseases, desmin mutations can trigger increased oxidative stress and cause abnormal protein aggregates. Moreover, inhibition of the clearance mechanisms during ageing might exacerbate protein accumulation and contribute to the progression of the disease (Schröder et al., 2007, Schröder and Schoser, 2009, Clemen et al., 2009, 2013, Winter et al., 2019, Herrmann et al., 2020, Spörrer et al., 2022).

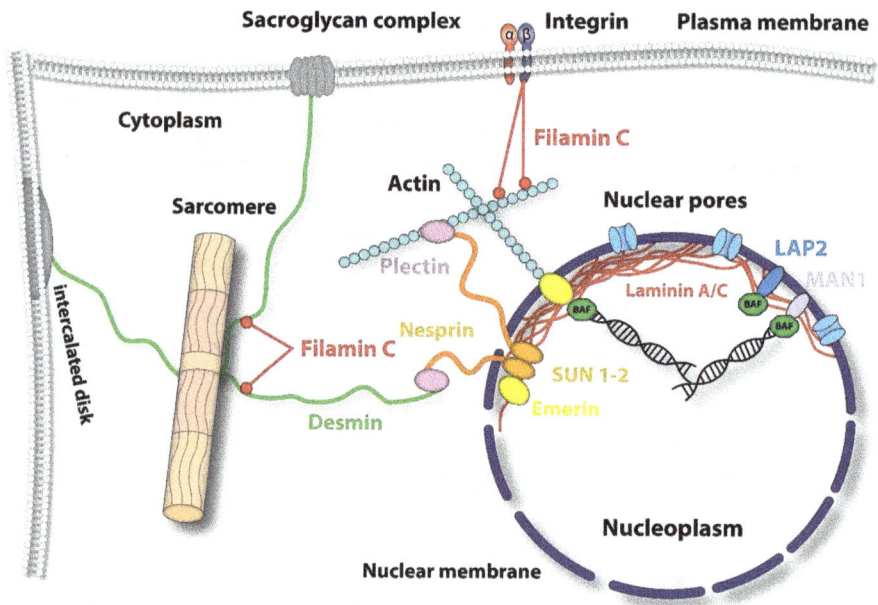

Figure 6.8.1: A representation of the intermediate filament proteins in the cell. These interact with the actin cytoskeleton connecting with integrins, the nuclear membrane, and intercalated disk, forming a tight cellular filament network. The different proteins are color-coded in the graph. Taken from Cell Biol. Int. (2018) **42**: 321 with permission.

The autosomal dominant missense p.Arg350Pro belongs to a subset of desmin mutations as the most commonly reported mutation. The exchange of arginine through proline affects the unfolding during desmin assembly that could lead to IF collapse. To date, the propensity of desmin mutants to modify the normal organization of myogenic cells has only been analyzed in murine C2C12 myoblast clones, expressing exogenously desmin mutants (Charrier et al., 2016). Disease mutations identified more recently showed skeletal and cardiac myopathies that correlate with pathological protein aggregation. Mücke et al. (2016) dissected the pathway and the kinetics of desmin assembly, in detail; and it was shown that its pathway deviates significantly from that of vimentin, another IF protein (Mücke et al., 2018, Schween et al., 2022). Further, comparing the assembly kinetics of mutant and wild-type desmin indicated how the interaction between the plakin family and cellular chaperones influence the assembly.

6.8.3 Desmin and Cell Mechanics

An important question raised within the present research community is the function of desmin and how its mutants exert their deleterious effects on human skeletal and cardiac muscle cells, with respect to their structure and function. Here, desmin is a key component of the 3D filamentous extra-sarcomeric cytoskeleton that interlinks neighboring myofibrils at the Z-disk, connecting the entire myofibrillar apparatus to costameres, intercalated discs, myotendinous, and neuromuscular junctions (Hnia et al., 2015). This network provides important anchorage points for the alignment of myofibrils and for the attachment to the sarcolemma, nuclei, and mitochondria by performing the important function of adapting striated muscle fibers to active and passive stresses. Studies in desminopathic patients showed that heterozygous/homozygous mutations affect the structure and function of the extra-sarcomeric network in different ways; however, nothing is known about the early disease stages that actually precede the clinical manifestation of muscle weakness in human desminopathies. To address this issue, Clemen et al. (2015) used hetero- and homozygous R349P knock-in mice, which possess the ortholog of the most frequently occurring human desmin missense mutation, R350P. The mice exhibited age-dependent skeletal muscle weaknesses, dilated cardiomyopathies, cardiac arrhythmias, and conduction defects. Further, as described in a mouse model, morphological and biomechanical alterations were evident in the early disease stages. Using nonlinear second harmonic generation (SHG) and 2-photon fluorescence morphometry analysis in combination with active and passive biomechanical recordings of muscle fibers, Diermeier et al. (2017a) unveiled an early disease pattern, in which mutant desmin showed aberrant myofibrillar alignment and orientation as the basis for compromised active force production. These authors showed altered passive and biomechanical properties, which made them more prone to fiber damage and provided initial insights into adaptive mechanisms that may compensate for force discrepancies in preclinical disease. Further, Diermeier et al. (2017b) used small fiber bundles from unfixed soleus mice muscles in multicellular biomechanics experiments. Since the morphological pathology of R349P desmin knock-in mice is most prominent in soleus muscle, these researchers restricted their biomechanical experiments to this mutation.

Fiber bundles were also used in a mechatronic device called "MyoRobot," which was custom-built, is automated, and mimics skeletal muscle (Haug et al., 2019). The fiber bundles, here, were measured by a force transducer pin and software-controlled voice coil actuator. An automated image-processing algorithm developed by Buttgereit et al. (2013) was used for the morphometric analysis of 3D SHG and multiphoton fluorescence images. A second morphometric parameter extracted from SHG microscopy called "verniers" described Y-shaped deviations that resulted from out-of-register deviations in the regular signal pattern of adjacent myofibrils (Friedrich et al., 2019).

These researchers tested the hypothesis that the mutated R349P desmin also exerts a detrimental effect on biomechanics by testing the steady-state axial elasticity of small fiber bundles. Two recordings from Des^{R349P} soleus fiber bundle experiments were carried out with simultaneous measurements on the individual, stretch-related, passive restoration force. Results from quasi-static passive biomechanics showed higher axial elastic stiffness in hetero- and homozygous Des^{R349P} soleus fiber bundles compared to the wildtype (Diermeier et al., 2017b). To determine the viscoelastic behavior of soleus fiber bundles from Des^{R349P} mice, stretch-jump experiments were also performed by stretching bundles successively; however, the relaxation kinetics proved inconsequential among the genotypes.

Subcellular morphological alterations detected by SHG provided a structural basis for explaining early alterations in biomechanical properties of slow-twitch muscle in Des^{R349P} desminopathy (Buttgereit et al., 2013, 2014). Since desmin is also known to link to the nuclear domain and the sarcoglykan complex of muscle fibers (Goldfarb and Dalakas, 2009), an impairment of mutant Des^{R349P} desmin as a means to contribute to lateral compliance was also suggested by Bonakdar et al. (2012) for human Des^{R350P} myoblasts. Further evidence for the effects of the Des^{R350P} mutation on the viscoelastic properties of IF-networks emerged from in vitro bulk assembly studies, where Des^{R350P} exhibited a merely weak increase in viscosity, when assembled on its own, but showed a marked hyperviscosity when co-assembled with equimolar amounts of wildtype desmin (Bär et al., 2006). Interestingly, mutations in the tail domain of desmin highlighted diminished stiffening in filament networks (Bär et al., 2010).

Each of the aforementioned studies provided initial insights into the detailed effects of the murine R349P desmin knock-in mutation on the passive and active biomechanical properties in preclinical stages of skeletal muscle, where desmin-positive protein aggregates are not yet present. The studies are supported by state-of-the-art multiphoton microscopy data that showed vast morphological alterations in the sub-cellular architecture of both fast- and slow-twitch muscle fibers, which point toward myofibrillar lattice disruptions that are more evident/accentuated in slow-twitch muscle (soleus). The lattice disruptions and less tightly oriented myofibrils suggest a compromise in biomechanical properties consistently observed in the passive quasi-static elasticity and for viscoelastic properties. Simply speaking, desminopathic muscle fiber bundles, myofibrillar bundles, as well as the membrane complex in myoblasts carrying the very same mutation were much stiffer compared to wildtype desmin. The severity of increased stiffness depended on the maturation level and was more pronounced in homozygous mutations in the preclinical adult stages (fiber and myofibrillar bundles). This might explain why affected muscles are prone to stretch-induced injury and aggravate subsequent protein aggregate formation, which is more pronounced in slow-twitch muscle. Interestingly, homozygous soleus muscle fibers show, by means of a mechanism not yet confirmed, a compensation of force over heterozygous preparations that otherwise reflect reduced myofibrillar Ca^{2+} sensitivity. Since the heterozygous Des^{R350P} genotype in humans is pathologically predominant, as reflected by the murine Des^{R349P} genotype, the specific result fully explains the detrimental effects of a single mutated desmin allele in affected patients: compromised passive extensibility of muscle, cellular architecture, and active force production (Diermeier et al., 2017b).

6.8.4 Lamin A/C and Diseases

Nuclear lamins are cytoskeletal proteins that belong to the family of IFs and are located on the inner nuclear membrane. There are two main classes of lamin proteins, A and B-type. B-type lamins are further classified into B1-lamins and B2-lamins encoded through the genes LMNB1 and LMNB2, respectively. As many mutations and particularly the lack of B-type lamins were found to be lethal to cells, no genetically inheritable disease is connected to mutations in the LMNB genes. In contrast to that, LMNA, the gene that encodes the A-type lamins A, AD10, C, and C2 is one of the most mutated genes in humans. The loss of A-type lamin function, however, can still lead to serious diseases, so-called laminopathies. The most prominent of these diseases are Emery-Dreifuss muscular dystrophy, cardiomyopathies, and premature ageing syndromes like Hutchinson-Gilford progeria syndrome. All the above laminopathies can be correlated to point mutations on the LMNA gene. Thus, the exact mechanism between the nanoscale punctual mutation and macroscopic changes of the tissue of diseased patients is largely unclear. It has repeatedly been suggested that laminopathies stem from a disturbance of a gene-regulating function in the LMNA gene. A general mechanical weakness, likely caused by diminished nucleocytoskeletal integrity after lamin A-loss, has previously been suggested to cause laminopathy (Bonne et al., 1999, Moir et al., 2000, Vignier et al., 2018, Pfeifer et al., 2019).

The nucleus is the most prominent cell organelle in all eukaryotic cells. It contains most of the cell's genetic material in the form of heterochromatin, euchromatin, and nucleoli, and protects and controls the genetic replication machinery. The nucleus regulates gene expression through various transmembrane nuclear pore complexes and channels and defines cell mechanical properties to a large extent, due to its dominating volume and higher stiffness. The deformability of the cell nucleus may likely be regulated by the state of chromatin, since chromatin condensation correlates with cell stiffness. The cell nucleus is surrounded by the nuclear membrane and stabilized by the nuclear skeleton, also called the nuclear lamina. The lamina is an organized meshwork of IFs, mainly lamin A/C and B, located at the interior boundary of the nuclear membrane, providing support and anchoring points for pores and channels. The nuclear lamina is connected to the cell cytoskeleton, e.g., to actin, microtubule, and IFs through LINC-complexes. These are assembled by the transmembrane proteins, SUN and nesprin, which interact with molecular motors such as dynein (Fatkin et al., 1999, Lloyd et al., 2002, Broers et al., 2004, Brull et al., 2018).

6.8.5 Laminin A/C and Cell Mechanics

IFs are a family of related cytoplasmic and nuclear proteins, which are, on average, 10 nm in diameter. They are categorized into six subfamilies according their similarities to amino acid and protein structures. Prominent examples are keratins (I and II), desmin and vimentin (III), neurofilaments (IV), lamins (V) and nestin (VI) (Mücke et al., 2018). The existence and number of certain IFs greatly depend on the cell type and their function. IFs are the least stiff of the three cytoskeletal proteins, having a Young's modulus of around 4×10^6 (Pa) (Charrier and Janmey, 2016). Moreover, they have a persistence length of only 1 micrometer, but are reported to counterbalance large strains. Lamins were found to play an important role in the cell's protection against nuclear stresses during the migration through confined spaces, thereby the nuclear lamina is assumed to function as protection against DNA compression and shear.

Local force generation, dynamic modification of stiffness, the viscosity of cells, and their responses to traction or compressional forces are general hallmarks of cellular and tissue mechanics (Dahl and Kalinowski, 2011). These parameters were examined by Lee et al. (2007) in lamin A-deficient mouse embryonic fibroblasts (MEFs). Either the disassembly of actin filaments or microtubule networks proved to lead to the decrease in cytoplasmic elasticity and viscosity. Further, studies by Lanzicher et al. (2015), using atomic force microscopy (AFM) on cardiomyocytes, which carry a lamin A/C mutation (D192G), showed increased maximum nuclear deformation load, nuclear stiffness, and fragility compared to control cells. They deduced from their experiments that a non-association of the cytoskeleton with lamins was the trigger for cellular morphological and adhesive changes that could lead to reported fatal cardiomyopathies (Chatzifrangkeskou et al., 2018). Aptke et al. (2017) investigated the mutation E145K on lamin A, which has been shown to cause Hutchinson-Gilford progeria syndrome (HGPS), by using the atomic force microscope. They found that this mutation dramatically increased nuclear stiffness compared to the wildtype in Xenopus oocytes.

Mechanical studies on lamina A-mutated in vitro systems have been conducted in a wide range of research facilities during the last years (Nikolova et al., 2004, Lammerding et al., 2004, 2006, Osmanagic-Myers et al., 2015b, Mitchell et al., 2015, Kolb et al., 2017). Pivotal results from Lange et al. (2015, 2017) showed that lamin A, in contrast to B-type lamins, is linked to cell mechanosensing, suggesting that lamin A is upregulated on stiffer matrix surroundings. Moreover, lamin A was found to hinder, but at the same time, to protect the cells against nuclear stresses during cell migration through confined spaces. Thereby, the nuclear lamina is assumed to function as protection against DNA compression and shear. Still, several questions remain unclear, one of which is, how the loss of lamin A results in a possible overall cell weakening, where the dose–response of lamin A-loss or overexpression on cell mechanics is concerned. In studies with K562 leukemia cells overexpressing lamin A, Lange et al. (2015) used the microconstriction methods and investigated lamin A fluorescence extension. Depending on the fluorescence expression levels after measurement, cells were sorted accordingly. This can be attributed to the fact that averaging the mechanical properties over the entire population would almost certainly lead to biased results. Such mechanical properties would, in turn, strongly depend on the transfection efficiency during the actual transfection process.

Expression levels of nuclear lamins have also widely been connected with overall cell stiffness and fluidity. All three network components are highly connected to each other, to the nucleus via LINC complexes, and to the cell membrane via focal adhesion sites and integrins. This poses a problem, when investigating the mechanical properties of reconstituted cytoskeletal networks in vitro and applying this knowledge to the in vivo complex system of a cell cytoskeleton.

Summarizing the above, cell mechanical measurements with a microconstriction setup showed that cell stiffness increases significantly in a dose–response manner with lamin A-overexpression level. At the same time, cell fluidity decreases significantly. The reason for this clear-cut correlation may be that lamin A supports the integrity of the nuclear lamina. The nuclear lamina, in turn, is connected to all other cell cytoskeletal components through so-called LINC complexes and might therefore provide stability for the actin cytoskeleton, as well. These results are in accordance with previous measurements on lamin A overexpressing adherent cells and nuclei. To the author's knowledge, a dose–response curve associating lamin A overexpression with cell stiffness and fluidity has not been explored.

Recently, Schürmann et al. (2016) examined the cellular mechanics of human fibrosarcoma (HT1080) cells in 2D under isotropic stretch in cells with overexpressed lamin A. From their results, they assumed stiffening of the nucleus membrane area and the cytoskeleton, as the cell area was smaller in these cells, compared to control cells for stretches up to 10%. The authors showed that, the increased stiffness of the mutant HT1080 cells resulted in complete detachment of cells from the extracellular matrix at 15% stretch, which confirmed the stiffening of the global cellular cytoskeleton through an isolated increase in nuclear stiffness in lamin A overexpressing cells.

To explain how mutations in lamin A of the nuclear envelope can affect the heart muscle, it has been proposed that nuclear envelope abnormalities can cause cellular fragility and decrease the mechanical resistance to stress. This could partially explain hypertrophic cardiac muscle disease, considering that the heart muscle is constantly subjected to mechanical force. It is believed that abnormal activation of stress-activated ERK1/2 signaling in mice hearts that carry lamin A mutations might be the cause. Administering drugs which inhibit ERK1/2 signaling could improve cardiac ejection fraction. Recent observations by Schwartz et al. (2017) also showed that pathogenic LMNA mutations in human muscle precursor cells, which are responsible for severe muscle dystrophies, exhibit accumulated contractile stress fibers, increased focal adhesions, and higher traction force, compared to control cells. Thus, deactivating the ROCK-dependent regulator, formin, responsible for remodeling actin, preserves the morphology of mutant cells. Further, the functional integrity of lamin/ nesprin-1 is necessary to modulate formin and cellular mechanical coupling. Previously, the role of cell and nuclear stiffness was investigated on multiple cell lines (the fibrosarcoma cell line HT-1080 and the breast cancer cell line MDA MB-231). These cell lines overexpressed lamin A that migrated through 3D devices consisting of a linear channel with a length of 630 μm, height of 3.7 μm, and decreasing channel width from 11.2 to 1.7 μm (Lautscham et al., 2015). All cell lines showed reduced cell migration, which was attributed to higher cell stiffness and lower adhesiveness. To separate the effect of cell stiffness from other invasion-modulating cell properties, the expression levels of lamin A were increased, which correlated with nuclear stiffness. The authors hypothesized that cells with higher lamin A levels experience higher resistance, when migrating through confined spaces due to the increased cell stiffness. In another study, the effect of lamin A by means of microconstriction method was investigated (Lange et al., 2015). To test how lamin A overexpression affects the overall cell mechanical properties, the stiffness and fluidity of various cells (leukemia cells, K562, and breast cancer cell line, MDA MB-231) were measured. Compared to wildtype cells, the stiffness cells that were overexpressed by lamin A increased significantly (Lange et al., 2017). This data confirms that lamin A contributes greatly to cell stiffness, but the method does not discriminate between the stiffness of the cell nucleus and the cytoskeleton. Lange et al. (2017) were ultimately unable to exclude the possibility that lamin A overexpression leads to altered cytoskeletal mechanics and structure.

More recently, we investigated the impact of A-type lamin (p.H222P) mutation on the mechanical properties of muscle cells by microconstriction rheology. We demonstrated that the expression of point mutation of lamin A in muscle cells increases cellular stiffness compared to cells expressing wild type lamin A, and that the chemical agent selumetinib, an inhibitor of the ERK1/2 signaling, reversed the mechanical alterations in mutated cells. These results highlight the interplay between A-type lamins and mechano-signaling, which are supported by cell biology measurements (Chatzifrangkeskou et al., 2020).

6.8.6 Plectin and Diseases

Plectin was first reported by Wiche et al. (1982), who found that plectin gene defects cause epidermolysis bullosa simplex with muscular dystrophy (EBS-MD). This, in turn, is characterized by severe skin blistering and muscular dystrophy. Using skeletal muscle, Wiche et al. (1982) showed that at least four plectin isoforms are responsible for targeting and linking desmin IF networks to Z-disks, costameres, mitochondria, and the nuclear/ER membrane system, severe skin blistering, and muscular dystrophy. Plectin deficiency leads to desmin aggregation and mitochondrial dysfunction. Further, they established numerous plectin isoform-specific knock-out mouse strains, elucidating the function of plectin in normal and EBS-MD muscles (Andrä et al., 1997). Moreover, Konieczny et al. (2008) established several plectin isoform-specific and conditional knock-out mouse strains, of which two closely mirror the human EBS-MD muscle pathology. Special focus was directed to plectin-mediated effects on the structure and function of the desmin cytoskeleton, mitochondrial positioning, and metabolism, as well as intracellular signaling events, including AMPK-mediated energy homeostasis, the mTOR pathway, and apoptosis.

6.8.7 Plectin and Cell Mechanics

Plectin is a prominent cytoskeletal linking protein based on IFs. It strengthens cells mechanically by interlinking, anchoring cytoskeletal filaments, and acting as scaffolding- and docking platform for signaling proteins. In this function, it controls the dynamics of the cytoskeleton; however, research results of its biomechanical effects in muscle are scarce. Hijikata et al. (1999) showed that plectin links desmin IFs to Z-disks and prevents individual myofibrils from disruptive contractions. Although Na et al. (2009) examined its role in setting cell stiffness, stress propagation, and traction generation in wildtype plectin and plectin-deficient skin fibroblasts, its influence on muscle biomechanics through the various organ scales was not known. Thus, Bonakdar et al. (2012) showed that pathogenic plectin mutations cause increased cell stiffness due to higher baseline contractile activation. This leads to higher intracellular stress during cyclic stretch and, consequently, to higher stress vulnerability in muscle. In related experiments, Winter et al. (2014) investigated the effect of a plectin knock-out in mouse myoblasts. These experiments are particularly relevant because the same procedure for obtaining immortalized myoblasts with the knock-out mutations of extra-sarcomeric cytoskeletal proteins was followed in all subsequent studies. Cell stiffness was decreased two times in the plectin knock-out cells. In agreement with lower stiffness, plectin knock-out cells showed a higher power-law exponent of the creep modulus, indicating a less stable cytoskeleton and a more fluid-like mechanical behavior of these cells. Furthermore, plectin knock-out cells were approximately 2.5 times less contractile, which indicates a diminished cytoskeletal prestress that is likely the primary cause for lower stiffness in these cells (Bonakdar et al., 2015). The hypothesis that cell death after stretching is caused by stretch-induced mechanical stress correlates with cell stiffness. Osmanagic-Myers et al. (2015a) confirmed that the softer plectin knock-out cells are approximately twice as less vulnerable to cyclic stretch, compared to wildtype cells. Almeida et al. (2015) also showed that plectin is an essential regulator of nuclear morphology and protects the nucleus from mechanical deformation.

6.8.8 Filamin C and Diseases

Filamin C is an actin-binding and regulatory protein that is closely associated in myofibril formation. The first mutation in the filamin C gene that caused myofibrillar myopathy (MFM) in humans was reported by Vorgerd et al. (2005). Studying the pathogenic consequences, these authors provided the biochemical evidence for altered filamin C properties that lead to protein aggregation. Further, in-depth studies on the pathogenesis of filamin C myopathy were carried out by the group of Dr. Fürst, who used ES cells stably transfected with wildtype and mutant filamin C as well as human samples (Fürst et al., 2013). More recently, compelling evidence of filamin C's involvement in human hypertrophic cardiomyopathy was shown with the help of SIFT and other screening algorithms (Gomez et al., 2017).

In cells, filamin C binds to both alpha-actinin and actin and can interact with the co-chaperone BAG3 and with the membrane fusion machinery containing the VPS protein (Selcen et al., 2009). These interactions are essential for chaperone-assisted selective autophagy (CASA) and found in muscle. Filamin C isoforms may also have degradation-independent functions in the regulation of mechanical-stress-related signaling pathways, thus necessitating the precise subcellular localization and dynamic behavior of all filamin C protein variants in muscle cells. Functional studies should therefore reveal their involvement in mechanically stress-induced degradation and signaling. Further, the impact of phosphorylation and the dynamics of complexes containing filamin C protein, in the context of contractile activity, is of importance; a similar study was conducted for desmin (Diermeier et al., 2017a).

6.8.9 Filamin C and Cell Mechanics

Filamin C is a key component of sarcomeric Z-disks and cell–matrix contacts, where it binds to a wide range of cytoskeletal and signaling proteins and to a large number of proteins, including aciculin, Xin, XIRP2, FILIP-1, myotilin, and podins. Further, many phosphorylation sites on filamin C have been identified, which could give the protein a regulatory function. It was shown that mechanical activity directly alters the dynamic behavior of filamin C and its interaction partners, and that protein complexes are immediately recruited to mechanically damaged areas. A lack or the functional impairment of components in this regulatory network has been reported to have severe muscle damage in human patients and animal models.

In our first experiments, we asked whether mechanical stress has a different effect on the dynamics of mutant filamin C than on wildtype cells (Winter and Goldmann, 2015). We were able to show that molecular processes contribute to a reduced mechanical stress resistance in diseased muscle cells, using live-cell confocal microscopy and protein expression studies on myoblasts derived from p.W2710X filamin C knock-in mice. Early, unpublished results have suggested that: (i) filamin C mutant cells detach at a higher percentage compared to wildtype cells after external stress application, (ii) the strain energy of mutant cells is lower compared to wildtype cells, and (iii) the stiffness of mutant cells is higher compared to wildtype cells. Some observations were confirmed by Chevessier et al. (2015), in that mutant filamin C in muscle interferes with the mechanical stability and strain resistance of myofibrillar Z-disks.

However, more in-depth studies are needed to unravel the biomechanical mechanisms responsible for muscle weakness in the filamin C myopathy (p.W2710X), using skeletal muscle preparations (whole muscles, fiber bundles, single fibers, myotubes) from heterozygous and homozygous mice. The role of mutated filamin C affecting the lateral versus the axial biomechanical properties should be elucidated. This can be accomplished using force transducer recordings in single myofibers with intact integrin–filamin C complexes and mechanically skinned fibers after removal of the sarcolemma, thereby leaving only the filamin C anchorage at the Z-disk. The lateral compliance can be determined through magnetic tweezer experiments in myotubes and in intact single fibers among genotypes. As a novel approach, the nonlinear tractions of the integrin–filamin C complex on the extracellular matrix should be considered with intact cells (myotubes, myofibers) embedded in a collagen hydrogel, and by applying traction force microscopy to quantify the traction forces in resting and field-stimulated hydrogels. The central question to be answered through these experiments is whether the filamin C-integrin or the filamin C-Z-disk anchorage is more crucial in determining the compromised biomechanical properties, for instance, in p.W2710X-filamin C myopathy. In addition, after answering this fundamental biological question, the efficacy of therapeutic approaches to filamin C treatment in the murine models, that is, mild exercise regimes and the use of chemical chaperones, should be addressed. Of relevance and importance is linking these methods to imaging projects so as to define whether and under which manipulations filamin C also acts as a mobile fraction in the mutated phenotype, that is, translocating from the Z-disk to the I-band region, and how this affects biomechanical properties of muscle (Leber et al., 2016).

More recently, Kathage et al. (2017) showed that the filamin C-associated protein, BAG3 regulates protein synthesis through mechanical strain, and Collier et al. (2019) reported that phosphorylation of another filamin C-associated protein (HspB1) is responsible for mechanosensitive chaperone interaction with filamin C. In conclusion, more in-depth studies are needed to elucidate the effect of mechanical stress on the localization and dynamic behavior of other filamin C-associated proteins and their variants.

References

Almeida, F. V., G. Walko, J. R. McMillan, J. A. McGrath, G. Wiche, A. H. Barber and J. T. Connelly (2015). "The cytolinker plectin regulates nuclear mechanotransduction in keratinocytes." Journal of Cell Science **128**: 4475–4486.

Andrä, K., H. Lassmann, R. Bittner, S. Shorny, R. Fässler, F. Propst and G. Wiche (1997). "Targeted inactivation of plectin reveals essential function in maintaining the integrity of skin, muscle, and heart cytoarchitecture." Genes & Development **11**: 3143–3156.

Apte, K., R. Stick and M. Radmacher (2017). "Mechanics in human fibroblasts and progeria: Lamin A mutation E145K results in stiffening of nuclei." Journal Molecular Recognition **30**. doi: 10.1002/jmr.2580.

Bär, H., N. Mücke, P. Ringler, S. A. Müller, L. Kreplak, H. A. Katus, U. Aebi and H. Herrmann (2006). "Impact of disease mutations on the desmin filament assembly process." Journal of Molecular Biology **360**: 1031–1042.

Bär, H., M. Schöpferer, S. Sharma, B. Hochstein, N. Mücke, H. Herrmann and N. Willenbacher (2010). "Mutations in desmin's carboxy-terminal "tail" domain severely modify filament and network mechanics." Journal of Molecular Biology **397**: 1188–1198.

Bonakdar, N., J. Luczak, L. A. Lautscham, M. Czonstke, T. M. Koch, A. Mainka, T. Jungbauer, W. H. Goldmann, R. Schröder and B. Fabry (2012). "Biomechanical characterization of a desminopathy in primary human myoblasts." Biochemical and Biophysical Research Communications **419**: 703–707.

Bonakdar, N., A. Schilling, M. Spörrer, P. Lennert, A. Mainka, L. Winter, G. Walko, G. Wiche, B. Fabry and W. H. Goldmann (2015). "Determining the mechanical properties of plectin in mouse myoblasts and keratinocytes." Experimental Cell Research **331**: 331–337.

Bonne, G., M. R. Di Barletta, S. Varnous, H. M. Be`cane, E. H. Hammouda, L. Merlini, F. Muntoni, C. R. Greenberg, F. Gary, J. A. Urtizberea, D. Duboc, M. Fardeau, D. Toniolo and K. Schwartz

(1999). "Mutations in the gene encoding lamin A/C cause autosomal dominant Emery-Dreifuss muscular dystrophy." Nature Genetics **21**: 285–288.

Brennich, M. E., U. Vainio, T. Wedig, S. Bauch, H. Herrmann and S. Köster (2019). "Mutation-induced alterations of intra-filament subunit organization in vimentin filaments revealed by SAXS." Soft Matter **15**: 1999–2008.

Broers, J. L. V., E. A. G. Peeters, H. J. H. Kuijpers, J. Endert, C. V. C. Bouten, C. W. J. Oomens, F. P. Baaijens and F. C. S. Ramaekers (2004). "Decreased mechanical stiffness in LMNA-/- cells is caused by defective nucleo-cytoskeletal integrity: Implications for the development of laminopathies." Human Molecular Genetics **13**: 2567–2580.

Brull, A., B. Morales-Rodriguez, G. Bonne, A. Muchir and A. T. Bertrand (2018). "The pathogenesis and therapies of striated muscle laminopathies." Frontiers in Physiology **9**: 1533.

Buttgereit, A., C. Weber, C. S. Garbe and O. Friedrich (2013). "From chaos to split-ups-SHG microscopy reveals a specific remodelling mechanism in ageing dystrophic muscle." The Journal of Pathology **229**: 477–485.

Buttgereit, A., C. Weber and O. Friedrich (2014). "A novel quantitative morphometry approach to assess regeneration in dystrophic skeletal muscle." Neuromuscular Disorders **24**: 596–603.

Capetanaki, Y., R. J. Bloch, A. Kouloumenta, M. Mavroidis and S. Psarras (2007). "Muscle intermediate filaments and their links to membranes and membranous organelles." Experimental Cell Research **313**: 2063–2076.

Charrier, E. E. and P. A. Janmey (2016). "Mechanical properties of intermediate filament proteins." Methods in Enzymology **568**: 35–57.

Chatzifrangkeskou, M., D. Yadin, T. Marais, S. Chardonnet, M. Cohen-Tannoudji, N. Mougenot, A. Schmitt, S. Crasto, E. Di Pasquale, C. Macquart, Y. Tanguy, I. Jebeniani, M. Pucéat, B. Morales-Rodriguez, W. H. Goldmann, M. Dal Ferro, M. G. Biferi, P. Knaus, G. Bonne, H. J. Worman and A. Muchir (2018). "Cofilin-1 phosphorylation catalyzed by ERK1/2 alters cardiac actin dynamics in dilated cardiomyopathy caused by lamin A/C gene mutation." Human Molecular Genetics **27**: 3060–3078.

Chatzifrangkeskou, M., D. Kah, J. R. Lange, W. H. Goldmann and A. Muchir (2020). "Mutated lamin A modulates stiffness in muscle cells." Biochemical and Biophysical Research Communications **529**: 861–867.

Chevessier, F., J. Schuld, Z. Orfanos, A. C. Plank, L. Wolf, A. Märkens, A. Unger, U. Schlötzer-Schrehardt, R. A. Kley, S. von Hörsten, K. Markus, W. A. Linke, M. Vorgerd, P. F. van der Ven, D. O. Fürst and R. Schröder (2015). "Myofibrillar instability exacerbated by acute exercise in filaminopathy." Human Molecular Genetics **24**: 7207–7220.

Clemen, C. S., D. Fischer, J. Reimann, L. Eichinger, C. R. Müller, H. D. Müller, H. H. Goebel and R. Schröder (2009). "How much mutant protein is needed to cause a protein aggregate myopathy in vivo? Lessons from an exceptional desminopathy." Human Mutation **30**: E490–E499.

Clemen, C. S., H. Herrmann, S. V. Strelkov and R. Schröder (2013). "Desminopathies: Pathology and mechanisms." Acta Neuropathologica **125**: 47–75.

Clemen, C. S., F. Stockigt, K. H. Strucksberg, F. Chevessier, L. Winter, J. Schütz, R. Bauer, J. M. Thorweihe, D. Wenzel, U. Schlötzer-Schrehardt, V. Rasche, P. Krsmanovic, H. A. Katus, W. Rottbauer, S. Just, O. J. Müller, O. Friedrich, R. Meyer, H. Herrmann, J. W. Schrickel and R. Schröder (2015). "The toxic effect of R350P mutant desmin in striated muscle of man and mouse." Acta Neuropathologica **129**: 297–315.

Collier, M. P., T. R. Alderson, C. P. de Villiers, D. Nicholls, H. Y. Gastall, T. M. Allison, M. T. Degiacomi, H. Jiang, G. Mlynek, D. O. Fürst, P. F. M. van der Ven, K. Djinovic-Carugo, A. J. Baldwin, H. Watkins, K. Gehmlich and J. L. P. Benesch (2019). "HspB1 phosphorylation regulates its intramolecular dynamics and mechanosensitive molecular chaperone interaction with filamin C." Science Advances **22**: 5. eaav8421.

Cooper, G. M. (2000). Intermediate filaments. The cell, a molecular approach." Chapter 11/IV. Sunderland. Boston (MA), Sinauer Associates, Inc. (Oxford University Press). ISBN-10: 0-87893-106-6.

Dahl, K. N. and A. Kalinowski (2011). "Nucleoskeleton mechanics at a glance." Journal of Cell Science **124**: 675–678.

Diermeier, S., J. Iberl, K. Vetter, M. Haug, C. Pollmann, B. Reischl, A. Buttgereit, S. Schürmann, M. Spörrer, W. H. Goldmann, B. Fabry, F. Elhamine, R. Stehle, G. Pfitzer, L. Winter, C. S. Clemen, H. Herrmann, R. Schröder and O. Friedrich (2017a). "Early signs of architectural and biomechanical failure in isolated myofibers and immortalized myoblasts from desmin-mutant knock-in mice." Scientific Report **7**: 1391.

Diermeier, S., A. Buttgereit, S. Schürmann, L. Winter, H. Xu, R. M. Murphy, C. S. Clemen, R. Schröder and O. Friedrich (2017b). "Preaged remodeling of myofibrillar cytoarchitecture in skeletal muscle expressing R349P mutant desmin." Neurobiology of Aging **58**: 77–87.

Etienne-Manneville, S. (2018). "Cytoplasmic intermediate filaments in cell biology." Annual Review of Cell and Developmental Biology **34**: 1–28.

Fatkin, D., C. Mac Rae, T. Sasaki, M. R. Wolff, M. Porcu, M. Frenneaux, J. Atherton, H. J. Vidaillet, S. Spudich, I. de Girolami, U. J. Seidman and C. E. Seidman (1999). "Missense mutations in the rod domain of the lamin A/C gene as causes of dilated cardiomyopathy and conduction-system disease D." New England Journal of Medicine **341**: 1715–1724.

Fletcher, D. A. and D. Mullins (2010). "Cell mechanics and the cytoskeleton." Nature **463**: 485–492.

Friedrich, O., M. Haug, B. Reischl, G. Prölß, L. Kiriaev, S. I. Head and M. B. Reid (2019). "Single muscle fibre biomechanics and biomechatronics – The challenges, the pitfall and the future." The International Journal of Biochemistry & Cell Biology **114**: 105563.

Fürst, D. O., L. G. Goldfarb, R. A. Kley, M. Vorgerd, M. Olive and P. F. van der Ven (2013). "Filamin C-related myopathies: Pathology and mechanisms." Acta Neuropathologica **125**: 33–46.

Goldfarb, L. G. and M. C. Dalakas (2009). "Tragedy in a heartbeat: Malfunctioning desmin causes skeletal and cardiac muscle disease." The Journal of Clinical Investigation **119**: 1806–1813.

Goldman, R. D., M. M. Cleland, S. N. P. Murthy, S. Mahammad and E. R. Kuczmarski (2011). "Inroads into the structure and function of intermediate filament networks." Journal of Structural Biology **177**: 14–23.

Gomez, J., R. Lorca, J. R. Reguero, C. Moris, M. Martin, S. Tranche, B. Alonso, S. Igleas, V. Alvarez, B. Diaz-Molina, P. Avanzas and E. Coto (2017). "Screening of the filamin C gene in a large cohort of hypertrophic cardiomyopathy patients." Circulation: Cardiovascular Genetics **10**: e001584.

Haug, M., C. Meyer, B. Reischl, G. Prölß, S. Nübler, S. Schürmann, D. Schneidereit, M. Heckel, T. Pöschel, S. J. Rupitisch and O. Friedrich (2019). "MyoRobot 2: Advanced biomechatronics platform for automated, environmentally-controlled muscle single fiber biomechanics assessment employing inbuilt real-time optical imaging." Biosensors & Bioelectronics **138**: 111284.

Herrmann, H., S. V. Strelkov, P. Burkhard and U. Aebi (2009). "Intermediate filaments: Primary determinants of cell architecture and plasticity." The Journal of Clinical Investigation **119**: 1772–1183.

Herrmann, H. and U. Aebi (2016). "Intermediate filaments: Structure and assembly." Cold Spring Harbor Perspectives in Biology **8**: pii: a018242.

Herrmann, H., E. Cabet, N. R. Chevalier, J. Moosmann, D. Schultheis, J. Haas, M. Schowalter, C. Berwanger, V. Weyerer, A. Agaimy, B. Meder, O. J. Müller, H. A. Katus, U. Schlötzer-Schrehardt, P. Vicart, A. Ferreiro, S. Dittrich, C. S. Clemen, A. Lilienbaum and R. Schröder

(2020). "Dual functional states of R406W-desmin assembly complexes cause cardiomyopathy with severe intercalated disc derangement in humans and knock-in mice." Circulation **142**: 2155–2171.

Hijikata, T., T. Murakami, M. Imamura, N. Fujimaki and H. Ishikawa (1999). "Plectin is a linker of intermediate filaments to Z-discs in skeletal muscle fibers." Journal of Cell Science **112**: 867–876.

Hnia, K., C. Ramspacher, J. Vermot and J. Laporte (2015). "Desmin in muscle and associated diseases: Beyond the structural function." Cell and Tissue Research **360**: 591–608.

Kathage, B., S. Gehlert, A. Ulbricht, L. Lüdecke, V. E. Tapia, Z. Orfanos, D. Wenzel, W. Bloch, R. Volkmer, B. K. Fleischmann, D. O. Fürst and J. Höhfeld (2017). "The co-chaperone BAG3 coordinates protein synthesis and autophagy under mechanical strain through spatial regulation of mTORC1." Biochimica Et Biophysica Acta – Molecular Cell Research **1864**: 62–75.

Köster, S., D. A. Weitz, R. D. Goldman, U. Aebi and H. Herrmann (2015). "Intermediate filament mechanics in vitro and in the cell: From coiled coils to filaments, fibers and networks." Current Opinion in Cell Biology **32**: 82–91.

Kolb, T., J. Kraxner, K. Skodzek, M. Haug, D. Crawford, K. K. Maaß, K. E. Aifantis and G. Whyte (2017). "Optomechanical measurement of the role of lamins in whole cell deformability." Journal of Biophotonics **10**: 1657–1664.

Konieczny, P., P. Fuchs, S. Reipert, K. S. Kunz, A. Zeöld, I. Fischer, D. Paulin, R. Schröder and G. Wiche (2008). "Myofiber integrity depends on desmin network targeting to Z-disks and costameres via distinct plectin isoforms." The Journal of Cell Biology **81**: 667–681.

Lammerding, J., P. C. Schulze, T. Takahashi, S. Kozlov, T. Sullivan, R. D. Kamm, C. L. Stewart and R. T. Lee (2004). "Lamin A/C deficiency causes defective nuclear mechanics and mechanotransduction." The Journal of Clinical Investigation **113**: 370–378.

Lammerding, J., L. G. Fong and R. T. Lee (2006). "Lamins A and C but not lamin beta1 regulate nuclear mechanics." Journal of Biological Chemistry **281**: 25768–25780.

Lange, J. R., J. Steinwachs, T. Kolb, L. A. Lautscham, I. Harder, G. Whyte and B. Fabry (2015). "Micro-constriction arrays for high throughput quantitative measurements of cell mechanical properties." Biophysical Journal **109**: 26–34.

Lange, J. R., C. Metzner, S. Richter, W. Schneider, M. Spermann, T. Kolb, G. Whyte and B. Fabry (2017). "Unbiased high-precision cell mechanical measurements with micro-constrictions." Biophysical Journal **112**: 1472–1480.

Lanzicher, T., V. Martinelli, L. Puzzi, G. Del Favero, B. Codan, C. S. Long, L. Mestroni, M. R. G. Taylor and O. Sbaizero (2015). "The cardiomyopathy lamin A/C D192G mutation disrupts whole-cell biomechanics in cardiomyocytes as measured by atomic force microscopy loading-unloading curve analysis." Scientific Report **5**: 13388.

Lautscham, L. A., C. Kämmerer, J. R. Lange, T. Kolb, C. Mark, A. Schilling, P. L. Strissel, R. Strick, C. Gluth, A. C. Rowat, C. Metzner and B. Fabry (2015). "Migration in confined 3D environments is determined by a combination of adhesiveness, nuclear volume, contractility, and cell stiffness." Biophysical Journal **109**: 900–913.

Leber, Y., A. A. Ruparelia, G. Kirfel, P. F. van der Ven, B. Hoffmann, R. Merkel, R. J. Bryson-Richardson and D. O. Fürst (2016). "Filamin C is a highly dynamic protein associated with fast repair of myofibrillar microdamage." Human Molecular Genetics **25**: 2776–2788.

Lee, J. S., C. M. Hale, P. Panorchan, S. B. Khatau, J. P. George, Y. Tseng, C. L. Stewart, D. Hodzic and D. Wirtz (2007). "Nuclear lamin A/C deficiency induces defects in cell mechanics, polarization, and migration." Biophysical Journal **93**: 2542–2552.

Lloyd, D. J., R. C. Trembath and S. Shackleton (2002). "A novel interaction between lamin A and SREBP1: Implications for partial lipodystrophy and other laminopathies." Human Molecular Genetics **11**: 769–777.

Mitchell, M. J., C. Denais, M. F. Chan, Z. Wang, J. Lammerding and M. R. King (2015). "Lamin/ A deficiency reduces circulating tumor cell resistance to fluid shear stress." American Journal of Physiology-Cell Physiology **309**: C736–746.

Moir, R. D., T. P. Spann, H. Herrmann and R. D. Goldman (2000). "Disruption of nuclear lamin organization blocks the elongation phase of DNA replication." The Journal of Cell Biology **149**: 1179–1191.

Mücke, N., S. Winheim, H. Merlitz, J. Buchholz, J. Langowski and H. Herrmann (2016). "In vitro assembly kinetics of cytoplasmic intermediate filaments: A correlative Monte Carlo simulation study." PloS One **11**: e0157451.

Mücke, N., L. Kämmerer, S. Winheim, R. Kirmse, J. Krieger, M. Mildenberger, J. Baßler, E. Hurt, W. H. Goldmann, U. Aebi, K. Toth, J. Langowski and H. Herrmann (2018). "Assembly kinetics of vimentin tetramers to unit-length filaments: A stopped flow study." Biophysical Journal **114**: 2408–2418.

Na, S., F. Chowdhury, B. Tay, M. Ouyang, M. Gregor, Y. Wang, G. Wiche and N. Wang (2009). "Plectin contributes to mechanical properties of living cells." American Journal of Physiology-Cell Physiology **296**: C868–C877.

Nikolova, V., C. Leimena, A. C. McMahon, L. C. Tan, S. Chandar, D. Jogia, S. H. Kesteven, J. Michalicek, R. Otway, F. Verheyen, S. Rainer, C. L. Stewart, D. Martin, M. P. Feneley and D. Fatkin (2004). "Defects in nuclear structure and function promote dilated cardiomyopathy in lamin A/C-deficient mice." The Journal of Clinical Investigation **113**: 357–369.

Osmanagic-Myers, S., S. Rus, M. Wolfram, D. Brunner, W. H. Goldmann, N. Bonakdar, I. Fischer, S. Reipert, A. Zuzuarregui, G. Walko and G. Wiche (2015a). "Plectin reinforces vascular integrity by mediating crosstalk between the vimentin and the actin networks." Journal of Cell Science **128**: 4138–4150.

Osmanagic-Myers, S., T. Dechat and R. Foisner (2015b). "Lamins at the crossroads of mechano-signaling." Genes & Development **29**: 225–237.

Pfeifer, C. R., M. Vashisth, Y. Xia and D. E. Discher (2019). "Nuclear failure, DNA damage, and cell cycle disruption after migration through small pores: A brief review." Essays in Biochemistry: EBC20190007.

Schröder, R. and B. Schoser (2009). "Myofibrillar myopathies: A clinical and myopathological guide." Brain Pathology (Zurich, Switzerland) **19**: 483–492.

Schröder, R., A. Vrabie and H. H. Goebel (2007). "Primary desminopathies." Journal of Cellular and Molecular Medicine **11**: 416–426.

Schwartz, C., M. Fischer, K. Mamchaoui, A. Bigot, T. Lok, C. Verdier, A. Duperray, R. Michel, I. Holt, T. Voit, S. Quijano-Roy, G. Bonne and C. Coirault (2017). "Lamins and nesprin-1 mediate inside-out mechanical coupling in muscle cell precursors through FHOD1." Scientific Report **7**: 1253. 10.1038/s41598-017-01324-z.

Schween, L., N. Mücke, S. Portet, W. H. Goldmann, H. Herrmann and B. Fabry (2022). "Dual-wavelength stopped-flow analysis of the lateral and longitudinal assembly kinetics of vimentin." Biophysical Journal 121: 1–12.

Schürmann, S., S. Wagner, S. Herlitze, C. Fischer, S. Gumbrecht, A. Wirth-Hücking, G. Prölß, L. A. Lautscham, B. Fabry, W. H. Goldmann, V. Nikolova-Krstevski, B. Martinac and O. Friedrich (2016). "The Iso-stretcher: An isotropic cell stretch device to study mechanical biosensor pathways in living cells." Biosensors & Bioelectronics **81**: 363–372.

Selcen, D., F. Muntoni, B. K. Burton, E. Pegoraro, C. Sewry, A. V. Bite and A. G. Engel (2009). "Mutation in BAG3 causes severe dominant childhood muscular dystrophy." Annals of Neurology **65**: 83–89.

Spörrer, M., D. Kah, R. C. Gerum, B. Reischl, D. Huraskin, C. A. Dessalles, W. Schneider, W. H. Goldmann, H. Herrmann, I. Thievessen, C. S. Clemen, O. Friedrich, S. Hashemolhosseini,

R. Schröder and B. Fabry, (2022). "The desmin mutation R349P increases contractility and fragility of stem cell-generated muscle micro-tissues." Neuropathology and Applied Neurobiology **48**: e12784.

Vignier, N., M. Chatzifrangkeskou, B. Morales-Rodriguez, M. Mericskay, N. Mougenot, K. Wahbi, G. Bonne and A. Muchir (2018). "Rescue of biosynthesis of nicotinamide adenide adenine dinucleotide protects the heart in cardiomyopathy caused by lamin A/C gene mutation." Human Molecular Genetics **27**: 3870–3880.

Vorgerd, M., P. F. van der Ven, V. Bruchertseifer, T. Löwe, R. A. Kley, R. Schröder, H. Lochmüller, M. Himmel, K. Köhler, D. O. Fürst and A. Hübner (2005). "A mutation in the dimerization domain of filamin C causes a novel type of autosomal dominant myofibrillar myopathy." American Journal of Human Genetics **77**: 297–304.

Wiche, G., H. Herrmann, F. Leichtfried and R. Pytela (1982). "Plectin: A high molecular weight cytoskeletal polypeptide component that copurifies with intermediate filaments of the vimentin type." Cold Spring Harbor Symposia Quantitative Biology **46**: 475–482.

Winter, L., I. Staszewska, E. Mihailovska, I. Fischer, W. H. Goldmann, R. Schröder and G. Wiche (2014). "Chemical chaperone ameliorates pathological protein aggregation in plectin-deficient muscle." The Journal of Clinical Investigation **124**: 1144–1157.

Winter, L. and W. H. Goldmann (2015). "Biomechanical characterization of myofibrillar myopathies." Cell Biology International **39**: 361–363.

Winter, L., A. Unger, C. Berwanger, M. Spörrer, M. Türk, F. Chevessier, K. H. Strucksberg, U. Schlötzer-Schrehardt, I. Wittig, W. H. Goldmann, K. Marcus, W. A. Linke, C. S. Clemen and R. Schröder (2019). "Imbalances in protein homeostasis caused by mutant desmin." Neuropathology and Applied Neurobiology **45**: 476–494.

Manfred Radmacher

6.9 Laminopathies

The primary function of the nucleus of eukaryotic cells is to securely store the cell chromosomes. A nucleus is surrounded by a double membrane: the outer and the inner nuclear membrane. The membranes are bridged with nuclear pore complexes, which facilitate the transport of mRNA into the cytosol. The inner membrane is underlined by the lamina, an elastic polymeric network, whose principal components are lamins, a class of intermediate filament proteins. The lamina offers a scaffold for binding transcription factors and provides mechanical stability to the nucleus (Sapra, 2020). Mutations in lamins result in laminopathies. One of these is the Hutchinson-Gilford progeria resulting in premature ageing. It is an inherited single point defect in the gene of lamin A, resulting in the exchange of amino acid 145 (E145K) (Broers, 2006). This disease is a prototype example, where a conformational change in a protein leads to a mechanical fingerprint: here, changed stiffness of the lamin layer and, as a consequence, a modified morphology of the nucleus. In addition, protein expression is hampered in a very general fashion, possibly induced by the changed properties of the lamina, which, as stated above, can serve as a scaffold for expression factors.

Despite the interest in the mechanobiology of the cell nucleus and particularly the lamin layer, there is only limited experimental knowledge on the mechanobiology of the nucleus (Sapra et al., 2020) and not many studies of the mechanics in laminopathies. The study of mechanical properties of the lamin layer, or even the nuclear membrane, is very complicated in situ, i.e., within the cell, due to the linkage between the actin cytoskeleton and the nucleus itself. Tension in the actin cytoskeleton will deform the nucleus such that tension is built up in the nucleus, and the mechanical properties of the nuclear region will be influenced by the tension of the cytoskeleton, and thus not reflect, or only reflect to some degree, the properties of the nucleus itself (Vishavkarma, 2014). Consequently, in an early study (Schäpe, 2009), the mechanical consequence of a laminopathy was studied in a model system: *Xenopus* oocytes. The nuclei of oocytes are easy to isolate due to their size of roughly 500 µm diameter; thus, they can be investigated by the AFM. It has been shown that, after the expression of human lamin A (wildtype and E145K mutant), the stiffness of the oocyte increased due to the increased thickness of the lamin layer. At comparable thicknesses of lamin layers, the networks formed from the mutant lamin were 2–3 times stiffer than the networks formed by the wildtype lamin. Mechanically speaking, the lamin layer resembles a thin, spherical shell, resulting in a linear force indentation curve if the tip is smaller than the radius of the curvature of the shell, i.e., the

Manfred Radmacher, University of Bremen, Bremen, Germany

https://doi.org/10.1515/9783110989380-019

nucleus (Boulbitch, 1998). Then, the response of the shell can be considered due to a point-like force with the consequence of a linear force law, where the apparent spring constant will depend on the elastic modulus of the shell, its radius, and most importantly, its thickness. Thus, it was essential, in this study, to characterize the thickness of the lamin layer by electron microscopy after measuring the apparent stiffness by AFM. A follow-up study showed that the lamin layer formed by the mutant protein is stiffer than lamin layers formed from the wild type protein (Kauffmann, 2011). This corroborates the idea that this mutation may have a mechanic implication. Finally, it could be demonstrated that nuclei from a progeria patient are much stiffer than nuclei from a young, healthy donor and have a similar stiffness as the nucleus of an older donor (Apte, 2017) (see Figure 6.9.1). Combining all these results, it can be argued that the mutation, E145K, which leads to the phenotype of Hutchinson-Guildford progeria in humans, is due to the different mechanical properties of the lamina layer. Distinct mechanics may change the binding of expression factors; it definitely changes the morphology of the nuclear lamina, thus resulting in the phenotype of this disease. So, there is a clear link between the mechanics of the nucleus and disease.

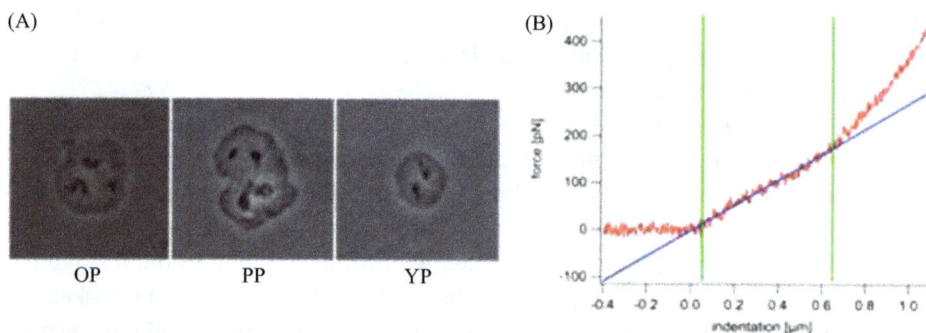

Figure 6.9.1: (A) Phase-contrast images of isolated nuclei from human skin fibroblasts from an old patient (OP), a 4-year old progeria patient (PP), and a 10-year old young patient (YP). (B) shows a typical force with a well-pronounced linear relation between force and indentation as expected from the response of a thin shell (lamin layer) to a point-like force (from Apte 2017).

References

Apte, K., R. Stick and M. Radmacher (2017). "Mechanics in human fibroblasts and progeria: Lamin A mutation E145K results in stiffening of nuclei." Journal of Molecular Recognition **30**(2): e2580–n/a.

Boulbitch, A. A. (1998). "Deflection of a cell membrane under application of a local force." Physical Review E **57**(2).

Broers, J. L. V., F. C. S. Ramaekers, G. Bonne, R. B. Yaou and C. J. Hutchison (2006). "Nuclear lamins: Laminopathies and their role in premature ageing." Physiological Reviews **86**(3): 967–1008.

Kaufmann, A., F. Heinemann, M. Radmacher and R. Stick (2011). "Amphibian oocyte nuclei expressing lamin A with the progeria mutation E145K exhibit an increased elastic modulus." Nucleus **2**(4): 310–319.

Sapra, K. T., Z. Qin, A. Dubrovsky-Gaupp, U. Aebi, D. J. Müller, M. J. Buehler and O. Medalia (2020). "Nonlinear mechanics of lamin filaments and the meshwork topology build an emergent nuclear lamina." Nature Communications **11**(1): 6205.

Schäpe, J., S. Prausse, R. Stick and M. Radmacher (2009). "Influence of lamin A on the mechanical properties of amphibian oocyte nuclei measured by atomic force microscopy." Biophysical Journal **96**(10): 4319–4325.

Vishavkarma, R., S. Raghavan, C. Kuyyamudi, A. Majumder, J. Dhawan and P. PA (2014). "Role of actin filaments in correlating nuclear shape and cell spreading." PLoS One **9**(9): e107895.

Lorena Redondo-Morata, Vincent Dupres, Sébastien Janel,
Frank Lafont

6.10 Integrative Approaches Using AFM to Study Pathogenic Bacteria and Infection

During infection, both the infectious agents (toxins, viruses, parasites, and bacteria) and the host cells act as sensors of their environment and respond to environmental conditions by using their surface constituents first. Therefore, understanding the biophysical properties of pathogens and hosts, especially their surfaces, is the key to understanding pathogenesis. To this end, the use of AFM has progressively increased over the years, as a stand-alone technique from which multiple parameters such as location, adhesion, binding, mechanics, and kinetics can be obtained. Equally emerging is the combination of this technique with others, such as fluorescence, to elucidate structure–activity relationships. Here we will review some of the major breakthroughs in the biophysics of infection by bacteria where the AFM technique has been paramount.

6.10.1 Bacteria

Pathogenic bacteria have the ability to adhere to a wide range of surfaces, leading to the formation of biofilms, and to interact and colonize the surface of host cells during infection. Bacteria have developed a great variety of mechanisms to promote or regulate their ability to adhere, including adhesion proteins (i.e., adhesins), surface appendages (pili, fimbriae), or hydrophobicity. Exploring the surface properties of bacteria and getting as much knowledge as possible about the mechanisms and strategies used during host cell colonization is of prime importance. This knowledge is also crucial in the face of increasing antibiotic resistance. Improving

Acknowledgments: We apologize to colleagues whose works were not cited herein owing to space limitation. This work was supported by a grant overseen by the French National Research Agency (ANR) as part of the "Investments d'Avenir" program (I-SITE ULNE/ANR-16-IDEX-1135 0004 ULNE) to L.R.-M, ANR (10-EQPX-04-01, 15-CE18-0016-01, 16-CE15-00003-03), FEDER (12001407), and EU-ITN Phys2Biomed 812772 to F.L.

Lorena Redondo-Morata, Vincent Dupres, Sébastien Janel, Frank Lafont, Cellular Microbiology and Physics of Infection Group, University of Lille, CNRS, Inserm, CHU Lille, Institut Pasteur Lille, U1019 – UMR 9017 – CIIL – Center for Infection and Immunity of Lille, Lille F-59000, France

https://doi.org/10.1515/9783110989380-020

understanding of the structure of the bacterial wall and the effect of antibacterial drugs on it is the key for developing new antimicrobial strategies.

AFM imaging is certainly, along with electron microscopy, the technique that has offered the greatest advances in our knowledge of the structure of bacterial cell walls over the last 20 years. It has the major advantage of being able to work in a liquid medium, therefore close to the physiological conditions, and thus keep the bacteria alive during the analysis. We will herein review the potentials offered by the AFM in this quest for knowledge of structural and biophysical properties of bacteria.

6.10.1.1 Bacteria Immobilization

Since the first applications of AFM in microbiology (Butt et al., 1990), cell immobilization on a solid substrate remains a prerequisite prior to performing AFM imaging or force measurements. For mammalian cells, a simple preparation method is to exploit the ability of the cells to spread and adhere to solid surfaces (e.g., glass or biocompatible polymers supports) (Matzke et al., 2001, Radmacher et al., 1992). For bacteria, cell-support contact area is significantly reduced and most of the cells do not spontaneously adhere to solid surfaces. Thus, a key challenge consists of developing procedures allowing imaging of the cells in their native state but without cell detachment by the scanning AFM probe. It then requires a firm immobilization of the microbial cells on the solid support.

Stronger attachment may be achieved through several approaches (Figure 6.10.1). Drying a droplet of a concentrated cell suspension on a mica or glass substrate (Amro et al., 2000) is probably the simplest way to proceed; nevertheless this approach may lead to significant alterations of the cell walls and thus limiting cell viability (Beckmann et al., 2006). In addition, a lack of stability is often observed, and cells can sweep away during scanning.

Physisorption has been intensively used to immobilize bacteria on flat surfaces based on electrostatic interactions. Thus, glass or mica can be coated with a poly-L-lysine (PLL) (Bolshakova et al., 2001, Fantner et al., 2010, Schaer-Zammaretti and Ubbink, 2003), PLL hydrobromide (Camesano and Logan, 2000), or poly(ethyleneimide) (Vadillo-Rodrigues et al., 2004, Velegol and Logan, 2002) to create a positively charged surface (Figure 6.10.1A). However, such polycations can have significant effects on bacteria physiology (Colville et al., 2010) and could easily contaminate modified AFM probes during force measurements. Gelatin-coated mica surfaces were also used to image gram-positive and gram-negative bacteria (Doktycz et al., 2003, Sullivan et al., 2005).

Chemical fixation (Figure 6.10.1B), including covalent binding and cross-linking, is commonly used to immobilize rod-shaped bacteria. For instance, one commonly used technique is to treat glass slides with aminosilane so that the bacteria could be coupled to the amino groups using the N-hydroxysuccinimide (NHS)/1-ethyl-3-(3-

dimethylaminopropyl)carbodiimide (EDC) treatment (Camesano and Logan, 2000, Liu et al., 2008). Glutaraldehyde may also be used to enhance immobilization (Razatos et al., 1998). Nevertheless, these two approaches are limited to the study of nonviable cells since bifunctional reagents used for cross-linking and reactive groups used in covalent binding may affect cell viability (D'Souza, 2001).

As an alternative approach, Kailas et al. (2009) introduced mechanical immobilization of bacteria using lithographically patterned surfaces (Figure 6.10.1C). Inert surfaces with hole arrays are used to immobilize and image *Staphylococcus aureus* cells. Due to the absence of chemical linkage and confinement effects, this approach appears very promising to follow dynamic events such as cell division (Kailas et al., 2009). In a similar way, Cerf et al. (2009) used chemically engineered template to probe the nanomechanical properties of *Escherichia coli* bacteria, a method that could be applied to various microbes (Formosa et al., 2015).

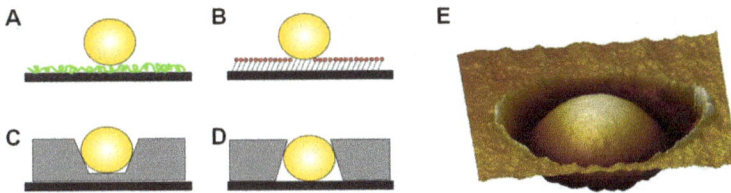

Figure 6.10.1: Different approaches for bacteria immobilization ((A) physisorption, (B) chemical fixation, (C) patterned surfaces, and (D) mechanical trapping) and AFM image of a *Lactobacillus plantarum* cell trapped in a porous membrane (E) (image scan size: 3 μm × 3 μm).

Another possibility is to immobilize the cells mechanically in a porous membrane or filters (Figure 6.10.1D and E). In this method, microbial cells are trapped in a polymer membrane with a pore size that is slightly smaller than the dimensions of the cell, allowing repeated imaging without cell detachment or cell damage (Dufrêne et al., 1999, Kasas and Ikai, 1995, Touhami et al., 2004). During the last two decades, this approach has been successfully used to immobilize various round-shaped microbial cells, including *S. aureus* (Touhami et al., 2004) or *Lactococcus lactis* (Gilbert et al., 2007), and in some occasions rod-shaped bacteria (Vadillo-Rodrigues et al., 2004, Francius et al., 2008). Only in very rare cases, bacteria spontaneously adhere to the membrane; it was observed for mycobacteria (Dupres et al., 2005), presumably because of the strong hydrophobic properties of the cells. Another advantage of the mechanical trapping approach is that it allows observing cell surface dynamics, such as cell growth and division (Touhami et al., 2004, Dague et al., 2008, Turner et al., 2010), as well as structural changes resulting from cell wall–drug interactions (Francius et al., 2008, Verbelen et al., 2006) and to probe mechanical properties of individual proteins (Alsteens et al., 2009, Dupres et al., 2009b). Nevertheless, several disadvantages must be pointed out: (i) the mechanical trapping could exert a pressure on the cell, (ii) only a small part of the cell appears accessible, and (iii) trapping may condition the selection of a bacterial population matching with the pore size. Limitation to commercially available pore sizes was another limitation since Turner et al. (2010) present an elegant study to modulate pore size using a chemical treatment.

Finally, bacterial cells have also been immobilized to glass-coated surfaces with highly adhesive polyphenolic proteins originating from the mussel *Mytilus edulis* (Louise Meyer et al., 2010).

To summarize, none of these immobilizing methods appear unambiguously applicable for all bacterial cells but each microbial specimen should be considered individually. While mechanical trapping in porous membrane is particularly suited for yeast cells (Alsteens et al., 2008a, Dupres et al., 2009a), fungi (Dague et al., 2008), or round-shaped bacteria (Touhami et al., 2004), chemical fixation and physical adsorption are widely used for rod-shaped bacteria (Schaer-Zammaretti and Ubbink, 2003) or gram-negative species (Gaboriaud et al., 2008, Greif et al., 2010). Moreover, in some occasions, imaging in air is required; this is the case when imaging of lateral bacterial appendices such as pili or flagella (Doktycz et al., 2003).

6.10.1.2 Imaging Bacteria

Already in the mid-1990s, the AFM demonstrated its ability to perform high-resolution imaging of many different bacterial protein crystals (Karrasch et al., 1994, Engel et al., 1985, Müller et al., 1995). It took a few more years and substantial improvements in the AFM instrumentation (both at the level of cantilever probes and imaging modes) for high-resolution imaging to be transposed to living microorganisms and enable subnanometer resolution to be achieved on such samples, thus revealing various cell surface features.

One of the first examples is the evidence of the presence of a crystalline bacterial cell surface layer (S-layer) on living *Corynebacterium glutamicum* bacteria (Dupres et al., 2009b). The presence of two highly ordered surface layers has then been demonstrated. The most external layer represents the hexagonal S-layer, and the inner layer displays regular patterns of nanogrooves that could act as a bimolecular template promoting the 2D assembly of S-layer monomers (Dupres et al., 2009b). More recent work has enabled to visualize S-layer lattices on *Lysinibacillus sphaericus* and *Viridibacillus arvi* (Günther et al., 2014) and on gram-negative *Tannerella forsythia* bacteria (Oh et al., 2013).

As a major component of both the gram-positive and gram-negative bacterial cell wall, peptidoglycan is the target of antibiotics such as β-lactams and glycopeptides that will impair its assembly. André et al. (2010) have used AFM to image the nanoscale organization of peptidoglycan in living *L. lactis* bacteria. They showed that wild-type cells display a smooth, featureless surface morphology, whereas mutant strains lacking cell wall exopolysaccharides feature 25-nm-wide periodic bands, made of peptidoglycan, running parallel to the short axis of the cell. Pasquina-Lemonche et al. successfully imaged at high-resolution the inner peptidoglycan surface, with glycan strand spacing typically less than 7 nm. They were able to discern morphological differences as a function of the location of the inner peptidoglycan layer; the cylinder of *Bacillus subtilis* showed dense circumferential orientation, while in *S. aureus* and division septa for both species, peptidoglycan was dense but randomly oriented (Pasquina-Lemonche et al., 2020).

But the main quality of the AFM is not so much high-resolution imaging itself, which could also be accessible by other techniques, but the possibility of imaging live bacteria and thus monitor dynamic events. Hence, Turner et al. (2010) have followed the division of *S. aureus* bacteria trapped in a porous polymer membrane with near molecular resolution. But one of the most common and interesting applications of AFM in the field of bacteriology is its ability to directly visualize the effect of drugs on the bacterial cell wall.

Kasas et al. (1994) were among the first research groups that used AFM imaging to study the effect of penicillin on the morphology of *B. subtilis* cells in air. In a similar approach, Braga and Ricci (1998) used AFM to visualize the action of cefodizime, a β-lactam antibiotic, on *E. coli*. Although the impact of these works was limited by the drying protocol applied during sample preparation, they were precursors and opened the door to numerous studies.

Continuous improvements in bacterial immobilization procedures, as previously discussed, made it possible to keep the cells alive throughout the analysis, thus ensuring the relevance of the results obtained. Liquid imaging of living cells was used by Francius et al. to observe the digestion of *S. aureus* cells walls by lysostaphin, a bacteriolytic enzyme targeting peptidoglycan (Francius et al., 2008). Time-lapse images collected following addition of lysostaphin revealed major structural changes in the form of cell swelling, splitting of the septum, and creation of nanoscale perforations.

The mycobacterial cell wall is overly complex and essential for growth and survival in the infected host. Due to this complexity, the molecular mechanisms by which antibiotics affect the cell wall structure and properties remain poorly understood. Verbelen et al. (2006) imaged the surface of hydrated *Mycobacterium bovis* bacillus Calmette Guérin (BCG) before and after exposure to ethambutol, an antimycobacterial drug that inhibits the synthesis of the polysaccharidic portion of the envelope. They found that the drug dramatically alters the fine surface architecture of the cells. Interestingly, the concentration of ethambutol causes the selective removal of the outer layers of the cell wall while concentric striations are observed on top of the remaining layers. In the same perspective, Alsteens et al. (2008b) imaged the surface of live *M. bovis* BCG cells prior and after incubation with four antimycobacterial drugs. If the overall integrity of the cells was maintained, all the four drugs induced substantial modifications of the cell surface architecture (increase of the surface roughness, layered structures, striations, and porous morphologies).

These modifications were suggested to reflect the inhibition of the synthesis of three major cell wall constituents (i.e., mycolic acids, arabinans, and proteins).

The use of antimicrobial peptides (AMPs) is an alternative to conventional antibiotics and antimicrobial drugs to fight pathogenic bacteria. These AMPs have been demonstrated to have their own advantages over the traditional antibiotics with a broad spectrum of antimicrobial activities, including gram-positive and gram-negative species (Lei et al., 2019). AMPs usually form a helix structure, act through the bacterial cell membrane, form ion channels or pores on the microbial membranes, leading to membrane permeability and causing leakage of intracellular substances to result in bacterial death (Lei et al., 2019). Nevertheless, how these peptides target and disrupt bacterial membranes is a key issue since a better knowledge of their mode of action could prove to be beneficial in the design of new antimicrobial strategies. Since their introduction a large number of AFM experiments have been attempted to study how these peptides disrupt membranes, mainly on reconstituted phospholipid bilayers (Hammond et al., 2021). These samples are flat enough so that subtle differences in topography can easily be seen, such as pore formation. Nevertheless, bacterial membranes differ greatly from reconstituted bilayers, both in their lipid composition and in the number of transmembrane proteins that they contain. It is therefore very tricky to know to what extent the mechanisms of action of peptides on bilayers can be transposed to live membranes. The topography (curvature) of bacterial pathogens makes high-resolution imaging challenging and does not currently allow detailed mechanistic studies of peptide-induced membrane disruption to be deciphered. Thus, the AFM studies on living pathogens mainly focused on modifications and alterations induced by AMPs across the membrane surfaces (Mortensen et al., 2009, Overton et al., 2020, Meincken et al., 2005).

In an original approach, Fantner et al. (2010) used high-speed AFM to measure the kinetics of a pore-forming, membrane-disrupting AMP (CM15) on individual live *E. coli* cells in an aqueous solution. With a rate of just few seconds per AFM image, it allows the characterization of the initial stages of the AMP action, revealing a two-step killing process: a time-variable incubation phase (which takes seconds to minutes to complete) followed by a more rapid (less than one minute) execution phase (Figure 6.10.2).

6.10.1.3 Stiffness Mapping of Bacteria

Unlike mammalian cells, bacteria are surrounded by thick, mechanically rigid cell walls, which resist the internal turgor pressure and play important roles in controlling cellular processes such as growth, division, and adhesion (Krieg et al., 2019). In many bacteria, these cell walls are composed of several layers of peptidoglycan. As previously mentioned, some antibiotics will directly affect the biochemical machinery that assembles the peptidoglycan structure, leading to mechanical cell wall

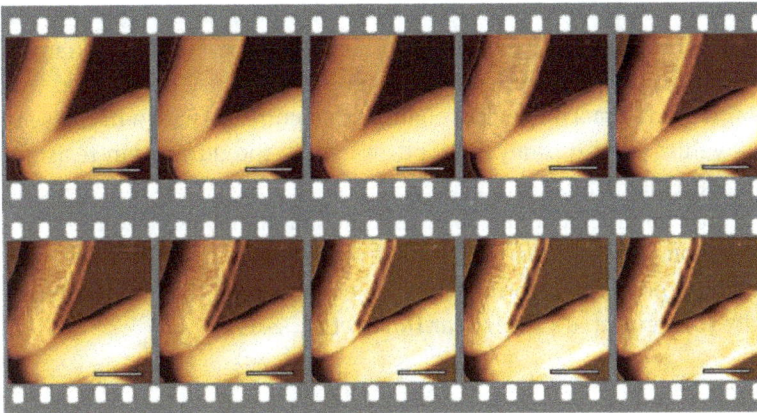

Figure 6.10.2: *Escherichia coli* cell disruption induced by CM15, imaged with fast AFM. Time series of CM15 antimicrobial action. CM15 injected at $t = -6$ s and images recorded every 13 s, with resolution of 1024 × 256 pixels and rate of 20 lines/s. The upper bacterium's surface starts changing within 13 s. The lower bacterium resists changing for 78 s. Scale bar = 1 µm. Reproduced from Fantner et al. (2010) with permission.

modifications. Besides imaging, measuring bacterial cell wall stiffness can then help elucidate the mechanism of antimicrobial compounds.

AFM has been extensively used to probe the mechanical properties of bacteria (Francius et al., 2008, Formosa et al., 2015, Kasas et al., 2018). In the same way as for experiments carried out on mammalian cells, force–distance curves are recorded on the surface of the bacteria. A fitting of the data associated with the contact part of the approach curve using the Hertz model is generally used to determine Young's modulus of the bacteria. The values found are in the order of MPa, much higher than those obtained for mammalian cells, highlighting the stiffness of the bacterial cell walls. The significant curvature associated with the topography of the bacteria nevertheless questions the relevance of the values obtained, as the contact area between the tip and the surface never remains constant with respect to the cell wall region analyzed.

In most experiments, mechanical measurements are correlated with imaging. Francius et al. (2008) studied the effect of lysostaphin on *S. aureus* cells. Treatment of the bacteria with this enzyme was found to decrease the bacterial spring constant and the cell wall stiffness, demonstrating that structural changes were correlated with major differences in cell wall nanomechanical properties. Young's modulus decreases, indicating a cell wall almost 10 times more elastic after treatment; it was attributed to the digestion of the peptidoglycan layer by the lysostaphin. This was confirmed by Loskill et al. (2014) with AFM experiments performed on *S. aureus* strains expressing a reduced peptidoglycan cross-linking and showing a reduction in cell wall stiffness.

Measuring cell mechanical properties can also participate to unravel the mechanisms by which AMPs act on the bacterial cell wall. Duramycin is an antibacterial peptide that acts specifically on certain gram-positive strains and very rarely on gram-negative strains. Hasim et al. used AFM to image and probe the mechanical properties of both susceptible and resistant gram-positive strains. They found that only sensitive strains exhibit morphological changes that correlate with a decrease in Young's modulus values after duramycin exposure while the duramycin-resistant strains retain their elasticity (Hasim et al., 2018).

Recently, multiparametric AFM has been used to simultaneously acquire high-resolution imaging and to probe the mechanical properties across the whole cell surface, thus making it possible to correlate the heterogeneity of the surface to variations in elasticity (Dufrêne et al., 2013; Saar Dover et al., 2015).

6.10.1.4 Single-Molecule Recognition and Molecular Imaging

Bacterial adhesion is the key step in the colonization of surfaces. It is a fundamental process that occurs both during biofilm formation and infection, leading to the invasion of host cells by pathogenic bacteria. The design of antiadhesive therapies can be beneficial in the light of increasingly problematic antibiotic resistance. However, the development of such therapies requires precise knowledge of the molecular mechanisms used by pathogenic bacteria to colonize the host surface and resist the mechanical stresses exerted by their environment. It is probably in this direction that the AFM, and more precisely single-molecule force spectroscopy (SMFS), has made it possible to make the greatest advances in recent years.

Specific interactions between receptors and related ligands play central role in bacterial adhesion. SMFS appears as a useful approach to probe the specific interaction between these complexes at the single-molecule level. This requires immobilizing the ligands, or other (bio)molecules, at the apex of the AFM tip (Hinterdorfer and Dufrêne, 2006). Several protocols can be followed to attach ligands. A simple method leading to the formation of a covalent bond between the probe and the ligand involves the use of self-assembled monolayers of alkanethiols (SAMs). The first step is to functionalize the tip surface with SAMs that terminate in carboxyl functions, which can then be reacted with the amino groups of proteins using EDC and NHS (Dammer et al., 1995). The investigations of antigen/antibody interactions require more complicated setup: both antigens and antibodies need firm attachment, their interactions usually depend on geometric factors (steric hindrance must be avoided) while nonspecific interactions must be reduced (expected specific forces are in the same piconewton range). To fulfill these requirements, different flexible cross-linkers were used to covalently attach a biomolecule to a silicon nitride tip (Dammer et al., 1996, Ebner et al., 2007a, 2007b, Hinterdorfer et al., 1996). As a cross-linker, Ebner and coworkers used poly(ethylene glycol) (PEG), a water-soluble polymer (6 nm extended length) which is known to prevent surface adsorption of proteins. Furthermore, the stretching of the spacer is not linear, providing characteristic force spectroscopy curves that allow discriminating between specific and unspecific interaction (Ebner et al., 2007a).

Using such biologically modified probes, AFM can be used to determine how cell-surface-bound molecules are organized, and how they interact with their environment (Dupres et al., 2009a).

Besides peptidoglycan, teichoic acids represent other major constituents of bacterial walls. André et al. (2011) used fluorescence microscopy and AFM to image their distribution in *Lactobacillus plantarum*, in relation with their functional roles. AFM topographic images of living bacteria showed a highly polarized surface morphology, the poles being much smoother than the side walls. Both AFM and fluorescence imaging with lectin probes revealed that the polarized surface structure correlates with a heterogeneous distribution of teichoic acids, the latter being absent from the surface of the poles. Based on control experiments performed on different mutant strains, this heterogeneity was found to play a key role in controlling cell morphogenesis (surface roughness, cell shape, elongation, and division).

Focusing more specifically on the adhesive properties of bacteria, molecular recognition based AFM is a valuable tool to detect and reveal the distribution of adhesion proteins, that is, adhesins (Alsteens et al., 2010). For instance, Dupres et al. (2005) investigated single heparin-binding hemagglutinin adhesins (HBHAs) on the surface of mycobacteria. Topographic images of the cells immobilized onto porous polymer membranes revealed a fairly smooth and homogeneous surface, consistent with both the expected cell wall architecture and earlier scanning electron microscopy observations. To address the question of the spatial distribution of the adhesins across the cell surface, adhesion force maps were recorded using a heparin-modified tip. These maps revealed that adhesion events were detected in about half of the locations. Together with blocking experiments, adhesion maps obtained on a mutant strain lacking HBHA and that did not show specific binding suggest the detection of single HBHA molecules. Interestingly, the HBHA distribution was not homogeneous over the mycobacterial surface, but rather concentrated into nanodomains that may have important biological functions (i.e., promoting adhesion to host cells).

In 2009, the same approach was used to investigate the binding strength and the surface distribution of fibronectin attachment proteins (FAPs) in mycobacteria (Verbelen and Dufrêne, 2009). The specific binding forces of FAPs were measured, and found to increase with the loading rate, as observed for other receptor–ligand systems. The distribution of FAPs was also mapped with a fibronectin-modified tip, revealing that the proteins were widely exposed on the mycobacterial surface. The authors then showed that treatment of the bacterial cells with enzymes or antibiotics led to a substantial reduction of the FAP surface density, confirming that the proteins localize specifically on the outermost cell surface.

The most recent developments in force spectroscopy make it possible to probe, on a molecular scale, the mechanisms that pathogenic bacteria display during the infection process. It is thus possible to specifically target various molecules of the bacterial cell wall by means of functionalized AFM probes, and to study the behavior of these molecules under stress. Different surface proteins have been identified to promote the adhesion of gram-positive bacteria to host cells and were then studied by AFM (Sullan et al., 2015, Herman et al., 2014). Mathelié-Guinlet et al. focused on the adhesin SpsD from *Staphylococcus pseudintermedius*, which is known to bind to fibrinogen. Using force-clamp spectroscopy, they showed that the interaction between SpsD and fibrinogen is strong and exhibits an unusual catch–slip transition in response to force (Mathelié-Guinlet et al., 2020). Mechanical force first prolongs the lifetime of the bonds up to a critical force around 1 nN, above which the bond lifetime decreases as an ordinary slip bond. This study is the first to reveal such a catch bond behavior in living gram-positive pathogens. It is thus suggested that such behavior could be used by pathogens to tightly control their adhesion properties during colonization and infection: catch binding to strengthen adhesion during colonization and biofilm formation while slip binding promoting cell dissemination.

Whatever the application of the AFM mentioned so far, one of the major drawbacks resides in the time-consuming operation-intensive process in contrast to the high-throughput approaches routinely applied in the omics fields. Consequently, experiments described in many publications are often done with too few samples to ensure proper statistical significance of the results. We therefore propose a solution which consists in automating the procedure on a sequential but multisample basis operating in fluid thus valid on biological cells (Dujardin et al., 2019). We demonstrated the potential of this approach to detect and scan both fixed and living bacteria before completion of data processing. Several hundred cells were thus detected and scanned automatically. To emphasize the relevance of this approach, the effect of two distinct treatments (gentamicin and heating) was then evidenced on physical parameters of fixed *Yersinia pseudotuberculosis* bacteria. If the system in its current state is directly applicable to image the topography of bacteria and probe their mechanical properties, its application to molecular imaging requires further development, including functionalized AFM probes management. The opening of this approach to more complex systems, such as mammalian cells or tissues, is currently under development.

6.10.1.5 Single-Cell Force Spectroscopy

As illustrated in the previous section, SMFS is a leading-edge approach to detect and map adhesive proteins directly on living bacteria surfaces. It also allows advances in the understanding of the molecular mechanisms involved in adhesion, at the scale of the single molecule. It is also fundamental to gain a more detailed understanding of the mechanisms that promote cell adhesion at the whole cell level to eventually understand how bacterial pathogens interact with surfaces. For this purpose, a single cell can be immobilized at the end of the AFM cantilever, in place of the probe, a so-called single-cell force spectroscopy (SCFS) mode (Helenius et al., 2008, Benoit et al., 2000).

The immobilization of individual bacteria at the end of an AFM cantilever, while ensuring that the cell remains alive, appeared challenging during the first attempts to study such systems. Indeed, chemical approaches commonly used have highlighted changes in the structure of the bacterial surface, which can lead to cell inactivation after immobilization (Vadillo-Rodríguez et al., 2004). More gentle approaches based on polydopamine or Cell-Tak, a commercial wet cell adhesive, were then proposed (Kang and Elimelech, 2009, Beaussart et al., 2014b, Zeng et al., 2014).

Studying the adhesion of pathogenic bacteria to medically relevant abiotic surfaces is one of the possibilities offered by SCFS (Thewes et al., 2014, Merghni et al., 2017). Probing surfaces of different natures (i.e., hydrophobic vs. hydrophilic, for instance) allow a better understanding of the mechanisms governing the formation of biofilms. In the context of host–pathogen interactions, SCFS can also be used to identify specific adhesins involved in cell adhesion (Formosa-Dague et al., 2016, Feuillie et al., 2018). Bacteria appendages such as pili and fimbriae also contribute to cell adhesion through nonspecific and specific interactions. SCFS also enabled to gain knowledge on how these appendages govern bacterial adhesion to abiotic surfaces or host cells (Beaussart et al., 2014a, Sullan et al., 2014, Beaussart et al., 2016).

Although extremely useful to better understand the adhesion mechanisms developed by pathogenic bacteria, the SCFS has one major limitation: the difficulty in having a sufficient panel of cell probes for each study, since each bacterium used as a probe involves the specific preparation of a cantilever, making this approach really time intensive. The fluidic force microscopy (FluidFM) technology is a valuable alternative which does not involve chemical fixing or glue (Dörig et al., 2010). It is based on microchanneled AFM cantilevers with nanosized apertures, connected to a pump through a microfluidic system, to easily capture and release individual cells on demand (Meister et al., 2009, Guillaume-Gentil et al., 2014, Amarouch, El Hilaly & Mazouzi, 2018). While this approach has proven to be very effective for yeast immobilization (Dörig et al., 2010), the small dimensions of the bacteria currently make their handling even more challenging (Potthoff et al., 2015).

6.10.2 Host–Pathogen Interactions

The host-bacteria system can be first envisioned from a reductionism perspective. In this manner, AFM allows studying the interactions between isolated molecules from both host and bacteria and among that latter, adhesins play a crucial role being the first front in the earliest interaction step. As an example, we can discuss the case of

Neisseria meningitidis, the causative agent of meningitis. *Neisseria meningitidis*, or meningococcus, uses signaling surface sensors from the endothelial cells, the beta2-adrenoreceptor and the beta-arrestin pathway, which, in turn, activates the host cell surface (Coureuil et al., 2010). After the knockdown of this receptor, bacteria are still able to bind the host cell surface thanks to pilins E and V adhesins-mediated interaction to CD147 as clearly established by several approaches (Bernard et al., 2014). Among those, AFM allowed to investigate interaction forces with CD147 linked to an inert substratum and pilins to the AFM tip (Bernard et al., 2014). Moreover, it has been possible to further investigate the rupture forces and adhesion work for the pilin/CD147 interaction when knocking down actinin 4, an actin-associated protein, demonstrating the importance of the anchorage of the surface machinery to the underlying cytoskeleton (Maïssa et al., 2017). This can be also examined in the case of the *Listeria monocytogenes* Internalin B adhesin interaction with the MET receptor. In that case, depending on the type of septin expression depletion, rupture forces and adhesion work were found affected (Mostowy et al., 2011). These two examples illustrate the added value of AFM when studying the specific interaction of adhesins on the tip and the cell surface receptors. Moreover, AFM can provide measurements of how disturbed an interaction is when playing with the level of expression of internal cellular molecular machineries directly or indirectly linked to plasma membrane receptors.

The next experimental configuration to consider is to link the bacterium to the cantilever. As in the previous paragraph, alteration of rupture forces and adhesion work can be investigated including when comparing heat-killed bacteria (or killed upon antibiotic treatment) versus live bacteria on the tip-less cantilever as shown for *Yersinia pseudotuberculosis* (Ciczora et al., 2019). Another aspect permitted using this experimental configuration is to study directly in real time the recruitment of fluorescent-labeled host cell proteins when the bacterium lands on the cell surface (Ciczora et al., 2019). These host proteins can be receptors, cytoskeleton (-associated) proteins, and signaling molecules. It is possible to compare on the same host cell the landing of living bacteria and killed bacteria – both coated on the cantilever (Ciczora et al., 2019). Also, while the bacterium on the cantilever lands on the host cell surface, one can monitor posttranslationally modified protein recruitment at the binding site (Ciczora et al., 2019). These are extremely valuable information for microbiologists to decipher the triggering of the cell response cascade.

The coating of the bacterium can be achieved by several ways. Physisorption is the most straightforward procedure (e.g., for *Yersinia pseudotuberculosis*; Ciczora et al., 2019)), but it is not possible for all bacteria species and this has to be tested in preliminary experiments. An alternative is to use chemical crosslinking, whenever the cell receptor is identified, so that it can be coated via a functionalized flexible linker (i.e., PEG). That is the case for the extracellular part of the MET receptor for Listeria (Mostowy et al., 2011). A third possibility is to use the FluidFM micropipette device that allows using a cantilever with a hollow tip and the microaperture (300 nm and a spring constant of 0.28–0.52 N/m) connected via a canal to a tubing/

pump system to apply a aspiration force in order to trap the bacterium, though only bacteria–glass interaction have been studied to date (Hofherr et al., 2020).

AFM could also be useful to study internalized bacteria (more detailed in the next section). With the introduction of the stiffness tomography approach (Roduit et al., 2008, 2009) and its validation (Roduit et al., 2009) it is now possible to sense bacteria within vacuolar compartment identified using fluorescence labeling of specific markers (Janel et al., 2017). For this, "correlative light AFM microscopy" (CLAM) and even "correlative light, atomic force, electron microscopy" (CLAFEM) are fundamental. In the latter case, electron microscopy allows to identify those intracellular structures of different stiffnesses that are not labeled with fluorescent probes (Janel et al., 2017). The interest of such an approach is to understand how the mechanical properties of vacuoles may influence clustering of signaling molecules that could affect the cell response. One can think about variation of lipid composition due to specific diet or diseases like diabetes, lysosomal storage disease for instance.

6.10.3 Intracellular Pathogens

Even though AFM seems most relevant for studying first steps of infections (binding, adhesion) due to its unique combination of sensitivity and versatility, it can also decipher later infection processes when pathogens reach inside the cell. Being the golden standard of surface analysis, the main added value of AFM is to monitor cellular surface changes upon infection. In this regard, infected erythrocytes led to many publications although more related to parasites than bacteria. *Plasmodium falciparum* hijacks the red blood cell to express proteins that progress to the cell membrane, allowing vascular cytoadhesion and further infection. These proteins are located in knobs-like membrane protrusions of nanometric size (Aikawa et al., 1996). AFM was employed to characterize the knobs height, diameter, number, and homogeneity for different parasite strains and we found a correlation between the genotype of the parasite and the phenotype of the cell membrane Figure 6.10.3A (Nacer et al., 2011). Kwon et al. (2019) first observed an increase in the curvature, then stiffness, and later height and volume at the different steps of *Plasmodium berghei* ANKA infection.

Virus-infected cells were also studied with the AFM (Häberle et al., 1992). Indeed, virus entry and/or budding can be observed in real time with good resolution. Surface roughness analysis of enveloped or nonenveloped virus-infected cells did not show any difference, while line profiles show minor, possibly because of different entry process (Moloney et al., 2004). Kuznetsov et al. (2004) observed Moloney murine leukemia virus (M-MuLV) in NIH 3T3 cells, and found out that the envelope glycoprotein greatly stabilizes the virion envelope, but not homogeneously throughout the viral population. Same individual retrovirus budding events were characterized in real time and classified into two populations: fast (<25 min) and slow (>45 min)

(Gladnikoff and Rousso, 2008). It demonstrated how human immunodeficiency and Moloney murine leukemia retroviruses infection reorganize the actin cytoskeleton from fibers to aggregates and how budding is actin driven (Gladnikoff et al., 2009). Correlating with fluorescence adds the unambiguous identification of what is seen in topography, for example, Figure 6.10.3B, the budding in HeLa cells expressing HIV-1 Gag-GFP and immuno-stained with anti-CD9 for STORM/AFM correlation (Dahmane et al., 2019).

After this digression from bacteria, remodeling of cell surface can be even more dramatic in the case of RhoA-inhibitory bacterial toxins effects on epithelial cells. *S. aureus* EDIN toxin or *Bacillus anthracis* edema toxin form large transient transendothelial cell macroaperture (TEM) tunnels by disrupting the contractile cytoskeletal network, allowing the bacteria to disseminate. AFM was key to monitor the height profile of TEM opening and to model it as liquid dewetting with an increased height at the cell border (Gonzalez-Rodriguez et al., 2012, Maddugoda et al., 2011). It was then further described how a stiff actin cable Figure 6.10.3C limits the TEM widening, following non-muscle myosin II increase of line tension and ezrin promotion of actomyosin cable structuration (Stefani et al., 2017).

However, AFM is not limited to surface characterization. It can assess elasticity of infected cells, and even intracellular stiffness changes. It has been demonstrated how discontinuities visible on the approach curve performed on cells with rather sharp tips are not instrumental artifacts, but intracellular compartments detected

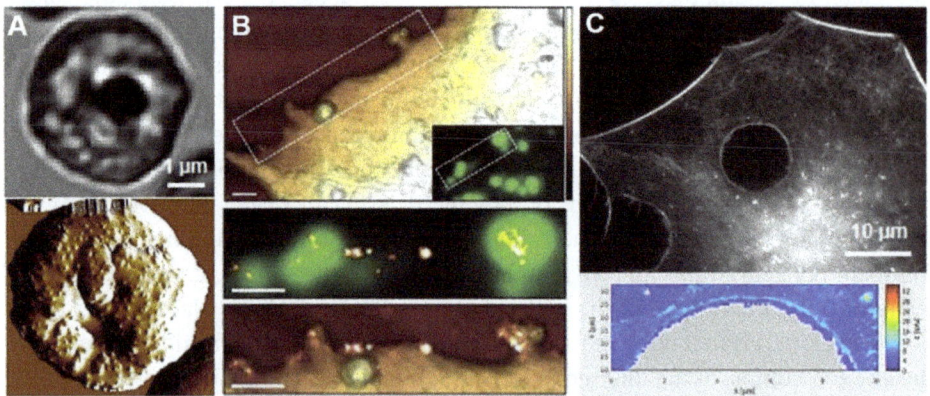

Figure 6.10.3: Cell surface remodeling upon infections with parasite, virus, and intoxication. (A) *Plasmodium falciparum* (3D7)-infected erythrocyte identified at the trophozoite step (top) and corresponding AFM error signal displaying parasite-derived knob-like structures responsible of adhesion in human endothelial cells (bottom). (B) CD9 recruitment at HIV-1 budding sites. HeLa cells expressing HIV-1 Gag-GFP were immuno-stained with anti-CD9 coupled to Alexa-647 and imaged by AFM (up), conventional fluorescence (middle), and dSTORM (bottom). Scale bars are 500 nm (up) and 200 nm (middle and bottom). Reproduced from (Dahmane et al., 2019) with permission. (C) Transendothelial cell macroaperture on HUVEC cells induced by edema toxin. Top: LifeAct-GFP fluorescence image, bottom: cell elasticity showing the stiff actin ring stabilizing the aperture opening.

because of their higher stiffnesses (Janel et al., 2019, Roduit et al., 2009). This has been applied to the detection of intracellular *Yersinia pseudotuberculosis* bacteria inside LC3-positive vacuole in *Potorous tridactylus* kidney (PtK2) cells, using correlative AFM, fluorescence, and electron microscopy (Janel et al., 2017).

6.10.4 In Vitro Reconstituted Systems to Study Toxins and Infection

In biophysics, to tackle the complexity of living systems, in vitro reconstituted systems are widely used. Whether from synthetic components or reconstituted from purified parts of the living systems, they have been used to understand the structure and function of viruses (Kuznetsov and McPherson, 2011), bacterial membranes, the bacterial outer membrane (Michel et al., 2017), host–cell interactions, and the effect of xenobiotics, like the effect of antimicrobial compounds (Domenech et al., 2009) or the kinetics of some antibiotics (Hammond et al., 2021). This section is not intended to be exhaustive but rather to survey some of the issues where the combination of a reconstituted system together with the use of AFM has been a key to advance in the understanding of the biophysics of infections.

6.10.4.1 Dynamics of Pore-Forming Toxins

Pore-forming toxins (PFTs) are soluble proteins that can oligomerize on the cell membrane and induce cell death by membrane insertion. PFTs can be secreted by bacteria but also by fungi, plants, and animals, constituting a defense mechanism. These toxins bind to the target membrane forming pores or channels through lytic capabilities that eventually will lead to the cell death. The structure of the membrane-spanning region defines the type of PFT: α-PFT (α-helical) or β-PFT (β-barrel) (Iacovache, Bischofberger, & van der, Goot 2010). β-PFTs oligomerize on the membrane surface before membrane insertion. These prepores are needed for the formation of the β-barrel, although the oligomerization does not have to be complete to form the transmembrane lytic pores. Arc-shaped oligomeric intermediates can also form functional pores (Hodel et al., 2016).

The first toxic proteins visualized by AFM were cholera toxin (Mou et al., 1995) and pertussis toxin (1994), produced by different types of bacteria. They exist as oligomers in solution, in the form of nonlytic prepores, that show tropism to specific components of the target membrane and hence is adsorbed. For instance, cholera toxin B-oligomers (CTX-B) are adsorbed on lipid bilayers containing monosialotetrahexosyl ganglioside (GM1), which is the receptor for CTX-B (Mou et al., 1995). Another bacterial PFT, α-hemolysin (αHL), exists as a water-soluble monomer that can self-assemble into well-defined structures upon oligomerization on a lipid membrane. One of the most studied models for this latter mechanisms has been α-hemolysin (αHL) (Cheley et al., 1997). For a review on assemblies of PFTs visualized by atomic force microscopy, see Yilmaz and Kobayashi (2016).

AFM probed to be useful to resolve the oligomeric structures of PFTs and the consequent induced membrane reorganization. However, the dynamics of the process remained unexplored until the AFM time resolution limit was pushed by the development of high-speed (HS)-AFM (Kodera et al., 2010). Improving the acquisition rate, molecular movies capturing the motion of membrane proteins were possible, and in 2013 HS-AFM was used for the first time to study the subsecond dynamics of PFT oligomers on a lipid membrane (Yilmaz et al., 2013). In this work, Yilmaz and coworkers studied lysenin that oligomerizes on the membrane forming hexagonal close-packed (hcp) structures. Hcp assembly was driven by reorganization of lysenin oligomers such as association/dissociation and rapid diffusion along the sphingomyelin/cholesterol containing membrane. Once the entire membrane was covered by the lysenin lattice, lysenin molecules were any longer diffusive neither dissociated. Later on, Yilmaz and Kobayashi (2015) studied the effect of lysenin binding on the dynamics of a phase separated membrane formed by liquid-ordered (L_o) cholesterol containing domains and a phosphatidylcholine-rich liquid-disordered (L_d) phase. Lysenin oligomerized on the sphingomyelin (SM)-rich L_o, keeping stable the phase boundary, which suggests that lysenin did not affect the line tension between L_o and L_d phases. After the full coverage of the SM-rich domain by oligomers, their hcp assembly gradually expanded into the L_d phase and eventually covered the entire membrane. This suggested that pore formation itself induced the exclusion of SM and cholesterol from the SM-rich domain, which was followed by further binding and oligomerization of lysenin. Other studies on PFTs by means of HS-AFM followed (Yilmaz and Kobayashi, 2016, Morante et al., 2016). The nonlytic oligomeric intermediate known as prepore plays indeed an essential role in the mechanism of insertion of the class of β-PFTs. However, in the class of α-PFTs, like the actinoporins produced by sea anemones, evidence of membrane-bound prepores is sometimes lacking. In Morante et al., they employed HS-AFM imaging together with single-particle cryo-electron microscopy (cryo-EM) to identify, for the first time, a prepore species of the actinoporin fragaceatoxin C bound to lipid vesicles (Morante et al., 2016). The size of the prepore coincided with that of the functional pore, except for the transmembrane region, which is absent in the prepore. Biochemical assays indicated that, in the prepore species, the N terminus is not inserted in the bilayer but is exposed to the aqueous solution. Our study reveals the structure of the prepore in actinoporins and highlights the role of structural intermediates for the formation of cytolytic pores by an α-PFT.

Munguira et al. (2016) used the PFT lysenin as a model to study anomalous diffusion of crowded membranes. This approach is unique and interesting because anomalous diffusion cannot be related with single-particle tracking approaches to local molecular details due to the lack of direct and unlabeled single-molecule observation capabilities. HS-AFM imaging was then used to analyze the lysenin prepore and pores on supported lipid membranes in highly crowded environment. They were able to describe the variance and kurtosis of the images as a function of the lag time and the area, ultimately describing the diffusion. They showed the formation of local glassy phases, where proteins are trapped in neighbor-formed cages for time scales up to 10 s, which had not been previously experimentally reported for biological membranes. Furthermore, around solid-like patches and immobile molecules a slower glasslike phase is detected, which traps the protein in a decreased diffusion area.

Finally, Ruan et al. (2016) used HS-AFM to investigate listeriolysin-O (LLO) activity. LLO is a virulence factor secreted during infection by *Listeria monocytogenes*. It enables the bacteria to escape from the phagocytic vacuole, which is necessary to the dissemination to other cells and tissues. HS-AFM was particularly used to analyze the assembly kinetics of the prepore-to-pore transition dynamics and the membrane disruption in real time. LLO toxin efficiency and disrupting mechanism depends on the membrane cholesterol concentration and the environmental pH. Moreover, LLO can form arc pores as well as damage lipid membranes as a lineactant, and this leads to large-scale membrane defects, as shown in Figure 6.10.4. Overall, the work of Ruan et al. provides a mechanistic basis of how large-scale membrane disruption leads to release of Listeria from the phagocytic vacuole in the cellular context.

6.10.4.2 Antimicrobial Peptides on Lipid Membranes

Quantitative characterization of membrane defects (pores) is important for elucidating the molecular basis of many membrane-active peptides. AFM observations on supported lipid bilayers are often complemented with fluorescence spectroscopy measurements to study the antimicrobial compound-induced time-dependent calcein leakage, as by Pan and Khadka (2016) where they studied melittin-induced defects or by Domenech et al. (2009) where they studied the interaction of oritavancin, semisynthetic antibiotic, with supported lipid bilayers mimicking gram-positive organisms.

Regarding AMPs, De Santis et al. (2017) developed an AMP capsid of de novo design. They introduced a de novo peptide topology that self-assembles into discrete antimicrobial capsids, which mimics the viral architecture. By means of AFM high-resolution and fast imaging, they could show that these artificial capsids assemble as 20-nm hollow shells that attack bacterial membranes and upon landing on phospholipid bilayers instantaneously (seconds) convert into rapidly expanding pores causing membrane lysis (minutes). This is shown in Figure 6.10.5. The designed capsids show several antimicrobial activities that will eventually lead to bacterial death.

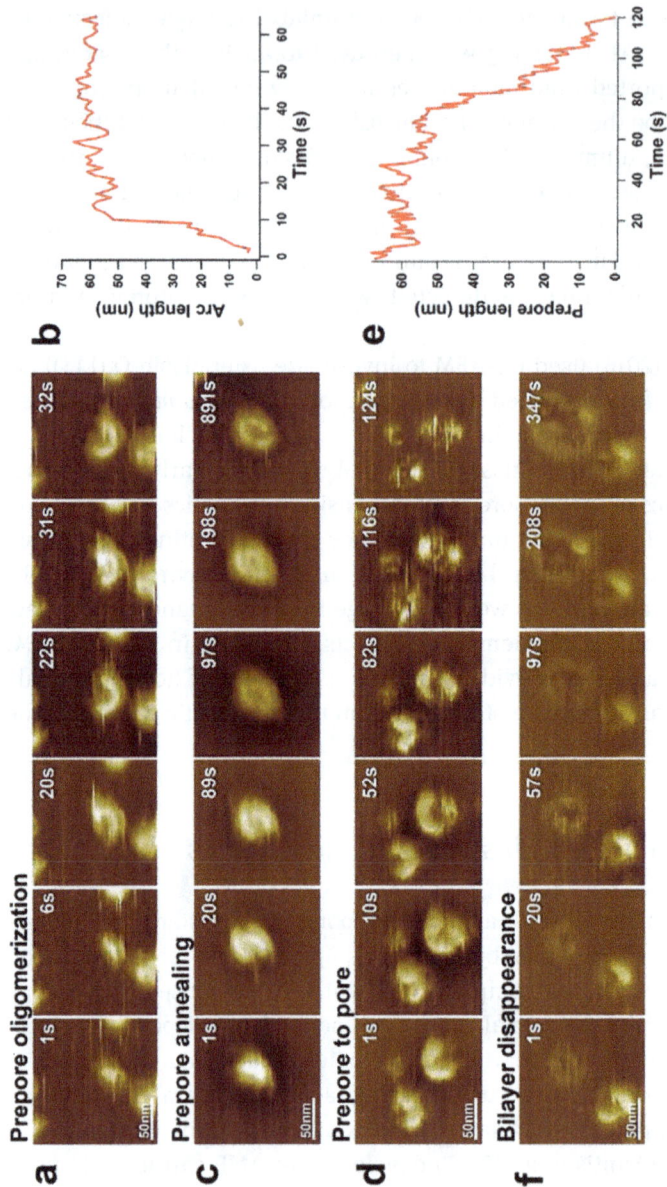

Figure 6.10.4: Dynamic characterization of LLO assembly and action. Buffer condition: 20 mM MES, pH5.6, 100 mM NaCl, 1 mM EDTA, LLO final concentration 500 nM. (a) Prepore oligomerization process. (b) The oligomerization rate is about 5 nm/s (about 2 subunits/s) until completion of an arc after about 10 s, oligomerization then stalls. (c) Annealing process of neighboring arcs: arc-shaped oligomers interlock to form a final ellipsoid LLO assembly that does not further evolve. (d) Membrane insertion process, that is, prepore-to-pore transition. (e) The prepore is stable on the bilayer for 60 s and then inserts rapidly and entirely with about 2 nm/s (about 1 subunit/s). (f) Following oligomerization and prepore-to-pore transition, bilayer disruption occurs at about 600 nm²/s. Reproduced from Ruan et al., (2016) (copyright from the Ruan et al., article distributed under the terms of the Creative Commons Attribution License).

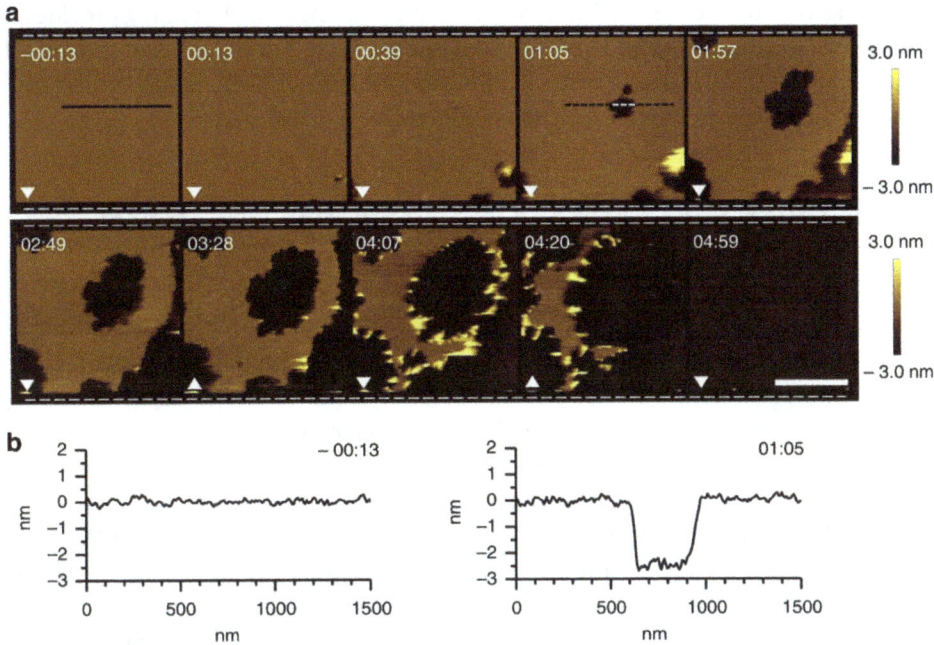

Figure 6.10.5: C 3-capsids disrupting phospholipid bilayers. (a) In-water AFM imaging of supported lipid bilayers treated with C 3-capsids (3 μM total peptide). Topography (height) images, captured at 13 s per frame, are shown. The time stamp corresponds to the middle line of each AFM scan, referenced to the time (00:00) of capsid injection. White arrowheads indicate the AFM scan direction. The scale bar is 1 μm. (b) Cross-sections along the lines marked in (a, – 00:13 and 01:57) before (left) and after (right) the addition of capsids. Reproduced from (De Santis et al., 2017), licensed under CC BY 4.0.

New approaches are continuously emerging in biophysics. A recent work from Kikuchi et al. (2020) focused on the study of bacterial extracellular vesicles (EVs). EVs are nanometer-sized lipid bilayer-based colloids naturally released from cells. In the case of bacteria, they are implicated in the transport of virulence factors, lateral gene transfer, interception of bacteriophages, antibiotic and eukaryotic defense factors, cell detoxification, and bacterial cell-to-cell communication (Schwechheimer and Kuehn, 2015, Kaparakis-Liaskos and Ferrero, 2015). There is an increasing interest in use EVs as biotechnological platforms, for example, in drug delivery or vaccines design (Gnopo et al., 2017). In this case, Kikuchi and coworkers used HS-AFM to analyze individual bacterial EVs. HS-AFM is typically operated in acoustic-modulation (tapping) mode, the authors used the phase image channel in this case, that correlates to different intrinsic viscoelastic properties of the material, to discern different EVs isolated from bacterial cultures. Interestingly, they demonstrated that a single bacterial species generates physically heterogeneous types of MVs. AFM phase imaging showed in this case a quantitative analysis of physical properties of individual EVs. The development of this new methodology provides a first step toward a better understanding of the biophysical properties of individual vesicles in their native state, which will lead to a better understanding of bacterial membrane vesicles biological functions (Cohen-Khait, 2020).

6.10.4.3 High-Speed AFM and Its Combination with Optical Microscopy

The combination of AFM and optical microscopies is a strong tool that can address long-standing questions about pathogens entry mechanisms, for example, as discussed earlier in Section 6.10.3, for revealing the membrane topography including viral budding sites (Dahmane et al., 2019). HS-AFM imaging allows one step further to decipher the molecular dynamics of pathogens or pathogenic compounds. Recently, Lim et al. used HS-AFM to decipher the dynamics of influenza A hemagglutinin, probing its interaction with exosomes (Lim et al., 2020). They showed conformational changes of the protein as a function of the acidity of the aqueous medium, and discussed it in the context of a possible mechanism of insertion of a fusion peptide into the exosomal layer and subsequently destabilizing it upon rupture of the exosome, releasing its content. The combination with optical microscopy would allow the great advantage of keeping the target localized during pathogen internalization while performing HS-AFM dynamic imaging.

However, there have not yet been any studies to our knowledge where HS-AFM coupled to optical microscopy is used to tackle the pathogen entry. Fukuda et al. (2013) developed the first tip-scan HS-AFM with total internal reflection fluorescence (TIRFM) microscopy. For the combination of HS-AFM with optical microscopy, the configuration of the AFM should be a tip-scanning setup rather than a sample scanning, to allow a configuration where the stand-alone HS-AFM can be mounted on an inverted optical microscope. They demonstrated the capability of their combined system by imaging with a HS-AFM wide-range scanner and TIRFM imaging chitinase A moving on a chitin crystalline fiber and myosin V walking on an actin filament. Later on, Umakoshi et al. (2020) achieved simultaneous, correlative HS-AFM/SNOM imaging in a setup where they accomplished a spatial and temporal resolution of 39 nm and 3 s per frame, respectively, for SNOM imaging.

The combination of HS-AFM with optics is still in its infancy but promising in the field of infection. Interestingly, Shibata et al. (2015) successfully imaged by means of HS-AFM the dynamics of mammalian living cells. Imaging high-aspect ratio objects such as living eukaryotic cells with HS-AFM is technically challenging. The authors overcame this difficulty by using customed microfabricated AFM cantilevers and attaching exceptionally long (~3 μm) and thin (~5 nm) tips of amorphous carbon to the cantilever. This allowed them to image for the first time the surface structure of live cells with the spatiotemporal resolution of nanometers and seconds, opening the avenue to study infected cells by the same means.

6.10.5 Conclusions

In the fields of microbiology and infection, AFM has proven to be particularly useful for several purposes, from structural and mechanical characteristics of bacterial cells, their adhesion properties to both surfaces and their hosts or the exploration of the effect or mechanism of action of antimicrobials or other therapeutic agents, and even following not only the first steps of infection but also deciphering later infection processes once the pathogens are inside the host cell. Both in cells and in vitro reconstituted systems, the contribution of AFM is unique in terms of its imaging capabilities, force sensitivity, and nanomanipulation, all in near-physiological conditions. In addition, at the single-molecule level, AFM also provides a tool to study the kinetics and thermodynamics of ligand-receptor interactions between pathogen and host factors. It is foreseeable that the combination of HS-AFM with advanced optical microscopy discussed in the previous section will be of great relevance for the study of pathogen–host cell interactions. In addition, one of the major avenues of development that will benefit this study of host–pathogen interactions is the improvement of acquisition rate during nanomechanical characterization. As discussed, herein nanomechanical mapping allows the assessment of the mechanical properties of heterogeneous surfaces such as the bacterial cell wall at high-spatial resolution. If the development of fast or high-speed mapping sees the light, it will serve to open a new avenue in the correlative studies between structure, dynamics, and mechanics in general and in particular to better understand thermodynamics and biophysics of infection processes.

References

Aikawa, M., K. Kamanura, S. Shiraishi, Y. Matsumoto, H. Arwati, M. Torii, Y. Ito, T. Takeuchi and B. Tandler (1996). "Membrane knobs of unfixed Plasmodium falciparum infected erythrocytes: New findings as revealed by atomic force microscopy and surface potential spectroscopy." Experimental Parasitology. Research Institute of Medical Sciences, Tokai University, Isehara, Japan **84**(3): 339–343. 10.1006/expr.1996.0122.

Alsteens, D., V. Dupres, S. A. Klotz, N. K. Gaur, P. N. Lipke and Y. F. Dufrêne (2009). "Unfolding individual Als5p adhesion proteins on live cells." ACS Nano **3**: 1677.

Alsteens, D., V. Dupres, K. M. Evoy, L. Wildling, H. J. Gruber and Y. F. Dufrêne (2008a). "Structure, cell wall elasticity and polysaccharide properties of living yeast cells, as probed by AFM." Nanotechnology. IOP Publishing **19**(38): 384005. 10.1088/0957-4484/19/38/384005.

Alsteens, D., M. C. Garcia, P. N. Lipke and Y. F. Dufrêne (2010). "Force-induced formation and propagation of adhesion nanodomains in living fungal cells." Proceedings of the National Academy of Sciences. National Academy of Sciences 107(48): 20744–20749. 10.1073/pnas.1013893107.

Alsteens, D., C. Verbelen, E. Dague, D. Raze, A. R. Baulard and Y. F. Dufrêne (2008b). "Organization of the mycobacterial cell wall: A nanoscale view." Pflügers Archiv – European Journal of Physiology 456(1): 117–125. 10.1007/s00424-007-0386-0.

Amarouch, M. Y., J. E. Hilaly and D. Mazouzi (2018). "AFM and FluidFM technologies: Recent applications in molecular and cellular biology." Scanning 2018. 10.1155/2018/7801274.

Amro, N. A., L. P. Kotra, K. Wadu-Mesthrige, A. Bulychev, S. Mobashery and G.-Y. Liu (2000). "High-resolution atomic force microscopy studies of the Escherichia coli outer membrane: Structural basis for permeability." Langmuir. American Chemical Society 16(6): 2789–2796. 10.1021/la991013x.

Andre, G., M. Deghorain, P. A. Bron, I. I. van Swam, M. Kleerebezem, P. Hols and Y. F. Dufrêne (2011). "Fluorescence and atomic force microscopy imaging of wall teichoic acids in lactobacillus plantarum." ACS Chemical Biology. American Chemical Society 6(4): 366–376. 10.1021/cb1003509.

Andre, G., S. Kulakauskas, M.-P. Chapot-Chartier, B. Navet, M. Deghorain, E. Bernard, P. Hols and Y. F. Dufrêne (2010). "Imaging the nanoscale organization of peptidoglycan in living Lactococcus lactis cells." Nature Communications 1(3): 1–8. 10.1038/ncomms1027.

Beaussart, A., M. Abellán-Flos, S. El-Kirat-Chatel, S. P. Vincent and Y. F. Dufrêne (2016). "Force nanoscopy as a versatile platform for quantifying the activity of antiadhesion compounds targeting bacterial pathogens." Nano Letters. American Chemical Society 16(2): 1299–1307. 10.1021/acs.nanolett.5b04689.

Beaussart, A., A. E. Baker, S. L. Kuchma, S. El-Kirat-Chatel, G. A. O'Toole and Y. F. Dufrêne (2014a). "Nanoscale adhesion forces of pseudomonas aeruginosa type IV Pili." ACS Nano 8(10): 10723–10733. 10.1021/nn5044383.

Beaussart, A., S. El-Kirat-Chatel, R. M. A. Sullan, D. Alsteens, P. Herman, S. Derclaye and Y. F. Dufrêne (2014b). "Quantifying the forces guiding microbial cell adhesion using single-cell force spectroscopy." Nature Protocols. Nature Publishing Group 9(5): 1049–1055. 10.1038/nprot.2014.066.

Beckmann, M. A., S. Venkataraman, M. J. Doktycz, J. P. Nataro, C. J. Sullivan, J. L. Morrell-Falvey and D. P. Allison (2006). "Measuring cell surface elasticity on enteroaggregative Escherichia coli wild type and dispersin mutant by AFM." Ultramicroscopy. Proceedings of the Seventh International Conference on Scanning Probe Microscopy, Sensors and Nanostructures 106(8): 695–702. 10.1016/j.ultramic.2006.02.006.

Benoit, M., D. Gabriel, G. Gerisch and H. E. Gaub (2000). "Discrete interactions in cell adhesion measured by single-molecule force spectroscopy." Nature Cell Biology 2(6): 313–317. 10.1038/35014000.

Bernard, S. C., N. Simpson, O. Join-Lambert, C. Federici, M.-P. Laran-Chich, N. Maïssa, H. Bouzinba-Ségard, et al. (2014). "Pathogenic Neisseria meningitidis utilizes CD147 for vascular colonization." Nature Medicine 20(7): 725–731. 10.1038/nm.3563.

Bolshakova, A. V., O. I. Kiselyova, A. S. Filonov, O. Yu Frolova, Y. L. Lyubchenko and I. V. Yaminsky (2001). "Comparative studies of bacteria with an atomic force microscopy operating in different modes." Ultramicroscopy. Proceedings of the Second International Conference on Scanning 86(1): 121–128. 10.1016/S0304-3991(00)00075-9.

Braga, P. C. and D. Ricci (1998). "Atomic force microscopy: Application to investigation of escherichia coli morphology before and after exposure to cefodizime." Antimicrobial Agents and Chemotherapy 42(1): 18–22.

Butt, H.-J., E. K. Wolff, S. A. C. Gould, B. Dixon Northern, C. M. Peterson and P. K. Hansma (1990). "Imaging cells with the atomic force microscope." Journal of Structural Biology **105**(1–3): 54–61. 10.1016/1047-8477(90)90098-W.

Camesano, T. A. and B. E. Logan (2000). "Probing bacterial electrosteric interactions using atomic force microscopy." Environmental Science & Technology. American Chemical Society **34**(16): 3354–3362. 10.1021/es9913176.

Cerf, A., J.-C. Cau, C. Vieu and E. Dague (2009). "Nanomechanical properties of dead or alive single-patterned bacteria." Langmuir. American Chemical Society **25**(10): 5731–5736. 10.1021/la9004642.

Cheley, S., M. S. Malghani, L. Z. Song, M. Hobaugh, J. E. Gouaux, J. Yang and H. Bayley (1997). "Spontaneous oligomerization of a staphylococcal alpha-hemolysin conformationally constrained by removal of residues that form the transmembrane beta-barrel." Protein Engineering **10**(12): 1433–1443. 10.1093/protein/10.12.1433.

Ciczora, Y., S. Janel, M. Soyer, M. Popoff, E. Werkmeister and F. Lafont (2019). "Blocking bacterial entry at the adhesion step reveals dynamic recruitment of membrane and cytosolic probes." Biology of the Cell **111**(3): 67–77. 10.1111/boc.201800070.

Cohen-Khait, R. (2020). "Imaging bacterial membrane vesicles with a delicate touch." Nature Reviews Microbiology. Nature Publishing Group **1–1**. 10.1038/s41579-020-00492-6.

Colville, K., N. Tompkins, A. D. Rutenberg and M. H. Jericho (2010). "Effects of Poly(l-lysine) substrates on attached escherichia coli bacteria." Langmuir. American Chemical Society **26**(4): 2639–2644. 10.1021/la902826n.

Coureuil, M., H. Lécuyer, M. G. H. Scott, C. Boularan, H. Enslen, M. Soyer, G. Mikaty, S. Bourdoulous, X. Nassif and S. Marullo (2010). "Meningococcus Hijacks a β2-adrenoceptor/β-Arrestin pathway to cross brain microvasculature endothelium." Cell **143**(7): 1149–1160. 10.1016/j.cell.2010.10.035.

Dague, E., D. Alsteens, J.-P. Latgé and Y. F. Dufrêne (2008). "High-resolution cell surface dynamics of germinating aspergillus fumigatus conidia." Biophysical Journal **94**(2): 656–660. 10.1529/biophysj.107.116491.

Dahmane, S., C. Doucet, A. Le Gall, C. Chamontin, P. Dosset, F. Murcy, L. Fernandez, et al. (2019). "Nanoscale organization of tetraspanins during HIV-1 budding by correlative dSTORM/AFM." Nanoscale **11**(13): 6036–6044. 10/ghmrrc.

Dammer, U., M. Hegner, D. Anselmetti, P. Wagner, M. Dreier, W. Huber and H. J. Güntherodt (1996). "Specific antigen/antibody interactions measured by force microscopy." Biophysical Journal **70**(5): 2437–2441.

Dammer, U., O. Popescu, P. Wagner, D. Anselmetti, H. J. Guntherodt and G. N. Misevic (1995). "Binding strength between cell adhesion proteoglycans measured by atomic force microscopy." Science. American Association for the Advancement of Science **267**(5201): 1173–1175. 10.1126/science.7855599.

Emiliana, D. S., H. Alkassem, B. Lamarre, N. Faruqui, A. Bella, J. E. Noble, N. Micale, et al. (2017). "Antimicrobial peptide capsids of de novo design." Nature Communications. Nature Publishing Group **8**(1): 2263. 10.1038/s41467-017-02475-3.

Doktycz, M. J., C. J. Sullivan, P. R. Hoyt, D. A. Pelletier, S. Wu and D. P. Allison (2003). "AFM imaging of bacteria in liquid media immobilized on gelatin coated mica surfaces." Ultramicroscopy. Proceedings of the Fourth International Conference on Scanning Probe Microscopy, Sensors and Nanostructures **97**(1): 209–216. 10.1016/S0304-3991(03)00045-7.

Domenech, O., G. Francius, P. M. Tulkens, F. Van Bambeke, Y. Dufrene and M.-P. Mingeot-Leclercq (2009). "Interactions of oritavancin, a new lipoglycopeptide derived from vancomycin, with phospholipid bilayers: Effect on membrane permeability and nanoscale lipid membrane organization." Biochimica Et Biophysica Acta-Biomembranes **1788**(9): 1832–1840. 10.1016/j.bbamem.2009.05.003.

Dörig, P., P. Stiefel, P. Behr, E. Sarajlic, D. Bijl, M. Gabi, J. Vörös, J. A. Vorholt and T. Zambelli (2010). "Force-controlled spatial manipulation of viable mammalian cells and micro-organisms by means of FluidFM technology." Applied Physics Letters. American Institute of Physics **97**(2): 023701. 10.1063/1.3462979.

D'Souza, S. F. (2001). "Immobilization and stabilization of biomaterials for biosensor applications." Applied Biochemistry and Biotechnology **96**(1): 225–238. 10.1385/ABAB:96:1-3: 225.

Dufrêne, Y. F., T. G. Marchal and P. G. Rouxhet (1999). "Influence of substratum surface properties on the organization of adsorbed collagen films: In situ characterization by atomic force microscopy." Langmuir. American Chemical Society **15**(8): 2871–2878. 10.1021/la981066z.

Dufrêne, Y. F., D. Martínez-Martín, I. Medalsy, D. Alsteens and D. J. Müller (2013). "Multiparametric imaging of biological systems by force-distance curve–based AFM." Nature Methods. Nature Publishing Group **10**(9): 847–854. 10.1038/nmeth.2602.

Dujardin, A., P. De Wolf, F. Lafont and V. Dupres (2019). "Automated multi-sample acquisition and analysis using atomic force microscopy for biomedical applications." PLOS ONE. Public Library of Science **14**(3): e0213853. 10.1371/journal.pone.0213853.

Dupres, V., D. Alsteens, G. Andre, C. Verbelen and Y. F. Dufrêne (2009a). "Fishing single molecules on live cells." Nano Today **4**(3): 262–268. 10.1016/j.nantod.2009.04.011.

Dupres, V., D. Alsteens, K. Pauwels and Y. F. Dufrêne (2009b). "In vivo imaging of S-layer nanoarrays on corynebacterium glutamicum." Langmuir. American Chemical Society **25**(17): 9653–9655. 10.1021/la902238q.

Dupres, V., F. D. Menozzi, C. Locht, B. H. Clare, N. L. Abbott, S. Cuenot, C. Bompard, D. Raze and Y. F. Dufrêne (2005). "Nanoscale mapping and functional analysis of individual adhesins on living bacteria." Nature Methods. Nature Publishing Group **2**(7): 515–520. 10.1038/nmeth769.

Ebner, A., P. Hinterdorfer and H. J. Gruber (2007a). "Comparison of different aminofunctionalization strategies for attachment of single antibodies to AFM cantilevers." Ultramicroscopy **107**(10–11): 922–927. 10.1016/j.ultramic.2007.02.035.

Ebner, A., L. Wildling, A. S. M. Kamruzzahan, C. Rankl, J. Wruss, C. D. Hahn, M. Hölzl, et al. (2007b). "A new, simple method for linking of antibodies to atomic force microscopy tips." Bioconjugate Chemistry. American Chemical Society **18**(4): 1176–1184. 10.1021/bc070030s.

Engel, A., A. Massalski, H. Schindler, D. L. Dorset and J. P. Rosenbusch (1985). "Porin channel triplets merge into single outlets in escherichia coli outer membranes." Nature **317**(6038): 643–645. 10.1038/317643a0.

Fantner, G. E., R. J. Barbero, D. S. Gray and A. M. Belcher (2010). "Kinetics of antimicrobial peptide activity measured on individual bacterial cells using high-speed atomic force microscopy." Nature Nanotechnology. Nature Publishing Group **5**(4): 280–285. 10.1038/nnano.2010.29.

Feuillie, C., P. Vitry, M. A. McAleer, S. Kezic, A. D. Irvine, J. A. Geoghegan and Y. F. Dufrêne (2018). "Adhesion of Staphylococcus aureus to corneocytes from atopic dermatitis patients is controlled by natural moisturizing factor levels." mBio. American Society for Microbiology **9**(4). 10.1128/mBio.01184-18.

Formosa, C., F. Pillet, M. Schiavone, R. E. Duval, L. Ressier and E. Dague (2015). "Generation of living cell arrays for atomic force microscopy studies." Nature Protocols **10**(1): 199–204. 10.1038/nprot.2015.004.

Formosa-Dague, C., Z.-H. Fu, C. Feuillie, S. Derclaye, T. J. Foster, J. A. Geoghegan and Y. F. Dufrêne (2016). "Forces between Staphylococcus aureus and human skin." Nanoscale Horizons. The Royal Society of Chemistry **1**(4): 298–303. 10.1039/C6NH00057F.

Francius, G., O. Domenech, M. P. Mingeot-Leclercq and Y. F. Dufrêne (2008). "Direct observation of Staphylococcus aureus cell wall digestion by lysostaphin." Journal of Bacteriology **190**(24): 7904–7909. 10.1128/JB.01116-08.

Fukuda, S., T. Uchihashi, R. Iino, Y. Okazaki, K. Yoshida, K. Igarashi and T. Ando (2013). "High-speed atomic force microscope combined with single-molecule fluorescence microscope." Review of Scientific Instruments **84**(7): 8–8. 10.1063/1.4813280.

Gaboriaud, F., B. S. Parcha, M. L. Gee, J. A. Holden and R. A. Strugnell (2008). "Spatially resolved force spectroscopy of bacterial surfaces using force-volume imaging." Colloids and Surfaces B: Biointerfaces **62**(2): 206–213. 10.1016/j.colsurfb.2007.10.004.

Gilbert, Y., M. Deghorain, L. Wang, B. Xu, P. D. Pollheimer, H. J. Gruber, J. Errington, et al. (2007). "Single-molecule force spectroscopy and imaging of the vancomycin/d-Ala-d-Ala interaction." Nano Letters. American Chemical Society **7**(3): 796–801. 10.1021/nl0700853.

Gladnikoff, M. and I. Rousso (2008). "Directly monitoring individual retrovirus budding events using atomic force microscopy." Biophysical Journal **94**(1): 320–326. 10/fkhc92.

Gladnikoff, M., E. Shimoni, N. S. Gov and I. Rousso (2009). "Retroviral assembly and budding occur through an actin-driven mechanism." Biophysical Journal **97**(9): 2419–2428. 10/ctxtcc.

Gnopo, Y. M. D., H. C. Watkins, T. C. Stevenson, M. P. DeLisa and D. Putnam (2017). "Designer outer membrane vesicles as immunomodulatory systems – Reprogramming bacteria for vaccine delivery." Advanced Drug Delivery Reviews **114**: 132–142. 10.1016/j.addr.2017.05.003.

Gonzalez-Rodriguez, D., M. P. Maddugoda, C. Stefani, S. Janel, F. Lafont, D. Cuvelier, E. Lemichez and F. Brochard-Wyart (2012). "Cellular dewetting: Opening of macroapertures in endothelial cells." Physical Review Letters **108**(21): 218105. 10.1103/PhysRevLett.108.218105.

Greif, D., D. Wesner, J. Regtmeier and D. Anselmetti (2010). "High resolution imaging of surface patterns of single bacterial cells." Ultramicroscopy **110**(10): 1290–1296. 10.1016/j.ultramic.2010.06.004.

Guillaume-Gentil, O., E. Potthoff, D. Ossola, C. M. Franz, T. Zambelli and J. A. Vorholt (2014). "Force-controlled manipulation of single cells: From AFM to FluidFM." Trends Biotechnol **32**: 381.

Günther, T. J., M. Suhr, J. Raff and K. Pollmann (2014). "Immobilization of microorganisms for AFM studies in liquids." RSC Advances. The Royal Society of Chemistry **4**(93): 51156–51164. 10.1039/C4RA03874F.

Häberle, W., J. K. H. Hörber, F. Ohnesorge, D. P. E. Smith and G. Binnig (1992). "In situ investigations of single living cells infected by viruses." Ultramicroscopy **42–44**: 1161–1167. 10/bvn8bg.

Hammond, K., M. G. Ryadnov and B. W. Hoogenboom (2021). "Atomic force microscopy to elucidate how peptides disrupt membranes." Biochimica et Biophysica Acta (BBA) – Biomembranes **1863**(1): 183447. 10.1016/j.bbamem.2020.183447.

Hasim, S., D. P. Allison, B. Mendez, A. T. Farmer, D. A. Pelletier, S. T. Retterer, S. R. Campagna, T. B. Reynolds and M. J. Doktycz (2018). "Elucidating Duramycin's bacterial selectivity and mode of action on the bacterial cell envelope." Frontiers in Microbiology. Frontiers **9**. 10.3389/fmicb.2018.00219.

Helenius, J., C.-P. Heisenberg, H. E. Gaub and D. J. Muller (2008). "Single-cell force spectroscopy." Journal of Cell Science. The Company of Biologists Ltd **121**(11): 1785–1791. 10.1242/jcs.030999.

Herman, P., S. El-Kirat-Chatel, A. Beaussart, J. A. Geoghegan, T. J. Foster and Y. F. Dufrêne (2014). "The binding force of the staphylococcal adhesin SdrG is remarkably strong." Molecular Microbiology **93**(2): 356–368. https://doi.org/10.1111/mmi.12663.

Hinterdorfer, P., W. Baumgartner, H. J. Gruber, K. Schilcher and H. Schindler (1996). "Detection and localization of individual antibody-antigen recognition events by atomic force microscopy."

Proceedings of the National Academy of Sciences. National Academy of Sciences **93**(8): 3477–3481. 10.1073/pnas.93.8.3477.

Hinterdorfer, P. and Y. F. Dufrêne (2006). "Detection and localization of single molecular recognition events using atomic force microscopy." Nature Methods. Nature Publishing Group **3**(5): 347–355. 10.1038/nmeth871.

Hodel, A. W., C. Leung, N. V. Dudkina, H. R. Saibil and B. W. Hoogenboom (2016). "Atomic force microscopy of membrane pore formation by cholesterol dependent cytolysins." Current Opinion in Structural Biology ((Engineering and Design • Membranes)) **39**: 8–15. 10.1016/j.sbi.2016.03.005.

Hofherr, L., C. Müller-Renno and C. Ziegler (2020). "FluidFM as a tool to study adhesion forces of bacteria – Optimization of parameters and comparison to conventional bacterial probe scanning force spectroscopy." (Ed.) Kerstin G. Blank PLOS ONE **15**(7): e0227395. 10.1371/journal.pone.0227395.

Iacovache, I., M. Bischofberger and F. Gisou van der Goot (2010). "Structure and assembly of pore-forming proteins." Current Opinion in Structural Biology . (Theory and Simulation / Macromolecular Assemblages) **20**(2): 241–246. 10.1016/j.sbi.2010.01.013.

Janel, S., E. Werkmeister, A. Bongiovanni, F. Lafont and N. Barois (2017). "CLAFEM: Correlative light atomic force electron microscopy." Methods Cell Biol **140**: 165–185. 10.1016/bs.mcb.2017.03.010.

Janel, S., M. Popoff, N. Barois, E. Werkmeister, S. Divoux, F. Perez and F. Lafont (2019). "Stiffness tomography of eukaryotic intracellular compartments by atomic force microscopy." Nanoscale **11**(21): 10320–10328. 10/ghmrpk.

Kailas, L., E. C. Ratcliffe, E. J. Hayhurst, M. G. Walker, S. J. Foster and J. K. Hobbs (2009). "Immobilizing live bacteria for AFM imaging of cellular processes." Ultramicroscopy **109**(7): 775–780. 10.1016/j.ultramic.2009.01.012.

Kang, S. and M. Elimelech (2009). "Bioinspired single bacterial cell force spectroscopy." Langmuir **25**(17): 9656–9659. 10.1021/la902247w.

Kaparakis-Liaskos, M. and R. L. Ferrero (2015). "Immune modulation by bacterial outer membrane vesicles." Nature Reviews. Immunology **15**(6): 375–387. 10.1038/nri3837.

Karrasch, S., R. Hegerl, J. H. Hoh, W. Baumeister and A. Engel (1994). "Atomic force microscopy produces faithful high-resolution images of protein surfaces in an aqueous environment." Proceedings of the National Academy of Sciences. National Academy of Sciences **91**(3): 836–838. 10.1073/pnas.91.3.836.

Kasas, S., B. Fellay and R. Cargnello (1994). "Observation of the action of penicillin on bacillus subtilis using atomic force microscopy: Technique for the preparation of bacteria." Surface and Interface Analysis **21**(6–7): 400–401. https://doi.org/10.1002/sia.740210613.

Kasas, S. and A. Ikai (1995). "A method for anchoring round shaped cells for atomic force microscope imaging." Biophysical Journal **68**(5): 1678–1680.

Kasas, S., P. Stupar and G. Dietler (2018). "AFM contribution to unveil pro- and eukaryotic cell mechanical properties." Seminars in Cell & Developmental Biology **73**: 177–187. 10.1016/j.semcdb.2017.08.032.

Kikuchi, Y., N. Obana, M. Toyofuku, N. Kodera, T. Soma, T. Ando, Y. Fukumori, N. Nomura and A. Taoka (2020). "Diversity of physical properties of bacterial extracellular membrane vesicles revealed through atomic force microscopy phase imaging." Nanoscale. The Royal Society of Chemistry **12**(14): 7950–7959. 10.1039/C9NR10850E.

Kodera, N., D. Yamamoto, R. Ishikawa and T. Ando (2010). "Video imaging of walking myosin V by high-speed atomic force microscopy." Nature **468**: 7320. 10.1038/nature09450.

Krieg, M., G. Fläschner, D. Alsteens, B. M. Gaub, W. H. Roos, G. J. L. Wuite, H. E. Gaub, C. Gerber, Y. F. Dufrêne and D. J. Müller (2019). "Atomic force microscopy-based mechanobiology." Nature Reviews Physics **1**(1): 41. 10.1038/s42254-018-0001-7.

Kuznetsov, Y. G., A. Low, H. Fan and A. McPherson (2004). "Atomic force microscopy investigation of wild-type Moloney murine leukemia virus particles and virus particles lacking the envelope protein." Virology **323**(2): 189–196. 10/fm597n.

Kuznetsov, Y. G. and A. McPherson (2011). "Atomic force microscopy in imaging of viruses and virus-infected cells." Microbiology and Molecular Biology Reviews : MMBR **75**(2): 268–285. 10.1128/MMBR.00041-10.

Kwon, S., D.-H. Lee, S.-J. Han, W. Yang, F.-S. Quan and K. S. Kim (2019). "Biomechanical properties of red blood cells infected by Plasmodium berghei ANKA." Journal of Cellular Physiology **234**(11): 20546–20553. 10/ghmz4p.

Lei, J., L. Sun, S. Huang, C. Zhu, P. Li, J. He, V. Mackey, D. H. Coy and Q. He (2019). "The antimicrobial peptides and their potential clinical applications." American Journal of Translational Research. e-Century Publishing Corporation **11**(7): 3919.

Lim, K., N. Kodera, H. Wang, M. S. Mohamed, M. Hazawa, A. Kobayashi, T. Yoshida, et al. (2020). "High-speed AFM reveals molecular dynamics of human influenza a hemagglutinin and its interaction with exosomes." Nano Letters. American Chemical Society **20**(9): 6320–6328. 10.1021/acs.nanolett.0c01755.

Liu, Y., J. Strauss and T. A. Camesano (2008). "Adhesion forces between Staphylococcus epidermidis and surfaces bearing self-assembled monolayers in the presence of model proteins." Biomaterials **29**(33): 4374–4382. 10.1016/j.biomaterials.2008.07.044.

Loskill, P., P. M. Pereira, P. Jung, M. Bischoff, M. Herrmann, M. G. Pinho and K. Jacobs (2014). "Reduction of the peptidoglycan crosslinking causes a decrease in stiffness of the staphylococcus aureus cell envelope." Biophysical Journal **107**(5): 1082–1089. 10.1016/j.bpj.2014.07.029.

Louise Meyer, R., X. Zhou, L. Tang, A. Arpanaei, P. Kingshott and F. Besenbacher (2010). "Immobilisation of living bacteria for AFM imaging under physiological conditions." Ultramicroscopy **110**(11): 1349–1357. 10.1016/j.ultramic.2010.06.010.

Maddugoda, M. P., C. Stefani, D. Gonzalez-Rodriguez, J. Saarikangas, S. Torrino, S. Janel, P. Munro, et al. (2011). "CAMP signaling by Anthrax edema toxin induces transendothelial cell tunnels, which Are resealed by MIM via Arp2/3-Driven actin polymerization." Cell Host and Microbe. Elsevier **10**(5): 464–474. 10.1016/j.chom.2011.09.014.

Maïssa, N., V. Covarelli, S. Janel, B. Durel, N. Simpson, S. C. Bernard, L. Pardo-Lopez, et al. (2017). "Strength of Neisseria meningitidis binding to endothelial cells requires highly-ordered CD147/β 2 -adrenoceptor clusters assembled by alpha-actinin-4." Nature Communications. Nature Publishing Group **8**(1): 15764. 10.1038/ncomms15764.

Mathelié-Guinlet, M., F. Viela, G. Pietrocola, P. Speziale, D. Alsteens and Y. F. Dufrêne (2020). "Force-clamp spectroscopy identifies a catch bond mechanism in a Gram-positive pathogen." Nature Communications. Nature Publishing Group **11**(1): 5431. 10.1038/s41467-020-19216-8.

Matzke, R., K. Jacobson and M. Radmacher (2001). "Direct, high-resolution measurement of furrow stiffening during division of adherent cells." Nature Cell Biology **3**(6): 607–610. 10.1038/35078583.

Meincken, M., D. L. Holroyd and M. Rautenbach (2005). "Atomic force microscopy study of the effect of antimicrobial peptides on the cell envelope of Escherichia coli." Antimicrobial Agents and Chemotherapy **49**(10): 4085–4092. 10.1128/AAC.49.10.4085-4092.2005.

Meister, A., M. Gabi, P. Behr, P. Studer, J. Vörös, P. Niedermann, J. Bitterli, et al. (2009). "FluidFM: Combining atomic force microscopy and nanofluidics in a universal liquid delivery system for single cell applications and beyond." Nano Letters. American Chemical Society **9**(6): 2501–2507. 10.1021/nl901384x.

Merghni, A., K. Bekir, Y. Kadmi, I. Dallel, S. Janel, S. Bovio, N. Barois, F. Lafont and M. Mastouri (2017). "Adhesiveness of opportunistic Staphylococcus aureus to materials used in dental office: In vitro study." Microbial Pathogenesis **103**: 129–134. 10.1016/j.micpath.2016.12.014.

Michel, J. P., Y. X. Wang, I. Kiesel, Y. Gerelli and V. Rosilio (2017). "Disruption of asymmetric lipid bilayer models mimicking the outer membrane of gram-negative bacteria by an active plasticin." Langmuir. American Chemical Society **33**(41): 11028–11039. 10.1021/acs.langmuir.7b02864.

Moloney, M., L. McDonnell and H. O'Shea (2004). "Atomic force microscopy analysis of enveloped and non-enveloped viral entry into, and egress from, cultured cells." Ultramicroscopy **100**(3–4): 163–169. 10/cmpsnh.

Morante, K., A. Bellomio, D. Gil-Carton, L. Redondo-Morata, J. Sot, S. Scheuring, M. Valle, J. M. Gonzalez-Manas, K. Tsumoto and J. M. M. Caaveiro (2016). "Identification of a membrane-bound prepore species clarifies the lytic mechanism of actinoporins." Journal of Biological Chemistry **291**(37): 19210–19219. 10.1074/jbc.M116.734053.

Mortensen, N. P., J. D. Fowlkes, C. J. Sullivan, D. P. Allison, N. B. Larsen, S. Molin and M. J. Doktycz (2009). "Effects of colistin on surface ultrastructure and nanomechanics of Pseudomonas aeruginosa cells." Langmuir: The ACS Journal of Surfaces and Colloids **25**(6): 3728–3733. 10.1021/la803898g.

Mostowy, S., S. Janel, C. Forestier, C. Roduit, S. Kasas, J. Pizarro-Cerda, P. Cossart and F. Lafont (2011). "A role for septins in the interaction between the listeria monocytogenes invasion protein InlB and the met receptor." Biophysical Journal **100**(8): 1949–1959. 10.1016/j.bpj.2011.02.040.

Mou, J., J. Yang and Z. Shao (1995). "Atomic force microscopy of cholera toxin B-oligomers bound to bilayers of biologically relevant lipids." Journal of Molecular Biology **248**(3): 507–512. 10.1006/jmbi.1995.0238.

Müller, D. J., F. A. Schabert, G. Büldt and A. Engel (1995). "Imaging purple membranes in aqueous solutions at sub-nanometer resolution by atomic force microscopy." Biophysical Journal **68**(5): 1681–1686.

Munguira, I., I. Casuso, H. Takahashi, F. Rico, A. Miyagi, M. Chami and S. Scheuring (2016). "Glasslike membrane protein diffusion in a crowded membrane." Acs Nano **10**(2): 2584–2590. 10.1021/acsnano.5b07595.

Nacer, A., E. Roux, S. Pomel, C. Scheidig-Benatar, H. Sakamoto, F. Lafont, A. Scherf and D. Mattei (2011). "Clag9 is not essential for PfEMP1 surface expression in non-cytoadherent Plasmodium falciparum parasites with a chromosome 9 deletion." PloS One **6**(12): e29039. 10/fzm2tp.

Oh, Y. J., G. Sekot, M. Duman, L. Chtcheglova, P. Messner, H. Peterlik, C. Schäffer and P. Hinterdorfer (2013). "Characterizing the S-layer structure and anti-S-layer antibody recognition on intact Tannerella forsythia cells by scanning probe microscopy and small angle X-ray scattering." Journal of Molecular Recognition: JMR **26**(11): 542–549. 10.1002/jmr.2298.

Overton, K., H. M. Greer, M. A. Ferguson, E. M. Spain, D. E. Elmore, M. E. Núñez and C. B. Volle (2020). "Qualitative and quantitative changes to escherichia coli during treatment with magainin 2 Observed in native conditions by atomic force microscopy." Langmuir. American Chemical Society **36**(2): 650–659. 10.1021/acs.langmuir.9b02726.

Pan, J. and N. K. Khadka (2016). "Kinetic defects induced by melittin in model lipid membranes: A solution atomic force microscopy study." The Journal of Physical Chemistry B. American Chemical Society **120**(20): 4625–4634. 10.1021/acs.jpcb.6b02332.

Pasquina-Lemonche, L., J. Burns, R. D. Turner, S. Kumar, R. Tank, N. Mullin, J. S. Wilson, et al. (2020). "The architecture of the Gram-positive bacterial cell wall." Nature. Nature Publishing Group **582**(7811): 294–297. 10.1038/s41586-020-2236-6.

Potthoff, E., D. Ossola, T. Zambelli and J. A. Vorholt (2015). "Bacterial adhesion force quantification by fluidic force microscopy." Nanoscale. The Royal Society of Chemistry 7(9): 4070–4079. 10.1039/C4NR06495J.

Radmacher, M., R. W. Tillamnn, M. Fritz and H. E. Gaub (1992). "From molecules to cells: Imaging soft samples with the atomic force microscope." Science (New York, N.Y.) 257(5078): 1900–1905. 10.1126/science.1411505.

Razatos, A., Y. L. Ong, M. M. Sharma and G. Georgiou (1998). "Molecular determinants of bacterial adhesion monitored by atomic force microscopy." Proceeding of National Academy of Sciences U.S.A. 95: 11059.

Roduit, C., F. Gisou van der Goot, P. De Los Rios, A. Yersin, P. Steiner, G. Dietler, S. Catsicas, F. Lafont and S. Kasas (2008). "Elastic membrane heterogeneity of living cells revealed by stiff nanoscale membrane domains." Biophysical Journal 94(4): 1521–1532. 10.1529/biophysj.107.112862.

Roduit, C., S. Sekatski, G. Dietler, S. Catsicas, F. Lafont and S. Kasas (2009). "Stiffness tomography by atomic force microscopy." Biophysical Journal 97(2): 674–677. 10.1016/j.bpj.2009.05.010.

Ruan, Y., S. Rezelj, A. B. Zavec, G. Anderluh and S. Scheuring (2016). "Listeriolysin O membrane damaging activity involves arc formation and lineaction – implication for listeria monocytogenes escape from phagocytic vacuole." Plos Pathogens 12: 4. 10.1371/journal.ppat.1005597.

Saar Dover, R., A. Bitler, E. Shimoni, P. Trieu-Cuot and Y. Shai (2015). "Multiparametric AFM reveals turgor-responsive net-like peptidoglycan architecture in live streptococci." Nature Communications. Nature Publishing Group 6(1): 7193. 10.1038/ncomms8193.

Schaer-Zammaretti, P. and J. Ubbink (2003). "Imaging of lactic acid bacteria with AFM – elasticity and adhesion maps and their relationship to biological and structural data." Ultramicroscopy. (Proceedings of the Fourth International Conference on Scanning Probe Microscopy, Sensors and Nanostructures) 97(1): 199–208. 10.1016/S0304-3991(03)00044-5.

Schwechheimer, C. and M. J. Kuehn (2015). "Outer-membrane vesicles from Gram-negative bacteria: Biogenesis and functions." Nature Reviews. Microbiology 13(10): 605–619. 10.1038/nrmicro3525.

Shibata, M., T. Uchihashi, T. Ando and R. Yasuda (2015). "Long-tip high-speed atomic force microscopy for nanometer-scale imaging in live cells." Scientific Reports 5: 7–7. 10.1038/srep08724.

Stefani, C., D. Gonzalez-Rodriguez, Y. Senju, A. Doye, N. Efimova, S. Janel, J. Lipuma, et al. (2017). "Ezrin enhances line tension along transcellular tunnel edges via NMIIa driven actomyosin cable formation." Nature Communications 8: 15839. 10.1038/ncomms15839.

Sullan, R. M. A., A. Beaussart, P. Tripathi, S. Derclaye, S. El-Kirat-Chatel, J. K. Li, Y.-J. Schneider, J. Vanderleyden, S. Lebeer and Y. F. Dufrêne (2014). "Single-cell force spectroscopy of pili-mediated adhesion." Nanoscale. The Royal Society of Chemistry 6(2): 1134–1143. 10.1039/C3NR05462D.

Sullan, R. M. A., J. K. Li, P. J. Crowley, L. Jeannine Brady and Y. F. Dufrêne (2015). "Binding forces of streptococcus mutans P1 adhesin." ACS Nano. American Chemical Society 9(2): 1448–1460. 10.1021/nn5058886.

Sullivan, C. J., J. L. Morrell, D. P. Allison and M. J. Doktycz (2005). "Mounting of Escherichia coli spheroplasts for AFM imaging." Ultramicroscopy. (Proceedings of the Sixth International Conference on Scanning Probe Microscopy, Sensors and Nanostructures) 105(1): 96–102. 10.1016/j.ultramic.2005.06.023.

Thewes, N., P. Loskill, P. Jung, H. Peisker, M. Bischoff, M. Herrmann and K. Jacobs (2014). "Hydrophobic interaction governs unspecific adhesion of staphylococci: A single cell force spectroscopy study." Beilstein Journal of Nanotechnology 5: 1501–1512. 10.3762/bjnano.5.163.

Touhami, A., M. H. Jericho and T. J. Beveridge (2004). "Atomic force microscopy of cell growth and division in staphylococcus aureus." Journal of Bacteriology **186**(11): 3286–3295. 10.1128/JB.186.10.3286-3295.2004.

Turner, R. D., N. H. Thomson, J. Kirkham and D. Devine (2010). "Improvement of the pore trapping method to immobilize vital coccoid bacteria for high-resolution AFM: A study of Staphylococcus aureus." Journal of Microscopy **238**(2): 102–110. 10.1111/j.1365-2818.2009.03333.x.

Umakoshi, T., S. Fukuda, R. Iino, T. Uchihashi and T. Ando (2020). "High-speed near-field fluorescence microscopy combined with high-speed atomic force microscopy for biological studies." Biochimica Et Biophysica Acta (BBA) – General Subjects. (Novel Measurement Techniques for Visualizing „live" Protein Molecules) **1864**(2): 129325. 10.1016/j.bbagen.2019.03.011.

Vadillo-Rodrigues, V., H. J. Busscher, W. Norde, J. De Vries, R. J. B. Dijkstra, I. Stokroos and H. C. van der Mei (2004). "Comparision of atomic force microscopy interaction forces between bacteria and silicon nitride substrata for three commonly used immobilization methods." Applied and Environmental Microbiology **70**: 5441.

Vadillo-Rodríguez, V., H. J. Busscher, W. Norde, J. de Vries, R. J. B. Dijkstra, I. Stokroos and H. C. van der Mei (2004). "Comparison of atomic force microscopy interaction forces between bacteria and silicon nitride substrata for three commonly used immobilization methods." Applied and Environmental Microbiology. American Society for Microbiology **70**(9): 5441–5446. 10.1128/AEM.70.9.5441-5446.2004.

Velegol, S. B. and B. E. Logan (2002). "Contributions of bacterial surface polymers, electrostatics, and cell elasticity to the shape of AFM force curves." Langmuir. American Chemical Society **18**(13): 5256–5262. 10.1021/la011818g.

Verbelen, C. and Y. F. Dufrêne (2009). "Direct measurement of Mycobacterium–fibronectin interactions." Integrative Biology. Oxford Academic **1**(4): 296–300. 10.1039/b901396b.

Verbelen, C., V. Dupres, F. D. Menozzi, D. Raze, A. R. Baulard, P. Hols and Y. F. Dufrêne (2006). "Ethambutol-induced alterations in Mycobacterium bovis BCG imaged by atomic force microscopy." FEMS Microbiology Letters **264**(2): 192–197. https://doi.org/10.1111/j.1574-6968.2006.00443.x.

Yilmaz, N. and T. Kobayashi (2016). "Assemblies of pore-forming toxins visualized by atomic force microscopy." Biochimica Et Biophysica Acta-Biomembranes **1858**(3): 500–511. 10.1016/j.bbamem.2015.10.005.

Yilmaz, N. and T. Kobayashi (2015). "Visualization of lipid membrane reorganization induced by a pore-forming toxin using high-speed atomic force microscopy." ACS nano **9**(8): 7960–7967. 10.1021/acsnano.5b01041.

Yilmaz, N., T. Yamada, P. Greimel, T. Uchihashi, T. Ando and T. Kobayashi (2013). "Real-time visualization of assembling of a sphingomyelin-specific toxin on planar lipid membranes." Biophysical Journal **105**(6): 1397–1405. 10.1016/j.bpj.2013.07.052.

Zeng, G., T. Müller and R. L. Meyer (2014). "Single-cell force spectroscopy of bacteria enabled by naturally derived proteins." Langmuir. American Chemical Society **30**(14): 4019–4025. 10.1021/la404673q.

Yang, J., J. Mou and Z. Shao (1994). "Structure and stability of pertussis toxin studied by in situ atomic force microscopy." FEBS Letters. No longer published by Elsevier **338**(1): 89–92. 10.1016/0014-5793(94)80122-3.

Christian Godon, Harinderbir Kaur, Jean-Marie Teulon,
Shu-wen W. Chen, Thierry Desnos, Jean-Luc Pellequer

6.11 Stiffening of the Plant Root Cell Wall Induced by a Metallic Stress

6.11.1 Introduction

Phosphate (Pi) is one of the main nutrients for plant growth, and Pi starvation has many physiological and developmental effects on crops. One of these responses is the inhibition of the primary root growth, correlated with a rapid decrease of cell elongation. We demonstrated that this inhibition occurs soon after the root-tip encounters substratum acidity, containing iron and low phosphate (–Pi). This condition is called low-Pi stress.

Genetic analysis in *Arabidopsis* unveiled several proteins whose mutation decrease or increase the root growth sensitivity to the low-Pi stress. These proteins belong to a putative functional pathway with two converging branches. In one branch of this pathway, the transcription factor STOP1 (SENSITIVE TO PROTON TOXICITY 1) directly activates the expression of *ALMT1* (ALUMINUM-ACTIVATED MALATE TRANSPORTER 1), coding a malate transporter of the plasma membrane. In seedlings exposed to a low external pH, Fe (as well as Al^{3+}) triggers the accumulation of STOP1 in the nucleus. RAE1 (RIBONUCLEIC ACID EXPORT protein 1), an E3 ubiquitin ligase, negatively regulates the stability of STOP1 in the nucleus. Thus, under iron deficiency, RAE1 promotes the degradation of STOP1, which can be prevented by the treatment with the proteasomal inhibitor, MG132. In the other branch of the pathway, *LPR1* (LOW PHOSPHATE ROOT 1) that codes for a cell wall located ferroxidase.

Acknowledgments: IBS acknowledges integration into the Interdisciplinary Research Institute of Grenoble (IRIG, CEA). This work was partly funded by the Agence Nationale de la Recherche (ANR-18-CE20-0023-03; ANR-09-BLAN-0118; ANR-12-ADAP-0019), CEA (APTTOX021401 and APTTOX021403), and Investissements d'avenir (DEMETERRES). This work acknowledges the AFM platform at the IBS. The authors acknowledge the support of the European Union's Horizon 2020 research and innovation program, under the Marie Skłodowska-Curie grant agreement no. 812772, project Phys2BioMed.

Christian Godon, Aix Marseille Université, CNRS, CEA, Institut de Biosciences et Biotechnologies Aix-Marseille, Laboratoire de Signalisation pour l'adaptation des végétaux à leur environnement, CEA Cadarache, Saint-Paul-lez-Durance 13108, France
Thierry Desnos, Aix Marseille Université, CNRS, CEA, Institut de Biosciences et Biotechnologies Aix-Marseille, Equipe Bioénergies et Microalgues, CEA Cadarache, Saint-Paul-lez-Durance, France
Harinderbir Kaur, Jean-Marie Teulon, Shu-wen W. Chen, Jean-Luc Pellequer, Univ. Grenoble Alpes, CEA, CNRS, IBS, Grenoble F-38000, France

https://doi.org/10.1515/9783110989380-021

Wild-type (WT, Col^{er105} [1]) seedlings exposed to a low external pH, Fe (as well as Al^{3+}) triggers the diminution of root elongation, while for the mutants *stop1* and *almt1*, the roots continue to grow. The current model postulates that exuded malate interacts with the apoplastic Fe. The mechanism inhibiting the cell elongation and cell wall modification is a matter of debate.

The primary cell wall (CW) is a 0.1–1 μm thick network of interconnected cellulose microfibrils and a matrix composed of hemicelluloses, pectins, and structural proteins (see Chapter 4.4). Located in between the two cell walls, the middle lamella primarily contains pectins. Cellulose microfibrils are made of highly crystalline domains (3–8 nm in diameter and several μm in length) linked together by amorphous regions with an interfibril spacing of 20–40 nm; they are very stable with negligible turnover (Cosgrove, 1997). Hemicelluloses, soluble only in strong alkali solutions, form a resilient and robust network with cellulose whereas pectins, soluble in hot aqueous buffer or diluted acids or with calcium chelators, form hydrated gels that push microfibrils apart, participate in wall thickness and wall porosity, and act as an adhesive layer between cells that are together in the middle lamella (Cosgrove, 2005). The major role of primary CW is to resist plant tensile stress while allowing a plant cell turgor-driven elongation. The cell wall is strong enough to support the cell turgor pressure (~1 MPa), which imposes a wall stress of about 10–100 MPa (Cosgrove, 1997).

The thickness, rigidity, and viscoelastic behavior of the cell wall determine the size, shape, morphology, and growth of a plant (Forouzesh et al., 2013). Studying plant biomechanics and mechanobiology extends our understanding of biological acclimation and adaptation of plants to changing physical environment (Moulia, 2013). Although biomechanical studies have been numerous for aerial parts of plants, there is much less study on roots due to the complexity of soils and the practical difficulties in visualizing roots in soils (Moulia, 2013).

One of the factors contributing to the complexity of the plant cell wall is the nonuniform spatial distribution of mechanical properties such as the elastic modulus (Yakubov et al., 2016). Several biophysical techniques have been used to study micro and nanomechanics on plant CW (Burgert and Keplinger, 2013) and several reviews on mechanical principles in plant growth, such as cell extension, growing CW, and CW architecture, can be found here (Cosgrove, 1997, Schopfer, 2006). Plant cell elongation theories are more than 100 years old, with a major account found in 1940 (Heyn, 1940), where it was emphasized that CW must be considered more as a living organ than a dead structure. To study the plant CW, the cellular force microscopy has been developed for large probes and high forces (up to mN), and revealed that

---

[1] Col is the accession name that stands for Columbia (the city where this *Arabidopsis thaliana* strain has been found). *er*, stands for *erecta*, a mutation in the *ERECTA* gene. 105 is the allele (i.e., a specific mutation) number 105. See Torii et al. (1996).

stiffness experiments provide a convoluted property of CW elasticity, turgor pressure, indenter geometry, and history in indentation stress (Routier-Kierzkowska et al., 2012). By using the conventional atomic force microscopy (AFM) with a nanosized tip (~10 nm in radius), it is possible to perform nanoindentation experiments, with forces ranging from 0.1 to 100 nN and with a cantilever of <3 N/m spring constant. It is commonly assumed that with standard AFM, nanomechanical experiments probe only the CW, with an average indentation below 500 nm. It is also commonly accepted that an indentation depth of <5% of the size of the investigated object is a reasonable depth target for AFM experiments on plants (Braybrook, 2015).

Most of mechanical properties of plant cell wall studies by AFM have been obtained on isolated cells (Zhao et al., 2005, Peaucelle et al., 2012, Zdunek and Kurenda, 2013, Yakubov et al., 2016) or on sectioned plant materials (Arnould et al., 2017, Torode et al., 2018, Kozlova et al., 2019). Only a few nanomaterial studies in plant tissues have been performed (Milani et al., 2011, Peaucelle et al., 2011, Milani et al., 2014, Balzergue et al., 2017). It has been proposed that strain-stiffening limits growth and restricts organ bulging (Kierzkowski et al., 2012), but it has also been recently found that plant roots can become stiffer as early as 30 min after exposition to iron stress (Balzergue et al., 2017, Godon et al., 2019).

To understand how the −Pi condition could rapidly inhibit root cell elongation, we reasoned that the mechanical properties of the CW could be modified. We explored the use of nanomechanical experiments with an AFM instrument to analyze the effect of the −Pi stress on the *Arabidopsis* root tip. We focused the probed region on the root epidermis cells located in the transition zone. In this region, cells have ceased their divisions and are ready to elongate rapidly. Results showed that a CW stiffening occurs around 30 min after the onset of the −Pi stress. This stiffening does not arise in the *almt1*, *stop1*, and *lpr1* mutants, and the WT grown on a medium with no added Fe. Using pharmacological drugs, we could show that peroxidases (probably the class III peroxidases that are abundant in the CW) are essential for this stiffening; they probably catalyze some covalent cross-links between macromolecular components of the cell wall. This is the first study showing plant CW stiffening, induced by stress. Most of the nanomechanical results with plant grown in −Pi condition have been published previously (Balzergue et al., 2017).

6.11.2 Elasticity Measurements in the Elongation Zone of Living Plant Roots

All the related materials and methods are located in Chapter 5.4. First, we asked how low Pi (−Pi) inhibits cell elongation. In plants, one mechanism that could alter cell expansion is the modification of the mechanical properties of their surrounding cell wall. Cross-links between some polysaccharides or proteins can be at the origin of this alteration of cell expansion. Thus, our hypothesis was that the cross-links in CW increase cell wall stiffness. To test the hypothesis that stiffness increases early after the onset of low Pi, we used a nanoindentation probe to measure cell surfaces stiffness on root plant seedlings. We measured the stiffness of the root surface in the transition zone, which is located between the root apical meristem (RAM) and the elongation zone (EZ) (Figure 6.11.1), where cells rapidly elongate when conditions are permissive. This region is localized at about 500 μm from the root tip.

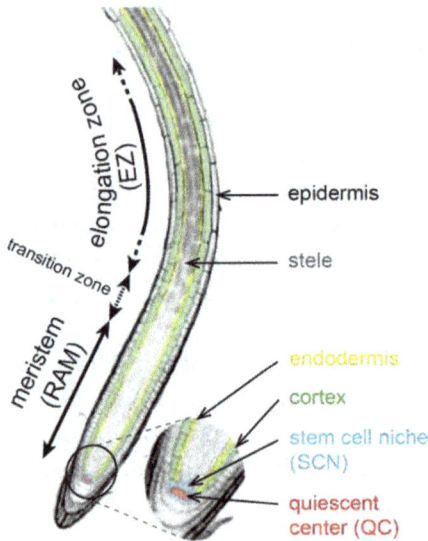

Figure 6.11.1: Scheme depicting different cell types of the primary root tip. Figure adapted from Balzergue et al. (2017).

AFM measurements were performed on WT seedlings that grew on a +Pi medium and transferred to a −Pi medium for up to 2 h. We discovered that cell wall stiffness increased as early as 30 min after transfer to −Pi and continued to increase in later time points (Figure 6.11.2). All the experimental details for these experiments could be found in Chapter 5.4.

Three different mutants *lpr1*, *stop1*, and *almt1*, in which root cell expansion is not restrained under −Pi and isolated in our laboratory (Balzergue et al., 2017), were tested under −Pi condition (Figure 6.11.3[2]).

---

2 Col[er105] is the wild-type seedling whereas *lpr1,lpr2* is a double mutant in the *lpr* genes. *Stop1*[48] and *almt1*[32] are mutants in *STOP1* and *ALMT1* genes. All the mutants *lpr*, *stop1*, and *almt1* are insensitive to the reduction of phosphate concentration in the environment (or the increase of iron concentration in the environment).

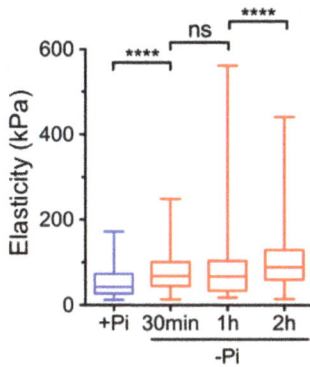

Figure 6.11.2: WT (Col^{er105}) seedlings were transferred to −Pi or +Pi medium for the indicated time, prior to measuring by AFM the stiffness of the cell surface in the transition zone of the primary root (See Methods, Chapter 5.4) (median +/− interquartile, Mann−Whitney's test. **** $P < 0.0001$; ns, not significant ($P > 0.05$)). −Pi indicates the absence of added Pi in the culture medium. Figure adapted from Balzergue et al. (2017).

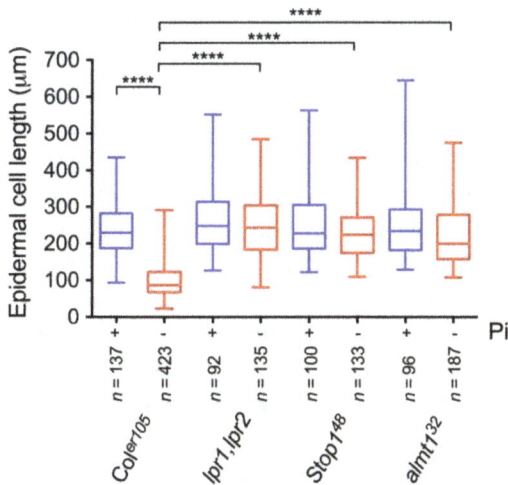

Figure 6.11.3: Seven-day-old seedlings of the indicated genotype were transferred to +Pi or −Pi medium for 24 h before measuring the final length of root epidermal cells (median +/− interquartile; Tukey's whiskers; Mann−Whitney's test: **** $P < 0.0001$; n, number of cells). Figure adapted from Balzergue et al. (2017).

We observed that the higher cell wall stiffness, measured for the WT, was decreased for the four *stop1*, *almt1*, and *lpr1,lpr2* mutants (Figure 6.11.4). For the first time, using AFM/nanoindentation probe, our results reveal a negative reciprocal relation between cell wall stiffness in the root transition zone and the final epidermal cell length. This suggests that low Pi triggers cell wall stiffening of pre-elongated cells to restrict their elongation.

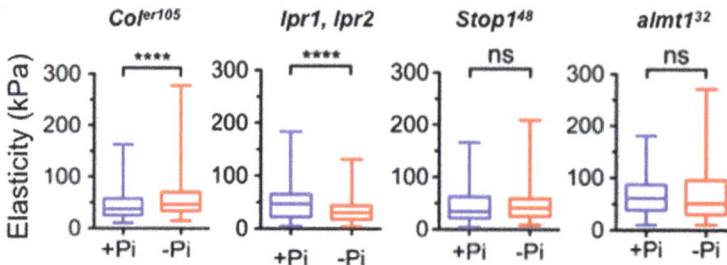

Figure 6.11.4: Seedlings of the indicated genotype were transferred to +Pi or −i medium for 30 min prior to measuring by AFM the stiffness of the cell surface in the transition zone of the primary root (median +/− interquartile, Mann–Whitney's test; **** $P < 0.0001$; ns, not significant ($P > 0.05$)). Figure adapted from Balzergue et al. (2017).

Class III peroxidases are thought to inhibit cell expansion by catalyzing cross-links between some polysaccharides or proteins, thereby tightening the cell wall (Passardi et al., 2004, Wolf et al., 2012). To test the hypothesis that −Pi induced peroxidase activity would cause cell wall stiffening at the root tips, we used a pharmacological approach to inhibit peroxidase activity. Salicylhydroxamic acid (SHAM), a peroxidase inhibitors (Rich et al., 1978, Balazs et al., 1986), treatment restored the WT root growth under −Pi condition and significantly increased the root epidermal cell length (Figure 6.11.5A). Consistent with the cell wall elongation restoration, we discovered that SHAM strongly decreased cell wall stiffness in −Pi (Figure 6.11.5B). These observations support the view that peroxidase activity catalyzes crosslinks in some cell wall components, thereby reducing cell expansion by chemical modification in the cell wall.

Recently, we demonstrated that in growth conditions with limited Pi allowing to distinguish the effect of Fe from the −Pi condition, Fe triggers the accumulation of STOP1 in the nucleus and increases the expression of *ALMT1* (Godon et al., 2019). In our previous work (Balzergue et al., 2017), we showed that in −Pi condition without Fe addition, root cell expansion is not restrained, and the root growth is comparable to seedlings growing on +Pi (Figure 6.11.6). Cell wall stiffness was measured in −Pi condition, with or without Fe addition, and results reveal a reciprocal relation between increased cell stiffness in the root transition zone and increased iron quantity in the medium. This suggests that Fe triggers cell wall stiffening to restrict cell elongation.

All the above experiments demonstrated that after the transfer of seedlings from the condition for which the Fe had no effect on root cell elongation (+Pi + Fe or −Pi–Fe) to a condition for which the Fe has an effect (−Pi + Fe), the cell stiffness increased rapidly (Balzergue et al., 2017). To evaluate the operational impact of our protocol (Azimzadeh et al., 1992, Godon et al., 2017, Teulon et al., 2019), it was decided to

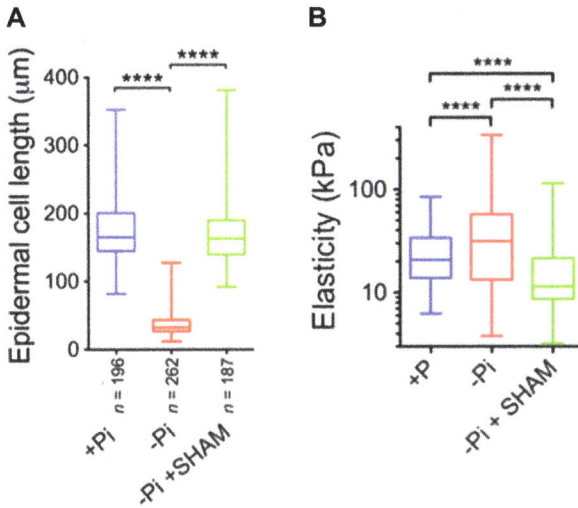

Figure 6.11.5: Effect of SHAM on root epidermal cell length. (A) Three-day-old WT seedlings were transferred for 7 days to + Pi or −Pi medium with or without 15 µM SHAM prior measuring the root epidermal cell length. (B) Three-day-old WT seedlings were transferred for 30 min to +Pi or −Pi medium with or without 15 µM SHAM prior to measuring by AFM the stiffness of the cell surface in the transition zone of the primary root (median +/− interquartile; Tukey's whiskers, Mann–Whitney's test: **** $P < 0.0001$, n = number of cells). Figure adapted from Balzergue et al. (2017).

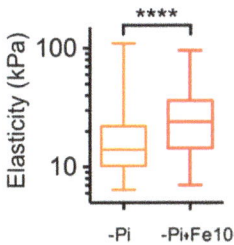

Figure 6.11.6: WT (Col^{er105}) seedlings were transferred to −Pi or −Pi + Fe (10 µM FeCl$_2$) medium for 30 min prior to measuring by AFM the stiffness of the cell surface in the transition zone of the primary root (median +/− interquartile, Mann–Whitney's test. **** $P < 0.0001$; ns, not significant ($P > 0.05$)). The experiment was performed twice with consistent results, and one representative experiment is shown. Fe10 means the presence of 10 µM FeCl$_2$. −Pi indicates the absence of added Pi to the culture medium. Figure adapted from Balzergue et al. (2017).

measure the stiffness on seedlings that have grown continuously in the (−Pi + Fe) medium, and not after a transfer. WT seeds were sown on −Pi plates containing 0, 8, 10, or 12 µM Fe, and they were grown for 4 or 7 days. We observed that after 4 days on plates containing 10 or 12 µM Fe, the primary root growth was strongly reduced, compared to lower concentrations of Fe, whereas the seedlings on 8 µM Fe were as long as those on the control plate without Fe (Figure 6.11.7A). Interestingly, all the seedlings grown for 4 days with Fe showed the same increased stiffness, compared to the control 0 Fe (Figure 6.11.7B). By letting seedlings grow for 3 more days (7-day-old) we observe that those under 8 µM Fe condition are shorter than the control 0 Fe, although not as short as those on 10 and 12 µM Fe. This shows that a Fe-triggered stiffening of root cell surface occurs even without an immediate reduced growth. These observations suggest that either the stiffening of internal cell walls (not accessible with our AFM setup) is lower at 8 µM Fe than at 10 and 12 µM, thus allowing the root to grow longer, or the stiffening is necessary to prevent root growth and there is another Fe-dependent reaction inhibiting the growth.

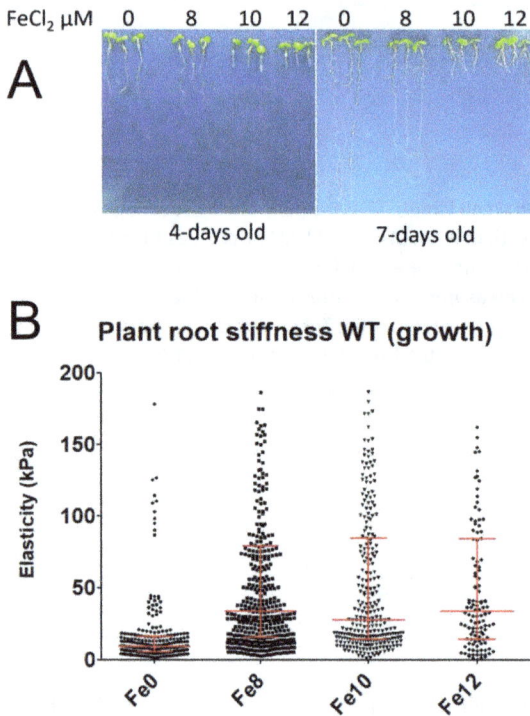

Figure 6.11.7: (A) WT seeds were sown on −Pi agar medium containing 0, 8, 10, or 12 µM Fe, and grown for four or seven days before taking the picture. (B) Stiffness of the root transition zone surface from seedlings grown 4 days in conditions as in (A). The minimum and maximum scale of red bar shows the interquartile region showing 50% of the data, whereas the middle red line signifies the median. Y-axis shows the stiffness (kPa) and X-axis shows the different iron concentrations. Sixty-one points are scattered outside the visible Y-scale.

In conclusion, a similar stiffening of plant roots has been observed, whether plants were growing on iron-free medium (then transferred to Fe-rich medium before nano-mechanical measurements) or on iron-containing medium until nanomechanical measurements. The impact of nanomechanical measurements (implying a probing at the nm scale) with AFM is demonstrated by the ability to detect very early events in plant root physiology, even before an observable phenotype. However, plants growing continuously in the presence of iron show a relatively small growth in time, whereas plants grown in the absence of Fe in agar and then transferred to a Fe-rich agar show a strong growth arrest (data not shown). This indicates that plants behave differently when they are sown under a stress condition, and they tend to counteract and develop their own signaling pathway to overcome the stress.

References

Arnould, O., D. Siniscalco, A. Bourmaud, A. Le Duigou and C. Baley (2017). "Better insight into the nano-mechanical properties of flax fibre cell walls." Industrial Crops and Products **97**: 224–228.

Azimzadeh, A., J. L. Pellequer and M. H. V. Van Regenmortel (1992). "Operational aspects of antibody affinity constants measured by liquid-phase and solid-phase assays." Journal of Molecular Recognition **5**: 9–18.

Balazs, C., E. Kiss, A. Leövey and N. R. Farid (1986). "The immunosuppressive effect of methimazole on cell-mediated immunity is mediated by its capacity to inhibit peroxidase and to scavenge free oxygen radicals." Clinical Endocrinology **25**: 7–16.

Balzergue, C., T. Dartevelle, C. Godon, E. Laugier, C. Meisrimler, J.-M. Teulon, A. Creff, M. Bissler, C. Brouchoud, A. Hagège, J. Müller, S. Chiarenza, H. Javot, N. Becuwe-Linka, P. David, B. Péret, E. Delannoy, M.-C. Thibaud, J. Armengaud, S. Abel, J.-L. Pellequer, L. Nussaume and T. Desnos (2017). "Low phosphate activates STOP1-ALMT1 to rapidly inhibit root cell elongation." Nature Communications **8**: 15300.

Braybrook, S. A. (2015). "Measuring the elasticity of plant cells with atomic force microscopy." Methods in Cell Biology **125**: 237–254.

Burgert, I. and T. Keplinger (2013). "Plant micro- and nanomechanics: Experimental techniques for plant cell-wall analysis." Journal of Experimental Botany **64**: 4635–4649.

Cosgrove, D. J. (1997). "Relaxation in a high-stress environment: The molecular bases of extensible cell walls and cell enlargement." The Plant Cell **9**: 1031–1041.

Cosgrove, D. J. (2005). "Growth of the plant cell wall." Nature Reviews. Molecular Cell Biology **6**: 850–861.

Forouzesh, E., A. Goel, S. A. Mackenzie and J. A. Turner (2013). "In vivo extraction of Arabidopsis cell turgor pressure using nanoindentation in conjunction with finite element modeling." The Plant Journal **73**: 509–520.

Godon, C., C. Mercier, X. Wang, P. David, P. Richaud, L. Nussaume, D. Liu and T. Desnos (2019). "Under phosphate starvation conditions, Fe and Al trigger accumulation of the transcription factor STOP1 in the nucleus of Arabidopsis root cells." The Plant Journal **99**: 937–949.

Godon, C., J.-M. Teulon, M. Odorico, C. Basset, M. Meillan, L. Vellutini, S.-W. W. Chen and J.-L. Pellequer (2017). "Conditions to minimize soft single biomolecule deformation when imaging with atomic force microscopy." Journal of Structural Biology **197**: 322–329.

Heyn, A. N. J. (1940). "The physiology of cell elongation." The Botanical Review **6**: 515–574.

Kierzkowski, D., N. Nakayama, A. L. Routier-Kierzkowska, A. Weber, E. Bayer, M. Schorderet, D. Reinhardt, C. Kuhlemeier and R. S. Smith (2012). "Elastic domains regulate growth and organogenesis in the plant shoot apical meristem." Science **335**: 1096–1099.

Kozlova, L., A. Petrova, B. Ananchenko and T. Gorshkova (2019). "Assessment of primary cell wall nanomechanical properties in internal cells of non-fixed maize roots." Plants **8**: 172.

Milani, P., M. Gholamirad, J. Traas, A. Arneodo, A. Boudaoud, F. Argoul and O. Hamant (2011). "In vivo analysis of local wall stiffness at the shoot apical meristem in Arabidopsis using atomic force microscopy." The Plant Journal **67**: 1116–1123.

Milani, P., V. Mirabet, C. Cellier, F. Rozier, O. Hamant, P. Das and A. Boudaoud (2014). "Matching patterns of gene expression to mechanical stiffness at cell resolution through quantitative tandem epifluorescence and nanoindentation." Plant Physiology **165**: 1399–1408.

Moulia, B. (2013). "Plant biomechanics and mechanobiology are convergent paths to flourishing interdisciplinary research." Journal of Experimental Botany **64**: 4617–4633.

Passardi, F., C. Penel and C. Dunand (2004). "Performing the paradoxical: How plant peroxidases modify the cell wall." Trends in Plant Science **9**: 534–540.

Peaucelle, A., S. Braybrook and H. Hofte (2012). "Cell wall mechanics and growth control in plants: The role of pectins revisited." Frontiers in Plant Science **3**: 121.

Peaucelle, A., S. A. Braybrook, L. Le Guillou, E. Bron, C. Kuhlemeier and H. Hofte (2011). "Pectin-induced changes in cell wall mechanics underlie organ initiation in Arabidopsis." Current Biology **21**: 1720–1726.

Rich, P. R., N. K. Wiegand, H. Blum, A. L. Moore and W. D. Bonner, Jr. (1978). "Studies on the mechanism of inhibition of redox enzymes by substituted hydroxamic acids." Biochimica et biophysica acta **525**: 325–337.

Routier-Kierzkowska, A. L., A. Weber, P. Kochova, D. Felekis, B. J. Nelson, C. Kuhlemeier and R. S. Smith (2012). "Cellular force microscopy for in vivo measurements of plant tissue mechanics." Plant Physiology **158**: 1514–1522.

Schopfer, P. (2006). "Biomechanics of plant growth." American Journal of Botany **93**: 1415–1425.

Teulon, J.-M., C. Godon, L. Chantalat, C. Moriscot, J. Cambedouzou, M. Odorico, J. Ravaux, R. Podor, A. Gerdil, A. Habert, N. Herlin-Boime, S.-W. W. Chen and J.-L. Pellequer (2019). "On the operational aspects of measuring nanoparticle sizes." Nanomaterials **9**: 18.

Torii, K. U., N. Mitsukawa, T. Oosumi, Y. Matsuura, R. Yokoyama, R. F. Whittier and Y. Komeda (1996). "The Arabidopsis ERECTA gene encodes a putative receptor protein kinase with extracellular leucine-rich repeats." The Plant Cell **8**: 735–746.

Torode, T. A., R. O'Neill, S. E. Marcus, V. Cornuault, S. Pose, R. P. Lauder, S. K. Kracun, M. G. Rydahl, M. C. F. Andersen, W. G. T. Willats, S. A. Braybrook, B. J. Townsend, M. H. Clausen and J. P. Knox (2018). "Branched pectic galactan in phloem-sieve-element cell walls: Implications for cell mechanics." Plant Physiology **176**: 1547–1558.

Wolf, S., K. Hematy and H. Hofte (2012). "Growth control and cell wall signaling in plants." Annual Review of Plant Biology **63**: 381–407.

Yakubov, G. E., M. R. Bonilla, H. Chen, M. S. Doblin, A. Bacic, M. J. Gidley and J. R. Stokes (2016). "Mapping nano-scale mechanical heterogeneity of primary plant cell walls." Journal of Experimental Botany **67**: 2799–2816.

Zdunek, A. and A. Kurenda (2013). "Determination of the elastic properties of tomato fruit cells with an atomic force microscope." Sensors **13**: 12175–12191.

Zhao, L., D. Schaefer, H. Xu, S. J. Modi, W. R. LaCourse and M. R. Marten (2005). "Elastic properties of the cell wall of Aspergillus nidulans studied with atomic force microscopy." Biotechnology Progress **21**: 292–299.

Outlook

Hans Oberleithner

7.1 Mechanics of Diseases: An Outlook

I would like to start this "Outlook" with a personal experience:

When, on a hot summer day in August 1992, I unleashed an "atomic force microscope" on living kidney cells in the Department of Cell and Molecular Physiology at Yale Medical School, entering the previously unknown world of nanostructures in living matter, I realized in a flash that I never wanted to leave that world again.

Almost 30 years have passed since then.

As the present work by nano-experts from all over the world impressively shows, physicists in particular have brought new innovative techniques to the nano-world and exported them to the life sciences, despite initial skepticism.

The skepticism existed on both sides. Here are the physicists who were used to working "with the highest precision" on inanimate matter, and there are the biologists, physiologists, and physicians who were used to fishing in the cloudy waters of animate matter.

Before I take a look into the future, I would like to turn the clock back half a century, with the intention of pointing out some significant "quantum leaps" in the life sciences that essentially took their starting point from the methods of physics and that were only made possible by the close linking of the two broad areas of science.

From my personal perspective, I would like to divide the last 60 years – please forgive the gross simplification – into four "quantum leaps":

Around 1960: The electrical concept of cells

The electrical cell voltage as a fundamental life expression of a cell can be predicted with almost physical/mathematical accuracy from the ion fluxes through the cell membrane, although the molecular structure of the membrane proteins was still completely unknown. The apparently "passive" protective shell of a cell turned out to be the "crucial player" of all vital processes.

Around 1980: Molecular imaging

Driven by the exciting developments in molecular biology/genetics, physicists develop methods for measuring/visualizing/manipulating cellular and macromolecular structures.

Nanotechnologies, originally reserved for the inanimate natural world, are invading the life sciences on silent soles. A hitherto unknown world of molecular images emerges.

Hans Oberleithner, Solegasse 14, Thaur 6065, Austria

https://doi.org/10.1515/9783110989380-022

Around 2000: Newton meets biology

Quantitative nanotechnologies discover the importance of *cell mechanics*. The possibilities of precise force measurements in the nano- and pico-newtons open up a completely new research perspective for bio-scientists – the possibility to explain cellular life processes on the basis of mechanical interactions between organic macromolecules.

Around 2020: Newton meets medicine

The nano-world of life sciences is expanding. Nanotechnologies have long since landed in the subfields of biomedical research. The mechanics of tissues, cells, and their substructures provide novel clues to the physiology/pathophysiology of any underlying life processes. As yet, it is still a "hunt and gather." Physicists learn from physicians, physicians learn from physicists; biologists learn from both.

Now comes my actual "outlook"

Around 2040: The mechanical concept of cells

As mentioned above, the decoding of the electrical cell voltage has led to fundamental insights into the life of a cell, and ultimately to a deep understanding of the complex organisms (man, animal, plant). The identification of myriad molecular structures inside and outside cells through the breakthrough advances in molecular biology/molecular genetics has made it possible to characterize nanomechanical properties of organ-specific tissues, single cells, cellular substructures, and even intermolecular forces. As can be seen from the various contributions in this volume, the ever-expanding arsenal of innovative nanomethods is currently being applied to many medical questions. It is becoming increasingly apparent that nanomechanics could possibly play a key role in attempting to explain the fundamental processes of life. Just as it was possible to unify the *electrical phenomena* of living cells into a generally valid concept (membrane concept) in the past, it may become possible in the upcoming years to derive from the *mechanical phenomena*, a generally valid principle for the life processes of cells.

In other words

While, in the course of many years, the analysis of individual components (membrane channels, ion pumps, etc.) had led to an overall concept of cell function, the basis of which was the biophysical measurement of electrical voltage (volts), another (alternative or complementary) overall concept of cell function may soon develop. It will emerge from the nanomechanical properties of individual components of a cell (plasma membrane, cytoskeleton, nucleus, etc.), and the basis of which will be the biophysical measurement of mechanical force (newtons).

From the various chapters of the present volume, it is clear that nanotechnologies – originally accessible mainly to physicists – have gained a foothold in the broad field of life sciences, so that a wealth of valuable data on the mechanics of individual components (tissues, extracellular and intracellular cell structures) already exists. The more information becomes available about the mechanical properties of whole cells and their substructures, the more rapidly a generally applicable concept for nanomechanics of living matter will emerge.

The focus will be on quantitative force measurements, in the order of nano- and pico-newtons. Due to the enormous progress in the development of high-end nanotechniques (atomic force microscopy, laser tweezers, fiber optics, etc.), intermolecular constellations can be simulated in vitro and systematically characterized via differentiated force measurements (viscoelasticity, frictional forces, electrostatic forces, tension forces, etc.).

It is likely that the current trend in the life sciences, particularly in biomedicine, will continue in looking at physiological as well as pathophysiological processes under well-defined, controlled conditions (in vitro) and only then cautiously increasing the complexity of experiments in tissues (in situ) to whole organism experiments (in vivo).

Despite all the euphoria about the great advances in molecular biology, the key role of electrolytes (inorganic ions) in generating cellular electricity should not be pushed to the background. Without inorganic ions, life would not exist. The physiological interaction of macromolecular structures can only function in a well-defined ionic environment. Therefore, it will be of great importance not only to have any conceivable macromolecule available, but also pay strict attention in experiments with living matter to whether force measurements in the broadest sense can be made under conditions that reflect the natural environment of a cell. This involves not only the quality/quantity of the electrolyte composition, but also parameters such as osmolality, molecular crowding, and many others.

However, if in the coming years we succeed in unifying the mechanical properties of specific cell structures – whether from humans, animals, or plants – in a generally valid concept, this could create fertile grounds that would enable cell research in a completely new light.

Subject Index (Volume 1 & Volume 2)

This is a merged subject index for volume 1 (Biomedical Methods) and volume 2 (Biomedical Applications). The bold number in front of the page references represents the volume number.

https://doi.org/10.1515/9783110989380-023

www.ingramcontent.com/pod-product-compliance
Lightning Source LLC
Chambersburg PA
CBHW080714220326
41598CB00033B/5420